D0151998

Polymers as Rheology Modifiers

ACS SYMPOSIUM SERIES 462

Polymers as Rheology Modifiers

Donald N. Schulz, EDITOR
Exxon Research and Engineering Company

J. Edward Glass, EDITOR
North Dakota State University

Developed from a symposium sponsored
by the Division of Polymeric Materials: Science and Engineering
at the 198th National Meeting
of the American Chemical Society,
Miami Beach, Florida,
September 10–15, 1989

American Chemical Society, Washington, DC 1991

Library of Congress Cataloging-in-Publication Data

Polymers as rheology modifiers / Donald N. Schulz, J. Edward Glass, editors

p. cm.—(ACS symposium series; 462)

Developed from a symposium sponsored by the Division of Polymeric Materials: Science and Engineering at the 198th national meeting of the American Chemical Society, Miami Beach, Florida, September 10–15, 1989.

Includes bibliographical references and index.

ISBN 0–8412–2009–3

1. Fluid dynamic measurements. 2. Rheology. 3. Polymer solutions.

I. Schulz, Donald N., 1943– . II. Glass, J. E. (J. Edward), 1937– . III. American Chemical Society. Division of Polymeric Materials: Science and engineering. IV. Series.

TA357.5.M43P65 1991
620.1'06—dc20 91–11687
CIP

The paper used in this publication meets the minimum requirements of American National Standard for Information Sciences—Permanence of Paper for Printed Library Materials, ANSI Z39.48–1984. ∞

PRINTED IN THE UNITED STATES OF AMERICA

ACS Symposium Series

M. Joan Comstock, *Series Editor*

1991 ACS Books Advisory Board

Foreword

THE ACS SYMPOSIUM SERIES was founded in 1974 to provide a medium for publishing symposia quickly in book form. The format of the Series parallels that of the continuing ADVANCES IN CHEMISTRY SERIES except that, in order to save time, the papers are not typeset, but are reproduced as they are submitted by the authors in camera-ready form. Papers are reviewed under the supervision of the editors with the assistance of the Advisory Board and are selected to maintain the integrity of the symposia. Both reviews and reports of research are acceptable, because symposia may embrace both types of presentation. However, verbatim reproductions of previously published papers are not accepted.

Contents

Preface

ADVANCED MATERIALS BRING TO MIND materials with high-performance solid-state properties such as mechanical strength, electrical conductivity, and nonlinear optical properties. However, advanced materials can also be fluids, and polymers can be the agents that convert them from simple fluids to high-performance fluids. Small amounts of polymer can have a profound effect on fluid rheological properties such as viscosification, drag reduction, and antimisting. Such complex fluids have found usefulness in applications as diverse as paints, coatings, fuels, lubricants, cosmetics, personal care products, and foods.

Numerous books cover rheology, and even more volumes cover polymers. However, few books are devoted primarily to polymers as materials for modifying or controlling fluid rheology. This volume aims to fill this void.

The first three chapters present basic rheological concepts. Chapter 1 contains a section on electrorheological fluids, that is, polymer suspensions whose viscosities change with electric fields. Chapters 4–7 emphasize gels and latices and describe synthesis, characterization, and fluid properties. The next seven chapters feature the special properties, such as shear thickening rheology, of associating polymer systems. These chapters give examples of ionic and H-bonding association groups on hydrocarbon backbones and hydrophobic associating groups on water-soluble backbones. Chapter 14 is the first report of the use of surfactants in combination with associative thickeners. The next four chapters describe rheology control based on polymer–polymer and polymer–solvent interactions. Chapters 19 and 20 examine deformation-related orientations in bulk and gelled polymer systems.

Acknowledgments

We wish to thank all the participants and contributors to this book and to the symposium on which it is based. We also acknowledge the support and encouragement of our families and our secretaries, Arlene Ozbun and

Penny Clancy. Finally, we thank A. Maureen Rouhi, the acquisitions editor at the ACS Books Department who kept us on task in a firm but patient manner.

DONALD N. SCHULZ
Exxon Research and Engineering Company
Annandale, NJ 08801

J. EDWARD GLASS
North Dakota State University
Fargo, ND 58105

February 13, 1991

RHEOLOGICAL CONCEPTS

Chapter 1

Polymers as Rheology Modifiers

An Overview

J. Edward Glass[1], Donald N. Schulz[2], and C. F. Zukoski[3]

[1]Polymers and Coatings Department, North Dakota State University, Fargo, ND 58105

[2]Corporate Research Laboratories, Exxon Research and Engineering Company, Annandale, NJ 08801

[3]Department of Chemical Engineering, University of Illinois, Champaign, IL 61820

The chapter summarizes the basis of the use of polymers to control or influence fluid rheology and outlines typical applications. Electrorheological fluids, a new class of polymeric rheology modifiers, are discussed.

Polymers are best known for their use as bulk materials (e.g., elastomers, plastics, and fibers). There is another world, however, where polymers are used to control solution and dispersion rheology. This world includes fields as diverse as fuels, lubricants, oil field chemicals, water treatment chemicals, coatings, and food applications. In these fields polymers affect the shear and elongational flow behavior (defined in Chapters 2 and 3) and thereby the performance of the fluid during and after application.

Polymers modify rheology by virtue of their high molecular weights, chain entanglements, and polymer–solvent interactions. Additional property control can be achieved by use of phase changes and associations. In special cases, polymer fluids also can be made to respond to external electrical fields. In the sections to follow, general flow behaviors are discussed, followed by the rheology requirements as they are desired or required in various applications. In the last section of this chapter, a relatively new class of polymeric rheology modifiers, electrorheological fluids, is discussed. These materials have not yet realized their full commercial acceptance but show promise for doing so.

Basic Concepts

The power of polymers to influence fluid rheology arises from the greater volume of a macromolecule in solution compared to the total of the molecular dimensions of the repeating units. The solution volume swept out by the polymer coil is known as the hydrodynamic volume (HDV),

0097–6156/91/0462–0002$06.00/0

which is determined by polymer structural parameters (e.g., chain length and chain stiffness) and polymer–solvent interactions, as well as polymer associations or repulsions. HDV also has a temperature, concentration, molecular weight, and deformation rate dependence.

For random-coil polymers, the effective HDV of a macromolecule is proportional to the cube of the root-mean-square end-to-end distance, $<\overline{r^2}>^{1/2}$, and is generally equated to the product $[n]M$, defined by the key equation in Flory's theory (*1*):

$$[n] = \frac{\phi <\overline{r^2}>^{3/2}}{M} \quad (1)$$

where M is molecular weight, $[n]$ is the intrinsic viscosity of the polymer, and ϕ is a universal constant for all polymer types. ϕ is influenced by the interaction between the polymer and solvent. This interaction is generally defined by an expansion coefficient, α, which is determined by comparing $<\overline{r^2}>^{1/2}$ or $[n]$ under solvent conditions where strong interaction forces are developed (e.g., a polar polymer in a polar solvent) to $<\overline{r^2}>_0^{1/2}$ or $[n]_0$ values under poor or theta solvent conditions. In the latter, the interaction is minimized, and the subscript is in reference to unperturbed dimensions.

Molecular Weight Effects

Polymer solutions exhibit a linear increase in viscosity with increasing molecular weight. At a certain molecular weight (M_c), the dependence of viscosity on molecular weight increases, generally by an exponential power of 3.4. For a polymer with a given molecular weight, the solution viscosity also increases monotonically with concentration, until a critical concentration is reached (Figure 1). Both observations reflect the onset of interchain associations. Data related to this behavior can be superimposed (*2*) by relating the solution viscosity dependence to a dimensionless parameter, $[n]c$, reflecting both the size, through $[n]$, the intrinsic viscosity, and its concentration.

Shear Thinning Behavior

Related phenomena also can be observed when polymer solutions are studied with increasing shear rate. The viscosity of a polymer solution will decrease with shear rate, a consequence, in part, of the disruption of overlapping chains faster than their ability to associate at higher deformation rates. The shear rate at which non-Newtonian behavior occurs increases with decreasing molecular weight (*3*), reflecting a shorter relaxation time of lower molecular weight polymers. This phenomenon also is apparent with increasing concentration for a polymer of constant molecular weight (Figure 2).

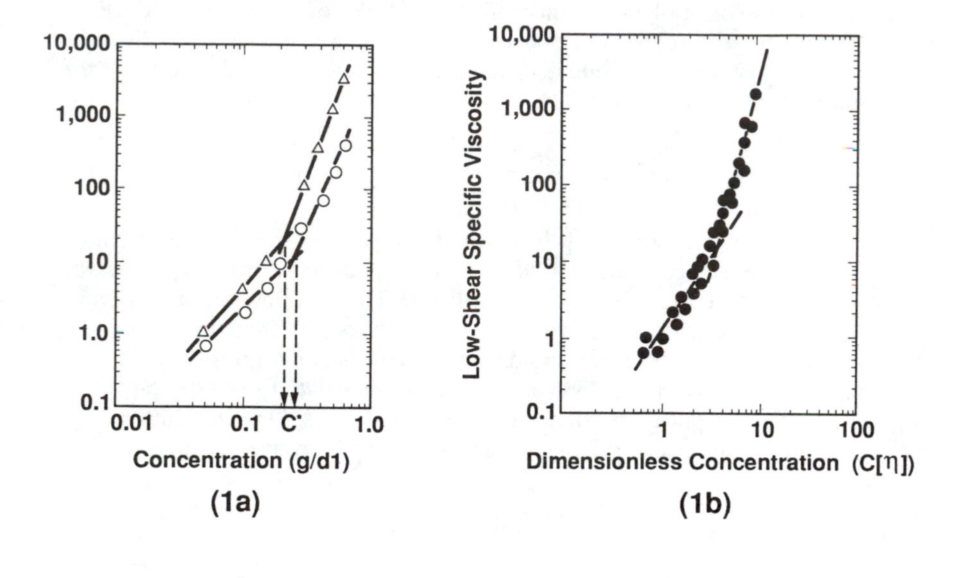

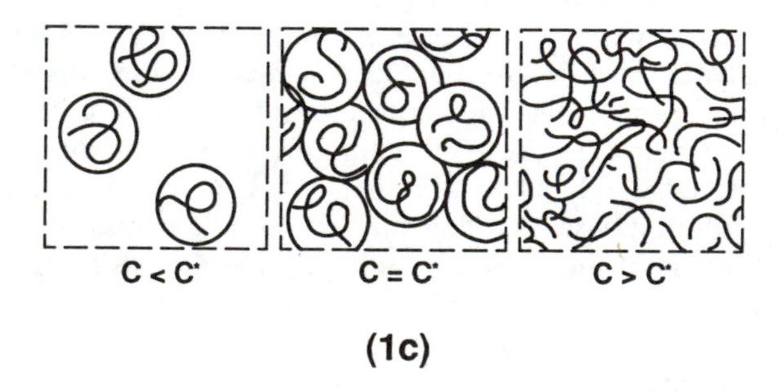

Figure 1. The relationship between low-shear specific viscosity and concentration for HP-guar. (1a) Plot of log shear specific viscosity vs concentration (g/dL); (1b) log shear specific viscosity vs dimensionless concentration $c[\eta]$. The break points in these curves occur where $c = c^*$ (1c). (Reproduced from reference 2. Copyright 1986 American Chemical Society.)

This behavior will be discussed in this volume, with a variety of uniquely different polymers, in both organic and aqueous media. This will occur at concentrations and molecular weights well below that noted with traditional structures; the phenomenon arises from units that effect atypical associations: mesogenic groups (liquid crystals), ionic interactions (ionomers) in organic media, and hydrophobic interactions in aqueous media. These interactions effect higher viscosities at low shear rates but without the high elastic response in high application deformation rate processes noted with truly high-molecular-weight polymers. This is a key factor in the acceptance of some of the novel rheology modifiers examined in this volume.

In many applications, disperse phases will be present. In the absence of polymers, they also exhibit shear thinning. Viscosities of such dispersions at low shear rates are greater than the viscosity of the individual disperse components; the phenomenon arises from multiplet rotations in a shear field and, with true agglomerates, entrapment of the continuous phase. As the shear deformation is increased, these multiple "associations" are disrupted, with an attendant drop in the viscosity. This phenomenon has been reported for a variety of disperse phase compositions (e.g., bentonite, coal particles, and titanium dioxide), but the mechanism for disperse systems has been defined (*4*) in greatest detail for monodisperse latices (Figure 3). Among disperse phases, at equal formulation volume fractions, smaller particles will impart greater viscosities because of greater surface area and the impact of hydration on the relative "effective volume fractions" of these dispersions.

Shear Thickening Behavior

Structure formation in polymer solutions under flow results in a number of experimental observations (*5*); one of these is shear thickening. Shear thickening occurs in both polymer solutions and in dispersions. In polymer solutions the phenomenon has been attributed to either intramolecular hydrodynamic interactions due to nonuniform changes of molecular distances during coil deformation of large chains, particularly in viscous solvents (*6*), or the transition of intra- to greater intermolecular associations of entanglement junctions.

Shear thickening in polymers was observed in ionomers (*7* and Chapter 9, this volume) and interpreted in terms of an increase in temporary associations among chains made possible by elongation under shear flow of the charged polymers. Established statistical properties of polymers have been used (*8*) to support such a mechanism. This theoretical work supports the original observations that the phenomenon is highly dependent on chemical composition and is maximized with only a few associating groups per chain. The shear thickening phenomenon also has been observed in aqueous systems utilizing hydrophobic bonding (Chapter 11, this volume), in ionic complexes (*9a,b* and Chapter 10, this volume), and

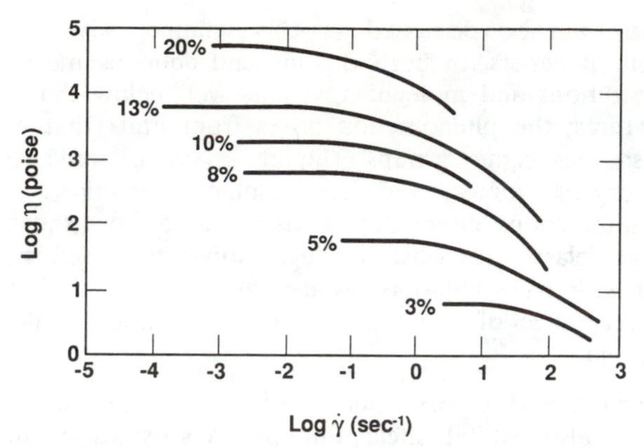

Figure 2. Viscosity data for solutions of high-molecular-weight poly(isobutylene) in decalin at 25 °C obtained on a cone and plate viscometer. (Reproduced with permission from reference 3. Copyright 1955 Academic)

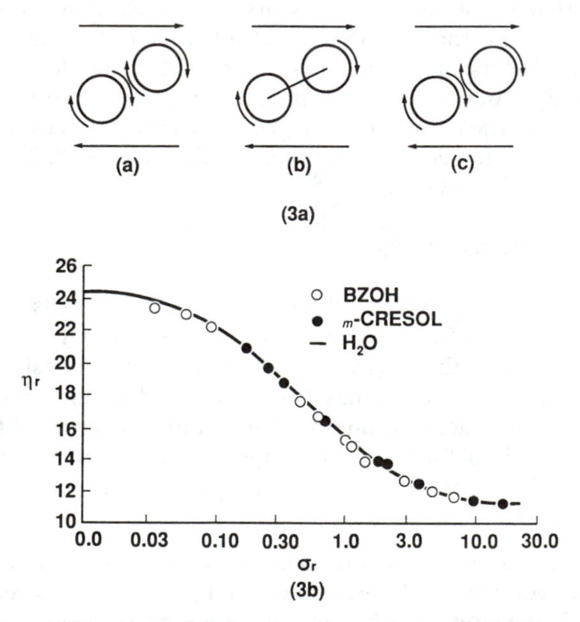

Figure 3. Shear thinning behavior for dispersions. (3a) Schematic drawing showing that proximity doublets, formed by Brownian diffusion, are destroyed by shear: (a) fluid motion if particles were to rotate independently, (b) rigid dumbbell rotation, and (c) a more realistic flow pattern, intermediate between (a) and (b). (3b) Relationship between relative viscosity and reduced shear stress for monodispersions of polystyrene spheres of various sizes in different media. (Reproduced with permission from reference 4. Copyright 1972 Elsevier.)

in surfactant-modified, water-soluble polymers (Figure 4; *10a,b* and Chapters 13 and 14, this volume).

One of the most complete experimental study of shear thickening has been in aqueous solutions with certain sodium borate/poly(vinyl alcohol) combinations (*11*), where the rheological behavior was correlated with ^{11}B NMR chemical shifts. The results are in agreement with the original mechanism proposed for ionomers. Many of the crosslinks in the system at rest were intramolecular. Shearing at sufficiently high rates was suggested to elongate the aggregates, and the equilibria of spontaneously breaking and reforming crosslinks shifted to more inter- rather than intramolecular associations. This creates a flow unit with a more effective hydrodynamic volume and higher viscosities. The mechanism is most significant at the point where the time scale of deformation (inverse shear rate) becomes shorter than the molecular relaxation time. At still higher shear rates the shear stresses become large enough to prevent interchain junctions reforming at a rate equal to breakage, and shear thinning is observed.

In high-molecular-weight polymer solutions at low concentrations, dilational flow occurs when individual high-molecular-weight chains are stretched in extensional flow with severe constrictions (e.g., in irregular channels related to subterranean media; *12*), leading to increased resistance factors (Figure 5) (pressure differential measurements taken during core testing that can be related to relative viscosities). The phenomenon has been studied as a function of polymer structure, molecular weight, solvent quality, and ionic environment (*13–15*). Shear thickening also is observed in concentrated dispersions (*16–22*). Most of the work has been on monodispersed latices (Figure 6), but the phenomenon also occurs in multimodal latices and pigments. Light scattering and light diffraction techniques, complementing rheological studies, have suggested that the dilatant flow behavior arises from an order–disorder transition (*21*).

Typical Applications

As Viscosifiers. Since polymers have an intrinsically large hydrodynamic volume (HDV), only low concentrations of polymer are needed to substantially increase the viscosity of the fluid. Viscosity index (VI) imprrovers, for automotive lubricant oils, are examples of a nonaqueous rheology modifier. Viscosity index is an empirical number that indicates the resistance of a lubricant viscosity to changes in temperature. High VI values indicate greater resistance to thinning at high temperatures (*23, 24*).

These polymers exist as compact coils in cold oil ("poor" solvent) and expand with increased temperature because of increased solvation. This inverse temperature response of polymer solutions enables the formulation of multigrade lubricating oils with flatter viscosity–temperature curves.

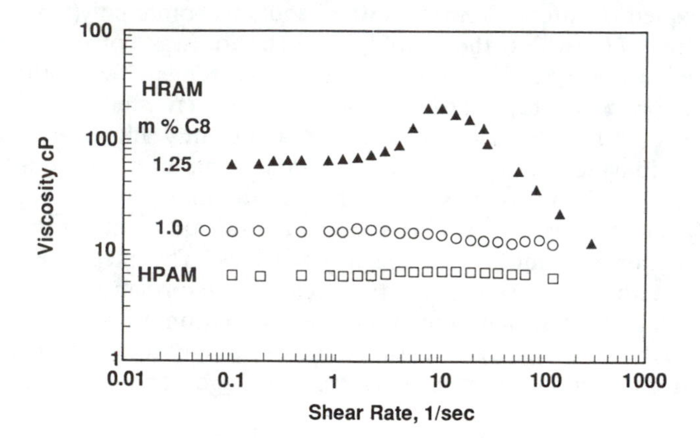

Figure 4. Shear thickening behavior for hydrophobically associating acrylamide polymers as a function of hydrophobe level. (Reproduced from reference 10a. Copyright 1989 American Chemical Society.)

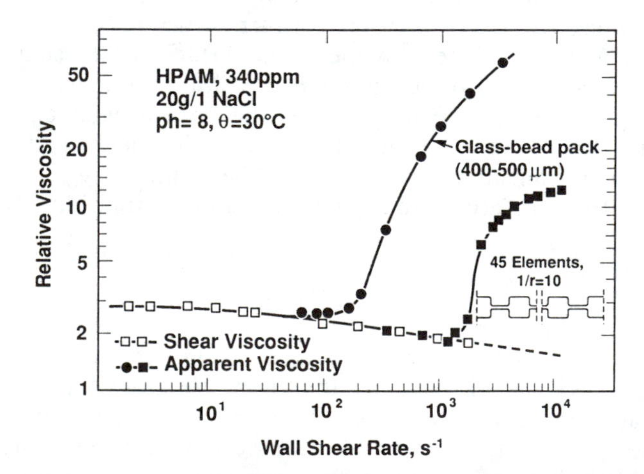

Figure 5. Comparison of rheological behavior in a glass-bead pack and in a model including successive constrictions; HPAM is hydrolyzed polyacrylamide. (Reproduced with permission from reference 53. Copyright 1981 Society of Petroleum Engineers.)

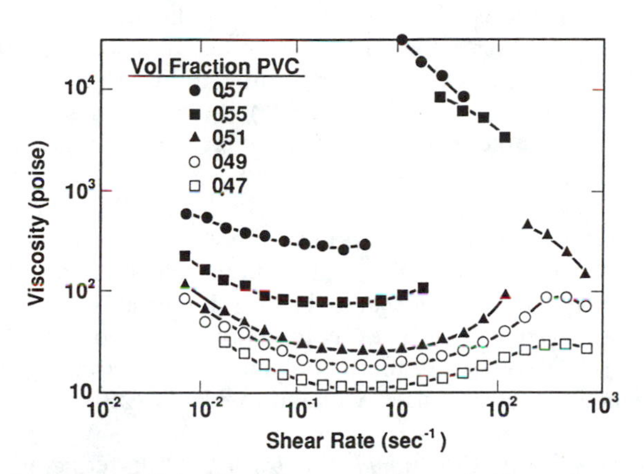

Figure 6. Shear thickening in concentrated dispersion of monodispersed latices. (Reproduced with permission from reference 18. Copyright 1972 Wiley.)

This effect results in easier engine starts in cold weather and retention of lubricating ability at high engine running temperatures.

Another needed property for VI improvers is shear (e.g., mechanical) stability. Shear stability is required for oils subjected to high deformation rates between the piston and cylinder wall or in gear pumps. Shear rates encountered in lubrication are as high as 10^5–3×10^6 s^{-1} (*24*). Engine oils also can be broken down by thermal oxidation at the high combustion temperatures. The main classes of VI improvers are saturated oil-soluble polymers, e.g., olefin copolymers (OCP), hydrogenated styrene–diene copolymers (HSD), and poly(alkyl methacrylates) (*23, 24*).

In either oil or aqueous systems, high viscosities are needed to inhibit sedimentation. Application areas include oil- or water-borne coatings, coal–water combustion slurries, drilling and completion fluids, and innumerable food applications. Except in food applications, the most cost-efficient method to achieve high low-shear-rate viscosities is through the use of clays. In oil-based coatings and drilling fluids, this is achieved generally by amine-treated montmorillonite (*25*). In aqueous systems, clay can be used alone or complemented by polymers. When clays are used, the viscosity has a significant elastic component (first normal stress [*26*] or storage modulus [*27*]; *see* Chapter 2 for definitions). This can impede production rates and performance. Although the use of polymers alone provides the best performance, they are often complemented with clay because of mechanical and thermal oxidation limitations of pure polymer systems. In clean systems, such as completion and workover fluids (*28*) and enhanced oil recovery systems, polymers that viscosify under high salinity and temperatures are used with antioxidants (*29*).

Fluids with a Significant Extensional Component. Drag-reducing agents are polymers that inhibit the development of bursts or turbulence and enhance the flow of a fluid through a pipe or the speed of a ship. The addition of only part-per-million levels of polymer can increase the flow by as much as 100%. Drag reduction occurs by the extension of the laminar flow regime to abnormally high Reynolds numbers (*23*). Both nonaqueous and aqueous polymers have been used. The mechanism (*30, 31*) of polymer-induced drag reduction is complex, and there is still no agreement on its details. One mechanism postulates that isolated polymer molecules are stretched beyond their random coil limit with a consequent increase in local viscosity and a dampening of turbulence. As such drag reduction (DR) is associated with transient elongational flow and elongational viscosity (*32*). On the other hand, those molecules that can form aggregates can presumably suppress small-scale turbulence by resisting rapid changes in alignment. Drag reduction scales with polymer coil volume. Thus, it tends to increase with increasing chain length and "goodness" of the solvent. Unfortunately, degradation also tends to increase with coil volume (*32*).

Many high-molecular-weight, hydrocarbon-soluble polymers have been screened as possible hydrocarbon drag reducers for crude-oil pipelines (*33*). One of the most successful crude-oil pipeline drag-reducing agents is ultrahigh-molecular-weight poly(1-octene). This material has been used to substantially increase the output of oil through the Alaska pipeline (*34, 35*). The limitation of this material is degradation when passing through each pumping station, which necessitates addition of more polymer.

Another area where extensional viscosity behavior is important is in antimisting agents. Certain polymers are very effective agents for reducing the mist of atomized liquids (*35, 36*, and Chapter 10, this volume). Some of the most effective antimisting agents for hydrocarbon solutions are those that exhibit high extensional viscosities at low polymer concentrations, based upon associations of functional groups (*37–40*). This may be a desirable feature in aviation fuel, but it is undesirable in most coating, petroleum recovery, and adhesion applications (*see* Chapter 20).

Disperse Phase Modifiers. Dispersions can provide very low cost means of achieving viscosity control. The technology associated with the low-cost modifiers are discussed in Chapter 7 (and in Chapters 25, 27, and 28 of reference 2). Disperse phases also can be used to control low-shear-rate viscosities without the extensional component that is undesirable in coating spray application. One established commercial example is high-solids coatings for automobile finishes, where EPA standards limit the use of high-molecular-weight polymers. The flocculation of microgels is used to obtain high viscosities at low shear rates until the oligomers have reacted to achieve interlinked macromolecular structures (*41*).

Electrorheology: Electrical Control of Suspension Viscosity

When large electric fields are applied to many nonaqueous suspensions, a reversible increase in suspension viscosity is observed. This change in stress-transfer ability, termed the electrorheological or ER response, was first observed by Winslow (*42*) in the late 1940s and is of growing interest from a technological and a scientific perspective. Potential applications of the ER response include shock absorbers, engine mounts, robotic actuators, and acoustic dampers. However, applications are still in development because of the limited understanding of how the electric field gives rise to changes in viscosity and how suspension properties can be optimized.

ER suspensions are typically composed of a solid particulate phase with a characteristic size of 0.5–100 microns suspended in a low-dielectric-constant oil. Large responses are noted when the particles have a polarizability much larger than that of the suspending oil (*43, 44*). Applications are limited by the level of stress that can be attained and the power requirements to achieve a given suspension viscosity. Understanding these phenomena requires an interdisciplinary program covering colloid science, suspension mechanics, and the physical chemistry of poorly conducting, low-dielectric-constant fluids exposed to field strengths on the order of 10^6 V/m. Recent progress has come by studying the suspensions from a colloid science perspective and through flow visualization experiments.

These studies have shown that upon application of the electric field, suspensions with initially random microstructures rearrange to form columns or fibers that span the electrode gap (*45, 46*). In his original work, Winslow noted the presence of these structures and associated the increased suspension viscosity with the degradation of the structures. Recent flow visualization experiments confirm that columnar structures are formed in the presence of the electrical field and that these structures are slowly degraded when the suspension is sheared perpendicular to the electrical field. Links between the suspension microstructure and suspension flow properties have been made by realizing that, in the presence of the electric field, particles polarize, At the first level of approximation, each particle interacts like a point dipole with all the other particles in suspension in a pairwise additive fashion (*47–49*). The force between two dipolar particles of radius *a* can be written as

$$F = 12\pi\epsilon_p\epsilon_c a^2\beta^2E^2\left[(a/R)^4(3\cos^2\theta - 1)\hat{r} + (\sin 2\theta)\hat{\theta}\right] \qquad (2)$$

where E is the magnitude of the applied field strength; ϵ_p and ϵ_c are the particle and continuous-phase dielectric constants, respectively; θ is the angle between the line of particle centers and the applied field; and R is the center-to-center pair separation. $\hat{\theta}$ and $\hat{r}$ are the unit vectors in the θ and r direction where $\hat{r}$ is directed along the line of particle centers. The relative polarizability of the particle is given as β [$= (\epsilon_p - \epsilon_c)/(\epsilon_p + 2\epsilon_c$]. This force law indicates that a pair of particles with their line of centers perpendicular with the electric field will experience a repulsion. If, however, the line of particle centers is parallel to the field, the pair feels an attraction. At all other angles the pair feels a torque that acts to align the pair with the field. This dipolar pair interaction is the origin of the fibrous or columnar structures observed in the flow visualization experiments.

When sheared, the particles are subjected to viscous forces that tend to degrade the structure. While critically dependent on the local structure, these viscous forces increase as $6\pi\eta_c a^2\dot{\gamma}$ where η_c is the solvent viscosity and $\dot{\gamma}$ is the macroscopic shear rate. At low shear rates or at large field strengths, the polarization forces that scale as $12\pi\epsilon_p\epsilon_c(a\beta E)^2$ dominate particle interactions, and the columnar structures are retained. At high shear rates or low field strengths, the viscous forces dominate over the polarization forces and degrade the structure. If the particles are large enough such that Brownian effects are negligible, the suspension viscosity should thus be a function of volume fraction and the dimensionless group, *Mn,* named after S. G. Mason for his early work on the interactions of particles in combined shear and electric fields (*48, 50*):

$$Mn = \eta_c \dot{\gamma} / 2\epsilon_p \epsilon_c \beta^2 E^2 \quad (3)$$

Extensive data taken on a variety of suspension indicate that, for a given suspension, all field strength, shear rate data can be reduced to a single curve when η/η_∞ is plotted against *Mn.* The data shown in Figures 7 and 8 show such a collapse. Here η/η_∞ is the suspension viscosity in the absence of the electric field.

As shown in Figure 7, the shear stress generated by an ER fluid becomes independent of shear rate as the shear rate is lowered and approaches that of the fluid with no field applied at elevated shear rates. In light of the above discussion, this behavior is the expected consequence of moving from polarization force control of suspension structure to viscous control as the shear rate is raised. One constitutive equation relating shear stress, τ, to the shear rate that has been successful in describing the continuous shear response of ER fluids is the Bingham equation

$$\tau = \tau_B + \eta_\infty \dot{\gamma} \quad (4)$$

where τ_B is the Bingham yield stress and η_∞ is the plastic viscosity. If the yield stress is assumed to depend on the polarization forces between particles and thus to scale as $(\beta E)^2$, this equation can be rewritten as (*51*)

$$\eta/\eta_\infty = \frac{Mn^*}{Mn} + 1 \quad (5)$$

where Mn^* is the Mason number that characterizes the switch from polarization to viscous control of suspension structure. The ability of this type of scaling to describe shear viscosities as a function of E and $\dot{\gamma}$ is clear from the fit to the data in Figure 8.

While these basic scaling concepts appear to hold for most ER suspensions, much needs to be understood before commercial applications of ER fluids will emerge. At elevated field strengths, the Bingham yield stress appears to scale linearly with E rather than E^2 as suggested by the polarization mechanism. Also the current levels required for many ER suspensions limit their applications. The origin of the current and how it can be minimized are problems of active research. Finally, particle physical properties that give rise to optimum ER responses are still debated. For example, from the polarization scaling given above, one expects the largest yield stress to occur when the particle dielectric constant is large in comparison with the continuous phase (i.e., $\epsilon_p/\epsilon_c >> 1$). However there are indication that an optimum particle conductivity may be required (*51*). The work of Gow and Zukoski (*52*) indicates that the ER response increases monotonically with ϵ_p/ϵ_c. However, in these experiments the particle conductivity also increased. Progress in understanding the links between electrical loss and stress transfer properties will come

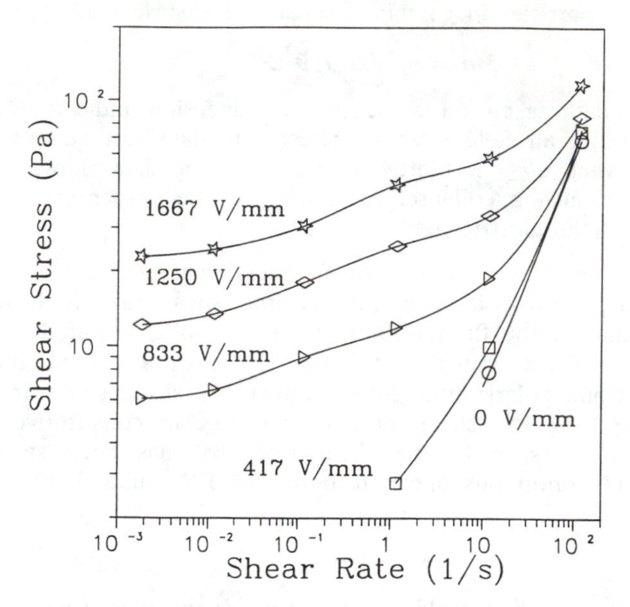

Figure 7. Shear stress vs. shear rate for a suspension of hollow silica spheres in corn oil at several electric field strengths. Volume fraction of spheres (ϕ) = 0.350.

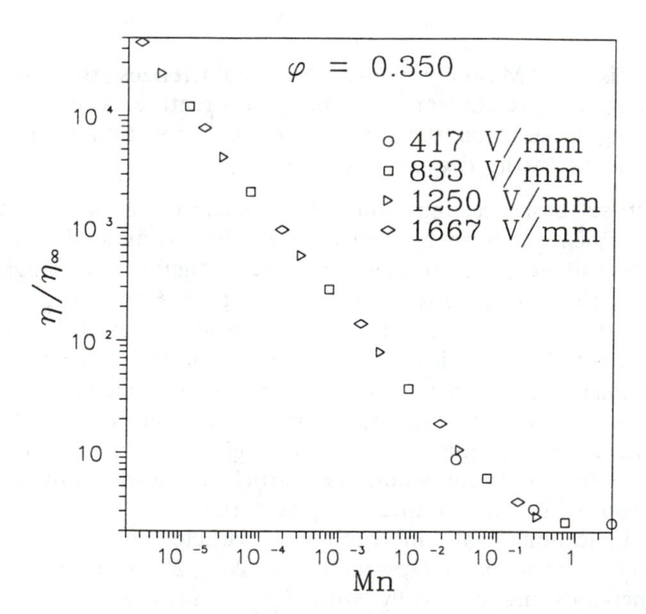

Figure 8. Relative suspension viscosity as a fraction of Mason numbers.

from careful rheological experiments and clever chemistry where particle dielectric properties can be varied in a systematic manner.

Despite our current level of understanding of the origin of ER response and how it can be optimized, interest in the phenomena is widespread. Active research programs into ER phenomena are underway at major automotive and chemical companies, as well as in many university laboratories. In addition, the use of ER fluids as acoustic dampers has drawn the attention of the defense community. While much research is oriented at developing engine mounts or clutches for automotive applications, which, if successful will result in tremendous markets for ER devices, there are many applications not requiring that the fluids operate under the extreme conditions required by the automotive industry. Robotic actuators, molding compounds, acoustic dampers, and electrically controlled valves are among some of the future applications of ER fluids. As a result, technological advancement as well as scientific interest in the ER response will continue.

Literature Cited

1. Elias, H. G. *Macromolecules,* Vol. 4; Plenum Press: New York, 1977.
2. Menjivar, J. A. In *Water-Soluble Polymers: Beauty with Performance;* Glass, J. E., Ed.; Advances in Chemistry Series 213; American Chemical Society: Washington, DC, 1986; Chapter 13.
3. DeWitt, T. W. *J. Colloid Interf. Sci.* **1955,** *10,* 174.
4. Kreiger, I. M. *Adv. Colloid Interf. Sci.* **1972,** *3,* 111.
5. Vrahopoulou, E. P., McHugh, A. J. *J. Non-Newtonian Fluid Mech.* **1987,** *25,* 157.
6. Burow, S., Peterlin, A., Turner, D. T. *Polym. Lett.* **1964,** *2,* 67.
7. Duvdevani, I., Agarwal, P. K., Lundberg, R. D. *Polym. Eng Sci.* **1982,** *22,* 499. Duvdevani, I.; Lundberg, R. D. U.S. Patent SIR H363 (November 3, 1987).
8. Witten, T. A., Jr.; Cohen, M. H. *Macromolecules* **1985,** *18,* 1915.
9. (a) Savins, J. G. *Rheol. Acta* **1967,** *7,* 87. (b) Schultz, R. K.; Myers, R. R. *Macromolecules* **1969,** *2,* 281.
10. (a) Bock, J.; Siano, D. B., Valaint, P. L.; Pace, S. J. In *Polymers in Aqueous Media: Performance through Association;* Glass, J. E., Ed.; Advances in Chemistry Series 223; American Chemical Society: Washington, DC, 1989; Chapter 22. (b) Jenkins, R. D.; Silebi, C. A.; El-Aasser, M. S. *Proc. Am. Chem. Soc. Div. Polym. Mater.: Sci. Eng.* **1989,** *61,* 629–633.

11. Maerker, J. M.; Sinton, S. W. *J. Rheol.* **1986,** *30*(1), 77.

12. Chauveteau, G. In reference 2, Chapter 14.

13. Durst, F.; Haas, R.; Kaczmar, B. U. *J. Polym. Sci.* **1981,** *26,* 3125.

14. Haas, R.; Kulicke, W. M. In *Proc. IUTAM Sympos.: Influence Polym. Addiv. Velocity Temp. Fields;* Gampert, B., Ed.; Springer: Berlin; 119.

15. Kulicke, W. M.; Haas, R. *Ind. Eng. Chem. Fundam.* **1984,** *23,* 316.

16. Metzner, A. B.; Whitlock, M. *Trans. Soc. Rheol.* **1958,** *2,* 239.

17. Morgan, R. J. *Trans. Soc. Rheol.* **1968,** *12,* 511.

18. Hoffman, R. L. *Trans. Soc. Rheol.* **1972,** *16,* 155.

19. Hoffman, R. L. *J. Colloid Interf. Sci.* **1974,** *46,* 491.

20. Collins, E. A., Hoffmann, D. J. *J. Colloid Interf. Sci.* **1979,** *71,* 21.

21. Tomita, M.; Van deVen, T. G. M. *J. Colloid Interf. Sci.* **1984,** *99,* 374.

22. Boersma, W. H.; Laven, J.; Stein, H. N. *AICHE J.* **1990,** *36*(3), 321.

23. Casale, A.; Porter, R. S. *Polymer Stress Reactions,* Vol. 2; Academic Press: New York; 548–558.

24. Benfaremo, N.; Liu, C. S. *Lubrication* **1990,** *76*(1), 1.

25. Kemnetz, S. J.; Still, A. L.; Cody, C. A.; Schwindt, R. *J. Coatings Technol.* **1989,** *61*(776), 47.

26. Heinle, S. A; Shah, S.; Glass, J. E. In reference 2, Chapter 11.

27. Glass, J. E. *J. Oil and Colour Chemists' Assoc.* **1975,** *58,* 169–177.

28. Glass, J. E. Society of Petroleum Engineers Publ. No. 7872, 1979.

29. Glass, J. E.; Ahmed, A.; Soules, D. A.; England-Jongewaard, S. K.; Fernando, R. H. Society of Petroleum Engineers Publ. No. 11691, 1983.

30. Peterlin, A. *Nature,* **1970,** *227,* 398.

31. Morgan, S. E.; McCormick, C. L. *Prog. Polym. Sci.* **1990,** *15,* 507.

32. Armstrong, R. C.; Gupta, S. K.; Basaran, O. *Polym. Eng. Sci.* **1980,** *20*(7), 466.

33. Holtmeyer, M.; Chatterji, J. *Polym. Eng. Sci.* **1980,** *20*(7), 473.

34. Mack, M. P. U.S. Patent 4,483,123 9; Feb. 21, 1984.

35. Burger, E. D.; Chorn, L. G.; Perkins, T. K. *J. Rheol.* **1980,** *24,* 603.

36. Peng, S. T. J.; Landell, R. F. In *Rheology,* Vol. 2: Fluids; Starita, G. A.; Marrucci, G.; Nicolais, L., Eds.; Plenum Press: New York, 1980.

37. Peng, S. T. J.; Landell, R. F. *J. Appl. Phys.* **1981,** *52,* 5988.

38. Ashmond, J. U.S. Patent 4,002,436; Jan. 1, 1977; assigned to ICI.

39. Duvdevani, I.; Eckert, J. A.; Schulz, D. N.; Kitano, K. U.S. Patent 4,523,939; June 10, 1985; assigned to Exxon.

40. Duvdevani, I.; Eckert, J. A.; Schulz, D. N.; Kitano, K. U.S. Patent 4,586,937; May 6, 1986; assigned to Exxon.

41. Bauer, D.; Briggs, L. M.; Dickie, R. A. *Ind. Engin. Chem. Prod. Res. Dev.* **1982,** *21*(4), 686.

42. Winslow, W. M. *J. Appl. Phys.* **1940,** *20,* 1137.

43. Gast, A. P.; Zukoski, C. F. *Adv. Colloid Interf. Sci.* **1989,** *30,* 153.

44. Block, H.; Kelly, J. P. *J. Appl. Phys.* **1988,** *21,* 1661.

45. Klingenberg, D. J.; Zukoski, C. F. *Langmuir* **1990,** *6,* 15.

46. Klingenberg, D. J.; Dierking, D.; Zukoski, C. F. *J. Chem. Soc. Faraday Transactions* in press.

47. Adriani, P. M.; Gast, A. P. *Phys. Fluids* **1988,** *31,* 2757 .

48. Allan, R. S.; Mason, S. G. *Proc. Roy. Soc. A* **1962,** *45,* 267.

49. Chaffey, C. E.; Mason, S. G. *J. Colloid Interf. Sci.* **1964,** *19,* 525; **1965,** *20,* 330; **1968,** *27,* 115.

50. Zis, I. Y. Z.; Cox, R. G.; Mason, S. G. *Proc. Roy. Soc. A* **1967,** *300,* 421.

51. Marshall, L.; Zukoski, C. F.; Goodwin, J. W. J. *J. Chem. Soc. Faraday Trans.* **1989,** *85,* 2785.

52. Gow, C. J.; Zukoski, C. F. *J. Colloid Interf. Sci.* **1990,** *136,* 175.

53. Chauveteau, G. *SPE Paper No 10060,* Society of Petroleum Engineers: San Antonio, TX, 1981.

RECEIVED February 1, 1991

Chapter 2

Rheological Measurements

Robert K. Prud'homme

Department of Chemical Engineering, Princeton University, Princeton, NJ 08544

The goal of this chapter is to provide a working knowledge of basic rheological terms, experimental results, and simple rheological models. The discussion is arranged in the following manner. In the first section the basic rheological flows and material functions (such as viscosity and dynamic moduli) are defined. In the second section the responses of general classes of polymeric fluids in these basic flow fields are presented. Emphasis is placed on the relationship between rheological response and material structure; that is "structure-property relationships" from a rheological point of view. In the final section simple rheological models are presented that can be used to represent the flow behavior of polymeric fluids.

Flows and Material Functions

Many industrial processes subject fluids to complex flow and temperature histories. To understand these complex flows, the response of the fluids in simple flow fields are studied to determine their "material functions" such as the viscosity, normal stress coefficients, or dynamic modul. In turn, these material functions are used to select appropriate mathematical models to describe the fluid rheology (called "constitutive equations" -- the subject of the final section) which can be used to predict the flow in complex geometries and flow fields. In addition, the material functions can be used to characterize materials, for example the weight-average molecular weight of a polymer can be determined from zero-shear rate viscosity measurements at high polymer concentrations. The definitions of several simple flows and associated material functions are given below. The measurement of material functions in these flows defines the practice of "rheometry", and several specialized references are available that cover this area (1-5).

0097–6156/91/0462–0018$08.50/0

Stress and Strain. The definition of stress requires the specification of the direction of the force and the orientation of the surface upon which the stress acts. Similarly, the definition of the rate of deformation or velocity gradient requires specification of the direction of the velocity and the direction in which the velocity varies. Figure 1 shows five flow fields extensively used in rheological measurements; steady shear, step shear rate, uniaxial extension, biaxial extension and oscillatory shear.

Steady-Shear Flows. Consider the flow shown in Fig. 1 where a fluid between two plates is sheared as the top plate moves with velocity U_x in the x-direction. The velocity gradient or shear rate is given by $\dot{\gamma}_{yx} = dv_x/dy = \dot{\gamma}$, and macroscopically is given by U_x/δ where δ is the plate separation. The stresses generated by the flow act parallel to the direction of shear (i.e., shear stresses) and perpendicular to the direction of shear (normal stresses). The experimentally observable stresses perpendicular to the direction of flow include the stress arising from fluid motion and the isotropic hydrostatic pressure. It is customary to eliminate the isotropic pressure by taking the difference between normal stresses, and it is in fact these differences that are experimentally measured:

τ_{yx} = *shear stress*, (1)
$\tau_{xx} - \tau_{yy} = N_1$ = *primary normal stress difference*, (2)
$\tau_{yy} - \tau_{zz} = N_2$ = *secondary normal stress difference*. (3)

These stresses are related to the velocity gradient, $\dot{\gamma}_{yx}$, thereby defining the material functions for steady shear flow:

$\tau_{yx} = -\eta\ \dot{\gamma}_{yx}$ *defines the viscosity* (4)

$\tau_{xx} - \tau_{yy} = -\Psi_1\ \dot{\gamma}_{yx}^2$ *defines the primary normal stress coefficient,* (5)

$\tau_{yy} - \tau_{zz} = -\Psi_2\ \dot{\gamma}_{yx}^2$ *defines the secondary normal stress coefficient* (6)

These material functions generally vary with shear rate. The normal stress coefficients are defined in terms of the square of the velocity gradient because the stress difference must be an even power of shear rate; that is, changing the direction of the shear (making $\dot{\gamma}_{yx}$ negative) does not change the direction or sign of the normal stress, whereas changing the direction of the velocity gradient does change the direction of the shear stress.

Uniaxial Extension/Compression. Consider the flow that either converges or diverges with respect to the z-axis as shown in Fig. 1. This flow is produced either by the stretching of a filament (extension) or biaxial stretching of a sheet as in the case of inflating a balloon (compression in the z-direction). The measurable stresses are the tensile normal stresses, and, again, to eliminate isotropic pressure terms, the difference in stresses are used to define the elongational viscosity material function:

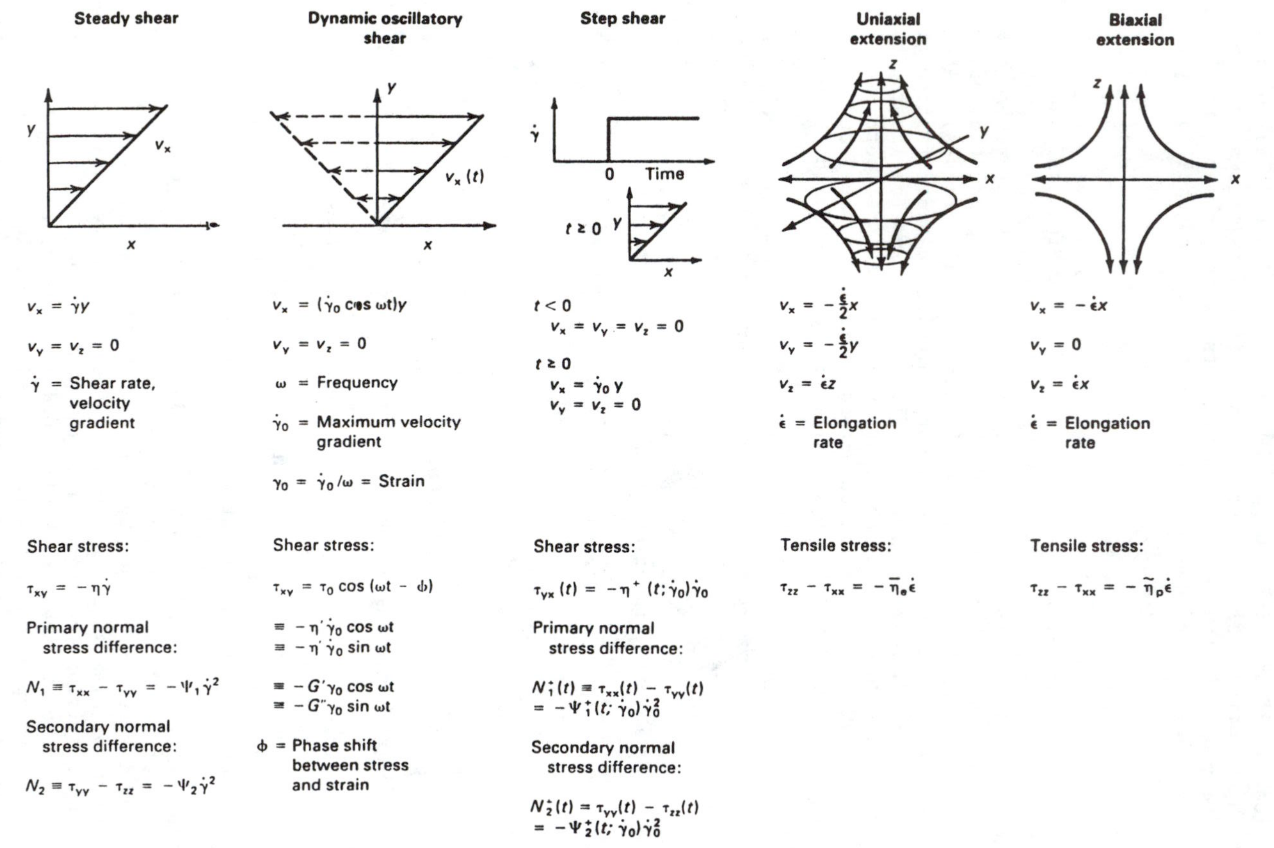

Figure 1. Common flow fields used to define material functions. (Reproduced with permission from *Electronic Materials Handbook,* Vol. 1, Packaging. Copyright 1989 ASM.)

$$\tau_{zz} - \tau_{xx} = \textit{tensile stress.} \quad (7)$$

Again these stresses are related to the velocity gradient, $\dot{\epsilon} = dv_z/dz$, thereby defining the elongational viscosity:

$$\tau_{zz} - \tau_{xx} = -\eta_e \dot{\epsilon} \quad \textit{defines the elongational viscosity} \quad (8)$$

Planar Extension. In planar extension the material is stretched in the z-direction, but it is constrained not to deform in the y-direction. This deformation arises when a sheet of material is stretched in one direction. The planar elongational viscosity is again defined by the tensile stress difference:

$$\tau_{zz} - \tau_{xx} = -\eta_p \dot{\epsilon} \quad \textit{defines the planar elongational function} \quad (9)$$

where again, $\dot{\epsilon} = dv_z/dz$, but $dv_y/dy = 0$ and $dv_x/dx = dv_z/dz$, unlike uniaxial extension where $dv_y/dy = -1/2(dv_z/dz)$, and $dv_x/dx = -1/2(dv_z/dz)$.

Dynamic Oscillatory Shear Flow. In oscillatory shear measurements or dynamic mechanical measurements a sinusoidally varying shear field is imposed on a fluid and the amplitude of the resulting shear stress and the phase angle between the imposed shear and the stress is measured. The test is said to be in the "linear viscoelastic" regime if the stress is linearly proportional to the imposed strain and the stress response is sinusoidal. A large body of literature exists on the relationship between dynamic oscillatory material functions and polymer molecular structure (6). Consider the oscillating velocity field shown in Fig. 1.

$$v_x = [\dot{\gamma}_{max} \cos(\omega t)]y \quad (10)$$

where ω is the frequency, $\dot{\gamma}_{max}$ is the maximum velocity gradient, and the maximum value of the strain is given by $\gamma_{max} = \dot{\gamma}_{max}/\omega$. The stress will also oscillate and will have some maximum value, τ_{max}, and some phase shift, ϕ, from the imposed shear.

$$\tau_{yx} = \tau_{max} \cos(\omega t - \phi). \quad (11)$$

The stress can be decomposed into two terms, one in-phase with the velocity and one 90 degrees out-of-phase. These can be written in terms of the maximum velocity gradient.

$$\tau_{yx} = -\eta' \dot{\gamma}_{max} \cos(\omega t) - \eta'' \dot{\gamma}_{max} \sin(\omega t) \quad (12)$$

where

$$\tau_{max} \cos\phi \equiv \eta' \dot{\gamma}_{max}, \text{ and } \tau_{max} \sin\phi \equiv \eta'' \dot{\gamma}_{max}. \quad (13)$$

This defines the two dynamic viscosity coefficients η' and η''. At low frequencies η' approaches the zero-shear-rate viscosity measured in steady shear. Alternately, coefficients can be defined in terms of the maximum strain instead of the strain rate.

$$\tau_{yx} = -G'\gamma_{max} \cos(\omega t) - G''\gamma_{max} \sin(\omega t) \tag{14}$$

where

$$\tau_{max} \cos\phi \equiv G'\gamma_{max}, \; \tau_{max} \sin\phi \equiv G''\gamma_{max}. \tag{15}$$

This defines the two functions G′ and G″ which are the storage and loss moduli, respectively. G′, proportional to the stress in-phase with strain, provides information about the elasticity of a material. For example, an ideal elastic rubber band would have all of its stress in-phase with strain or displacement. G″, the loss modulus, is proportional to stress out-of-phase with displacement and, therefore, in-phase with rate-of-displacement or shear-rate. For a purely viscous liquid, all of the stress would be out-of-phase with displacement. It should be kept in mind that linear viscoelasticity assumes that the stress is linearly proportional to strain and that the stress response involves only the first harmonic and not higher harmonics in frequency (i.e., the stress is a sinusoidal). Experimentally,both of these conditions should be verified.

These linear viscoelastic dynamic moduli are functions of frequency. They have proven to be sensitive probes of the structure of polymer solutions and gels. Figure 2 shows the dynamic moduli for a polymer solution during gelation (7). The material begins as a solution in Fig. 2a and ends as a solid gel in Fig. 2d. For a polymer solution at low frequency, elastic stresses relax and viscous stresses dominate with the result that the loss modulus, G″, is higher than the storage modulus, G′. Both decrease with decreasing frequency, but G′ decreases more quickly. For a gel the stress cannot relax and, therefore, is independent of frequency. Also, because the gel is highly elastic the storage modulus, G′ is higher than the loss modulus, G″.

Linear viscoelastic measurements can also be used in conjunction with classical polymer kinetic theory to relate the storage modulus of a gel to the number density of crosslinks. By following the storage modulus with time, the chemical kinetics of gel formation can be measured (8,9). Polymer kinetic theory (6) shows that the frequency independent, low frequency limit of the storage modulus for a gel is given by

$$G' = G^{\circ} = \nu kT + G_{en} \tag{16}$$

where G° is the equilibrium shear modulus, ν is the number density of network strands, k is Boltzmann's constant, T is the absolute temperature, and G_{en} is a contribution arising from entanglements that are not covalent crosslinks.

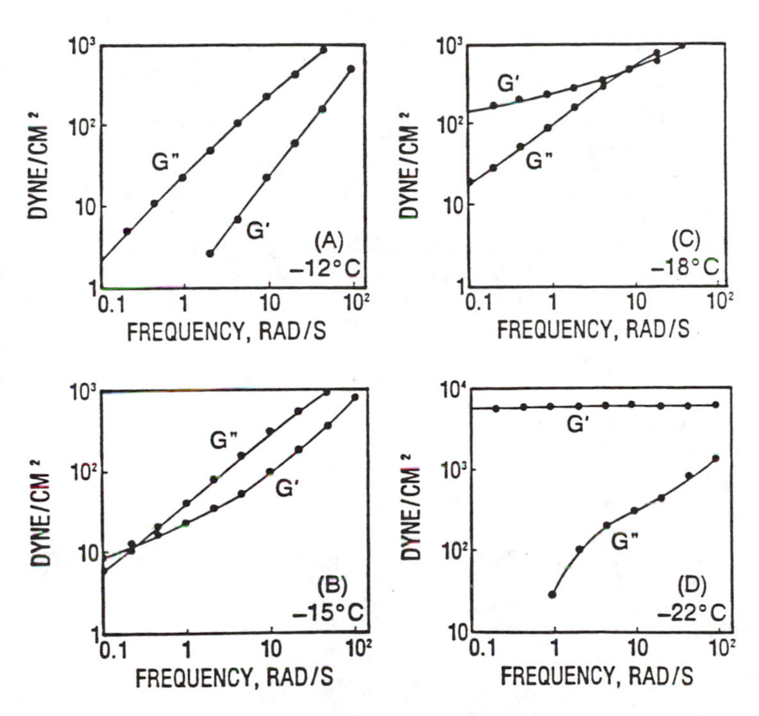

Figure 2. Dynamic moduli versus frequency during the process of gelation. The material is polystyrene in carbon disulfide that gels upon cooling. (Reproduced from reference 7. Copyright 1983 American Chemical Society.)

Step Shear Rate

In this experiment, the shear rate is changed instantaneously from one value to another and the shear stress is monitored with time. Most commonly, the initial shear rate is zero and the resulting material function is the "stress growth function":

$$\tau_{xy}(t) = \bar{\eta}(\dot{\gamma}_{xyi}, t)\dot{\gamma}_{xy} \quad (17)$$

which defines the growth from an initial shear rate $\dot{\gamma}_{xyi}$ to the final shear rate $\dot{\gamma}_{xy}$. There are corresponding functions for the transient normal stresses:

$$\tau_{xx} - \tau_{yy} = N_1(t) \equiv \bar{\Psi}_1(\dot{\gamma}_{xyi}, t)\, \dot{\gamma}^2_{xy} \quad (18)$$

$$\tau_{yy} - \tau_{zz} = N_2(t) \equiv \bar{\Psi}_2(\dot{\gamma}_{xyi}, t)\, \dot{\gamma}^2_{xy} \quad (19)$$

Step Strain. In step strain experiments, an instantaneous strain is applied to the material and the decay in the stress is monitored with time. This defines the shear modulus, G(t), for an applied shear strain of magnitude γ_0 :

$$\tau_{xy}(t) = G(t)\,\gamma_0 \qquad \textit{defines shear modulus} \quad (20)$$

and it defines the Young's modulus, E(t), if an elongational strain of magnitude ϵ_0 is applied.

$$\tau_{xx}(t) - \tau_{yy}(t) = E(t)\,\epsilon_0 \qquad \textit{defines Young's modulus} \quad (21)$$

Constant Stress Creep Experiments. The creep test is the inverse of the step shear rate experiment; a constant stress τ^0_{xy} is applied to the material, and the strain is monitored with time. This defines the compliance J:

$$\gamma_{xy}(t) = J(t)\,\tau^0_{xy} \qquad \textit{defines the compliance } J(t) \quad (22)$$

Thixotropic Loop. A deformation history that provides qualitative information about time-dependent fluid rheology is the thixotropic loop where the shear rate is continuously ramped from zero to a higher value over a prescribed time period. The resulting shear stress is measured. This test is sensitive to the kinetics of structure evolution which can be important in aggregated colloidal dispersions. If the structures in the material are broken apart by shear and cannot reform during the time of the shear rate ramp, then the stresses during the decreasing shear rate ramp will be lower than the stresses during the increasing leg.

The region between the stress and shear rate curves in the increasing and decreasing ramps is known as the "thixotropic loop". It is difficult to quantify the results observed in a thixotropic loop experiment because the response is a complex convolution of the kinetics of shear induced structure breakdown and the kinetics of aggregation. But the thixotropic loop often provides a useful "finger print" of the material.

Experimental Geometries and Simple Flows. Although the material functions are defined for the flows specified in the previous section, it is often most convenient to measure the material functions using alternate geometries or experiments that approximate the ideal flow geometry. Table I gives several examples of geometries from which material functions can be determined for low viscosity fluids (10).

Examples of Material Behavior

Since rheology is frequently used as a probe of molecular structure and interactions, it is helpful to have a general idea of what the material functions look like for commonly encountered fluids.

Steady Shear Viscosity and Normal Stresses. Low molecular weight fluids and resins generally are Newtonian fluids, which means that they have a constant viscosity independent of shear rate. They also display no elastic normal stresses. For polymer melts, filled polymers, polymer solutions and dispersions, the viscosity is not constant, but decreases at higher shear rates as shown in Fig. 3. Whereas many systems display shear thinning viscosities, only long chain polymeric molecules exhibit high values of elastic normal stresses. These elastic stresses are responsible for phenomena such as die swell, where a polymer extruded through an orifice swells to a diameter greater than the orifice diameter. Solid fillers reduce the level of normal stresses as shown in Fig. 4, which shows the viscosity and normal stress of a polypropylene melt filled with 50% wt calcium carbonate filler (11,12,13). Two points should be noted, the viscosity increases with filler concentration, and the primary normal stress difference decreases. For dispersions of solids in non-polymeric media, the viscosity may also show shear thinning (14), exactly like the polymeric analog; but there will be virtually no normal stresses. For both Newtonian and polymeric continuous phases, viscosity increases with increasing volume fraction up to a critical volume fraction above which the viscosity diverges to infinity and the material will not flow. This critical volume fraction is about 63% for non-interacting, monodisperse, spheres and decreases markedly when the particles have long aspect ratios (e.g., chopped glass fibers) or when the particles are strongly interacting as is the case when particle sizes are below 1 μm and surface forces become dominant. It can be increased to about 80% for a broad distribution of sphere sizes. The effect of particle aspect ratio is seen in Fig. 5, which is the viscosity versus shear rate for polyamide 6 melts filled with glass fibers of different aspect ratios, all at 30% by weight loading (15). The higher aspect ratio fillers produce higher viscosities. The effect of particle surface interactions is seen for the polystyrene calcium carbonate system in Fig. 6, where decreasing particle size from 17 μm to 0.07 μm at the same volume fraction of particles (ϕ = 30% by volume) results in a ten-fold increase in viscosity (16). The effect of surface

Table I. Experimental Geometries for Measuring Fluid Rheology

Experimental Geometry	Measured Quantities
Flow in a tube (capillary viscometer) 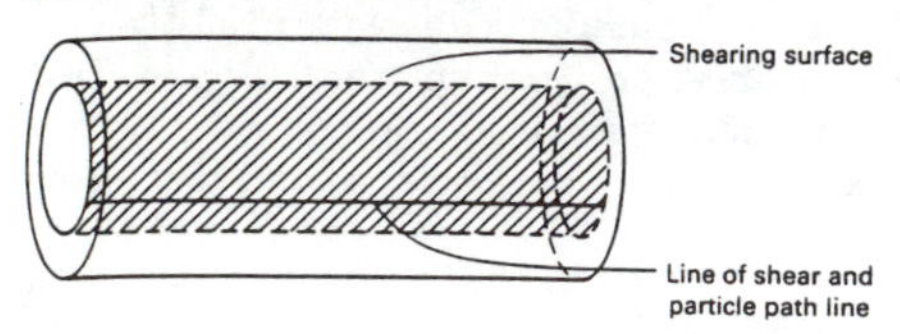	Q = Volume rate of flow ΔP = Pressure drop through tube R = Tube radius L = Tube length $\dot{\gamma}_R$ = Shear rate at tube wall τ_R = Shear stress at tube wall
Torsional flow between a cone and disk 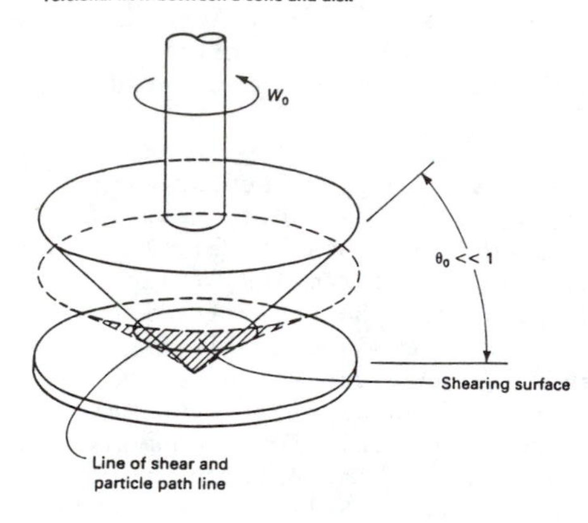	R = Radius of circular plate θ_0 = Angle between cone and plate (usually less than 100 mm) W_0 = Angular velocity of cone T = Torque on plate F = Force required to keep tip of cone in contact with circular plate
Torsional flow between parallel plates 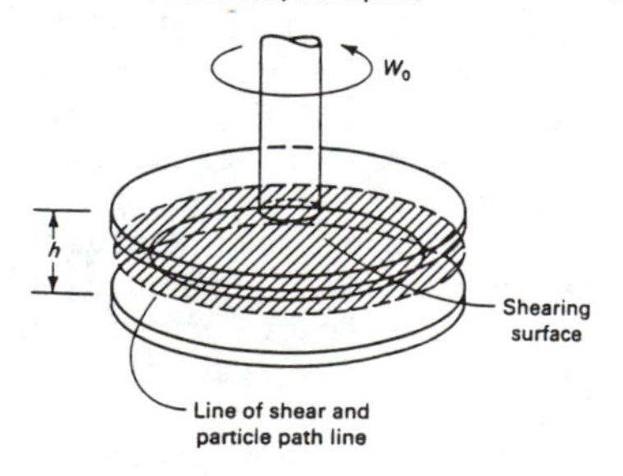	R = Radius of disks H = Separation of disks W_0 = Angular velocity of upper disk T = Torque required to rotate upper disk F = Force required to keep separation of two disks constant
Torsional flow between concentric cylinders (Couette geometry) 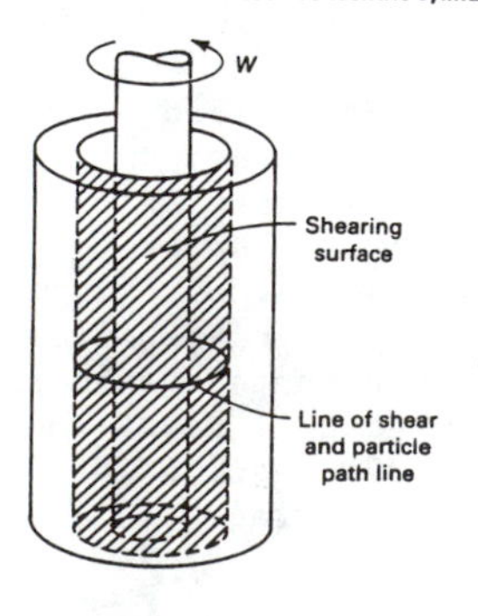	R_1, R_2 = Radii of inner and outer cylinders H = Height of cylinders W_1, W_2 = Angular velocities of inner and outer cylinders T = Torque on inner cylinder

Table I. Continued

Material Function Determination

Flow in a tube (capillary viscometer)

$$\eta(\dot{\gamma}_R) = \frac{\tau_R}{(Q/\pi R^3)}\left[3 + \frac{d \ln (Q/\pi R^3)}{d \ln \tau_R}\right]$$

$$\dot{\gamma}_R = \frac{1}{\tau_R^2}\frac{d}{d\tau_R}(\tau_R^3 Q/\pi R^3)$$

$$\tau_R = \Delta PR/2L$$

Torsional flow between parallel plates

$$\eta(\dot{\gamma}_R) = \frac{(T/2\pi R^3)}{\dot{\gamma}_R^2}\left[3 + \frac{d \ln (T/2\pi R^3)}{d \ln \dot{\gamma}_R}\right]$$

$$\Psi_1(\dot{\gamma}_R) - \Psi_2(\dot{\gamma}_R) = \frac{(F/\pi R^2)}{\dot{\gamma}_R^2}\left[2 + \frac{d \ln (F/\pi R^2)}{d \ln \dot{\gamma}_R}\right]$$

$$\dot{\gamma}_R = \frac{W_0 R}{H}$$

Torsional flow between a cone and disk

$$\eta(\dot{\gamma}) = \frac{3T}{2\pi R^3 \dot{\gamma}}$$

$$\Psi_1(\dot{\gamma}) = \frac{2F}{\pi R^2 \dot{\gamma}^2}$$

$$\dot{\gamma} = W_0/\theta_0$$

Torsional flow between concentric cylinders

$$\eta(\dot{\gamma}) = \frac{T(R_2 - R_1)}{2\pi R_1^3 H |W_2 - W_1|}$$

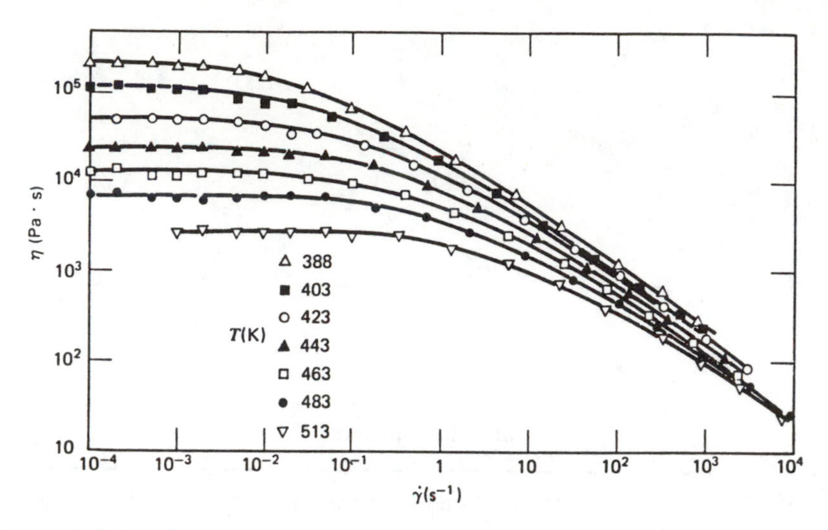

Figure 3. Viscosity versus shear rate for a low-density polyethylene melt at several temperatures. Data at shear rates below 5×10^{-2} s^{-1} were taken on a rotational viscometer, and viscosities at higher shear rates were taken on a capillary viscometer. (Reproduced with permission from reference 23. Copyright 1971 Hanser.)

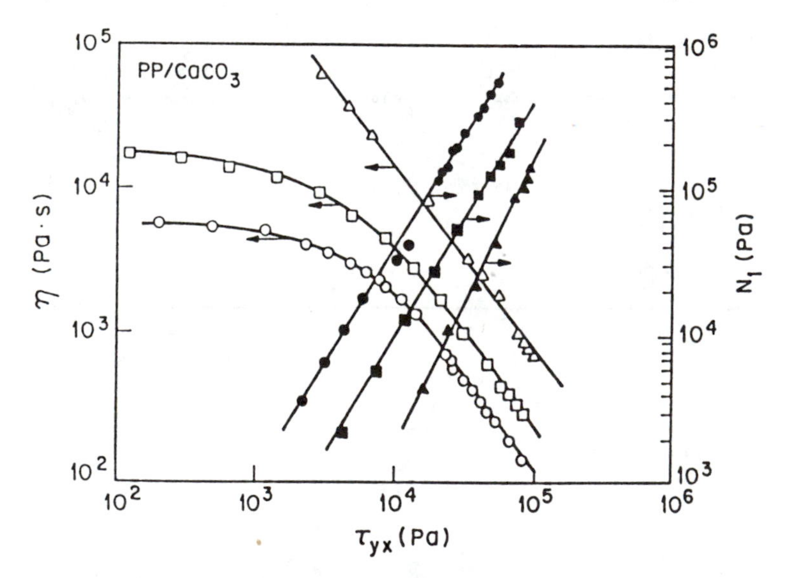

Figure 4. Viscosity, η (open symbols), and first normal stress difference, N_1 (closed circles), as a function of shear stress for polypropylene melts filled with $CaCO_3$ (50% wt) with and without a titanate coupling agent: (□, ■) pure propylene, (○, ●) with titanate treatment, and (Δ, ▲) without titanate treatment. (Reproduced with permission from reference 11. Copyright 1981 Society of Plastics Engineers.)

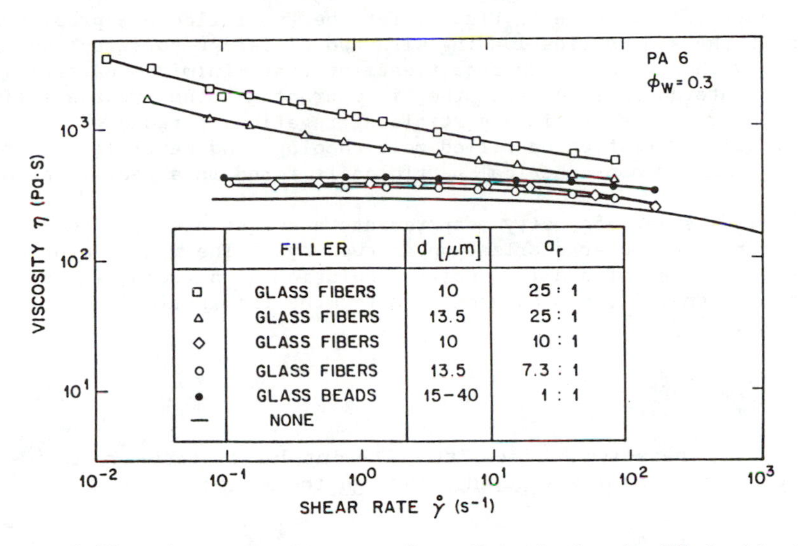

Figure 5. Viscosity versus shear rate for polyamide 6 melts filled with glass fibers of different aspect ratios (a_r) and diameters (d) at a constant mass fraction of 30% fibers. (Reproduced with permission from reference 15. Copyright 1984 Steinkopff.)

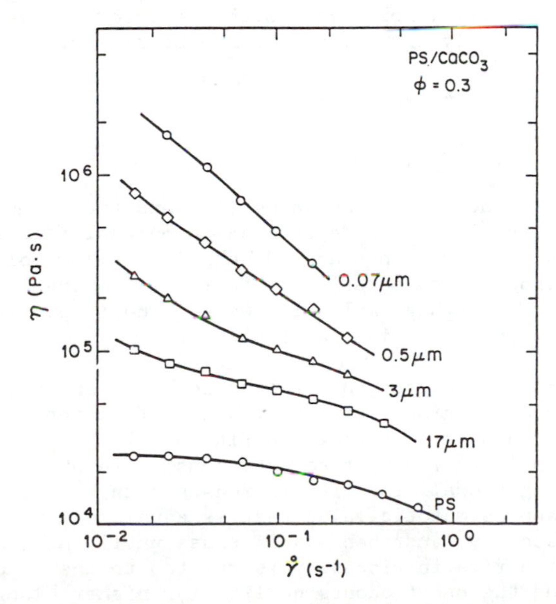

Figure 6. Viscosity versus shear rate for polystyrene melts with $CaCO_3$ fillers of various particle sizes shown on the figure. The filler loading is constant at 30% vol. (Reproduced with permission from reference 16. Copyright 1983 Wiley.)

treatments is also seen in Fig. 4, for the two filled polypropylene melts at the same solids loading with two different surfaces: one is treated with a titanate surface treatment that minimizes particle surface interactions and the other is untreated. The treated filler has a lower viscosity since particle aggregation is reduced. Additional information on filled melt rheology and particle orientation for non-spherical particles is found in a recent review (17).

Steady shear viscosity measurements are also used for polymer molecular weight characterization in two ways. The measurement of the viscosity η, of a dilute polymer solution at a succession of concentrations, C_p, can be used to determine the intrinsic viscosity, $[\eta]$:

$$[\eta] \equiv \lim_{C_p \to 0} \frac{\eta - \eta_s}{C_p \eta_s} \tag{23}$$

where η_s is the solvent viscosity. The intrinsic viscosity is related to molecular weight, M_w, through the Mark-Houwink expression:

$$[\eta] = K\ M_w^a \tag{24}$$

where K and a are constants tabulated for each polymer in standard references (18).

Also, the zero shear viscosity, η_0, of a polymer solution or melt can be used to determine its molecular weight. Figure 7 (6) shows similar viscosity molecular weight behavior for several amorphous, linear polymers. The data can be represented by

$$\eta_0 = K_1 (M_w) \qquad \textit{for } M_w < M_c \tag{25}$$

$$\eta_0 = K_2 (M_w)^{3.4} \qquad \textit{for } M_w < M_c \tag{26}$$

The constants K_1 and K_2 are tabulated (18) and the transition from the first order to 3.4 order dependence on molecular weight comes from the transition from unentangled behavior at low molecular weight [corresponding to material II in the following section] to entangled behavior at high molecular weight [corresponding to material III in the following section].

Uniaxial Extension/Compression. Polymeric melts and viscoelastic fluids display elongational viscosities as a function of elongational strain rate as shown in Fig. 8. The data are shown for a polystyrene melt at 170°C at several constant elongation rates (24). The elongational viscosity increases with time initially, may reach a constant asymptotic value that is equal to three times the zero shear viscosity, and then may increase until the fiber breaks. The onset of the rise in viscosity is related to the rate of elongation, and the onset occurs earlier for higher elongation rates. This final "strain hardening" is one of the mechanisms stabilizing the stretching of polymer fibers and is therefore very desirable. However, it makes reliable "steady elongational viscosity" data very difficult to obtain. Most data on fiber

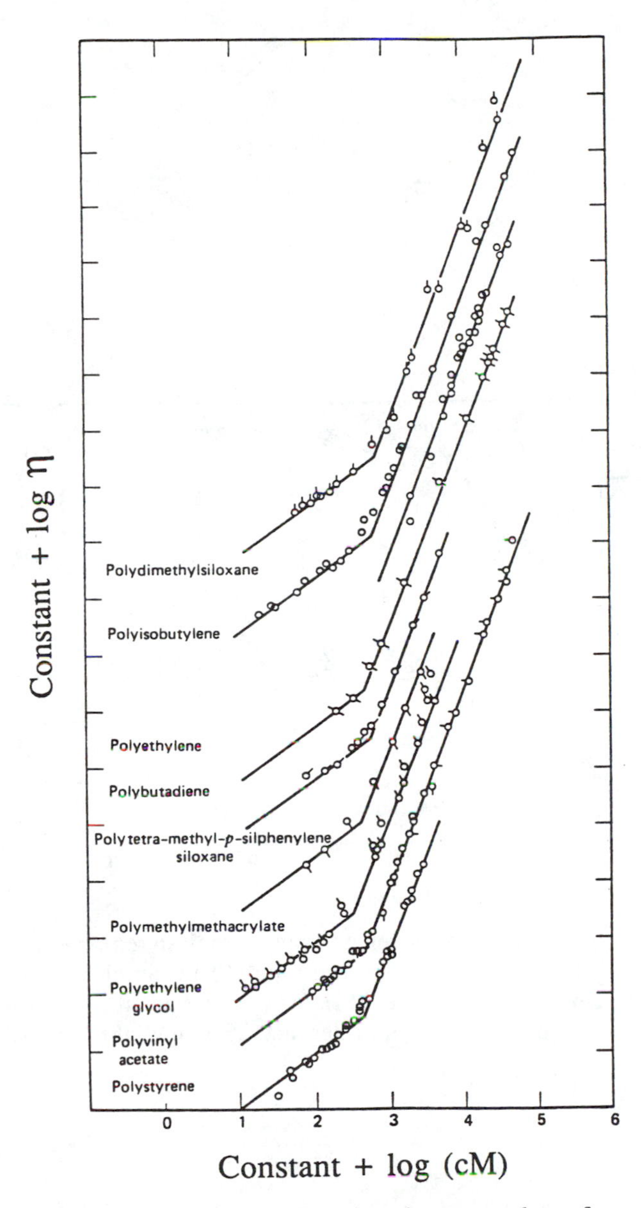

Figure 7. Log of the zero-shear-rate viscosity versus log of concentration times molecular weight. The data are shifted along the vertical and horizontal axes; the two shift factors are tabulated for many polymer systems. The molecular weight at the break point between the region of slope 1 and the region of slope 3.4 at higher molecular weight defines the critical molecular weight for entanglement, Mc. (Reproduced with permission from reference 24. Copyright 1968 Springer.)

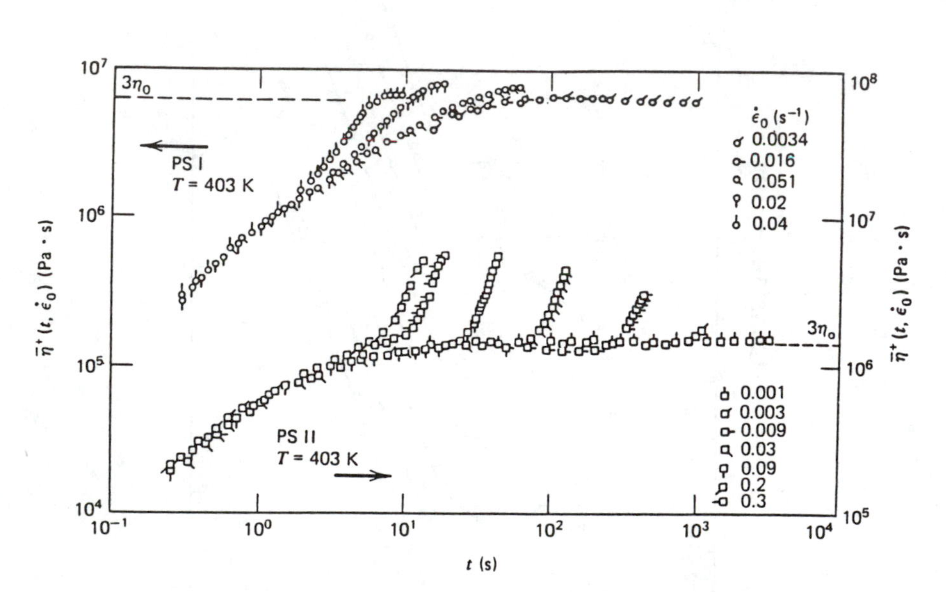

Figure 8. Elongational viscosity versus time for polystyrene melts at various constant elongation rates shown on the figure. The samples were PSI (Mw = 7.4×10^4, Mw/Mn = 1.2) and PSII (Mw = 3.9×10^4, Mw/Mn = 1.1). (Reproduced with permission from reference 25. Copyright 1980 Wiley.)

elongation is not at steady state conditions and is, therefore, only qualitative in nature.

Dynamic Oscillatory Shear. Dynamic oscillatory measurements are sensitive probes of molecular structure and interactions in melts and solutions. The frequency dependence of the storage and loss moduli, G′ and G" respectively, are shown in Figs. 9 and 10 (adapted from Ferry (6)) for several types of polymeric materials. Additional details on the materials can be found in the original citation. The materials are designated on the figure by the numerals:

I. Dilute Polymer Solution: Data for a 860,000 Mw narrow molecular weight distribution polystyrene in a chlorinated biphenyl solvent at 25°C (the data are taken at several temperatures and then normalized to 25°C using time/temperature superposition as described in ref. 6).

II. Low Molecular Weight Amorphous Polymer: Data for a 10,500 MW poly(vinyl acetate) at 75°C.

III. High Molecular Weight Amorphous Polymer: Data for a 600,000 narrow molecular weight distribution polystyrene at 100°C.

IV. Lightly Crosslinked Amorphous Polymer: Data for a lightly vulcanized Hevea rubber with an equilibrium modulus of $E = 7x10^5$ Pa at 25°C.

Figures 9 and 10 show the response of the dilute solution and the low molecular weight polymer, I and II respectively, where both moduli decrease with decreasing frequency. The stresses relax with increasing time or decreasing frequency, with the elastic stresses, G′, relaxing more quickly than the viscous stresses, G". In the limit of low frequencies G" will attain a constant slope of -1 on a log-log plot and in this region: $G''/\omega = \eta_0$, i.e., the zero shear viscosity is given by the loss modulus divided by frequency.

For the high molecular weight polymer, III, there is a high frequency plateau in the glassy regime. In this region the frequency is faster than the relaxation processes of the polymer chains and the material behaves as an amorphous solid. Another plateau appears at a lower frequency corresponding to a "pseudo network" composed of temporary chain entanglements that do not have sufficient time to relax during the time of an oscillation at frequency ω. In this region the material behaves like a crosslinked rubber. Finally, at low frequencies the moduli drop -- stresses can relax during the time scale of the imposed oscillation. The zero shear viscosity is given by $\eta_0 = G''/\omega$ in this region.

For the crosslinked rubber, IV, at high frequencies the moduli approach the same glassy plateau as the uncrosslinked polymer. At these high frequencies the polymer does not "feel" the constraints of the chemical crosslinks since the spacing between crosslinks is longer than the lengths of the polymer chain that can relax over the time-scale of an oscillation. The"rubbery" plateau now extends over

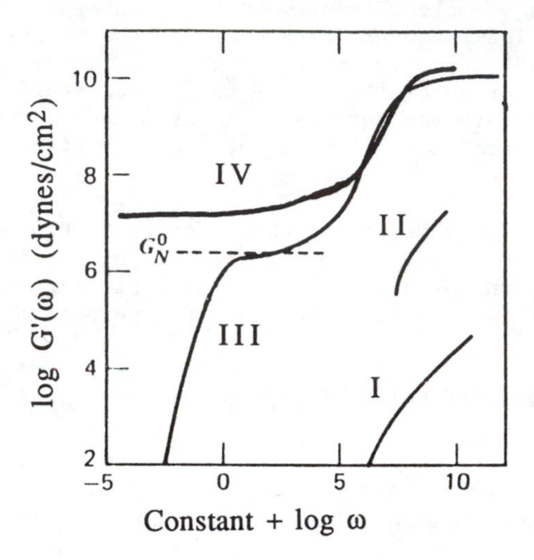

Figure 9. Storage moduli versus frequency for several classes of materials. Materials are identified in the text.

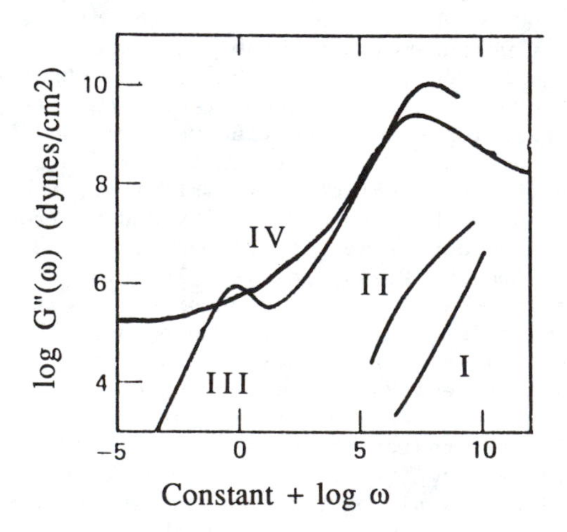

Figure 10. Loss moduli versus frequency for several classes of materials. Materials are identified in the text.

the entire low frequency range since elastic stresses cannot relax no matter how slowly the deformation is applied. The loss moduli, G", decreases with decreasing frequency and is two orders of magnitude below G'.

Step Shear Rate. If a steady shear rate is imposed on a sample initially at rest, the stress will rise with time and ultimately reach a steady shear viscosity characteristic of that shear rate. However, at high shear rates the viscosity will overshoot the equilibrium viscosity as is shown in Fig. 11 for a 2% wt polyisobutylene solution in Primol (20). The maximum in the stress value occurs at increasingly shorter times as the shear rate increases. It is found experimentally that the maximum occurs at a constant strain value (i.e. shear rates times time). The demarcation between shear rates that are slow enough so as to produce monotonic behavior and shear rates that produce overshoot is given by the dimensionless Weissenberg Number (10):

$$Wi = \lambda \cdot \dot{\gamma}_{xy} \qquad (27)$$

where λ is the characteristic fluid relaxation time and γ_{xy} is the shear rate. For Weissenberg numbers greater than one, flow time is faster than polymer relaxation time and non-linear effects such as stress overshoot are observed. Similarly, the onset of shear thinning in steady shear viscosity is governed by the Weissenberg number -- for Wi greater than one, shear thinning is observed. The concept of characteristic relaxation times and flow time scales arises repeatedly in rheology.

Another interesting observation comes from plotting the viscosity versus time (and not normalizing it by the equilibrium viscosity) as shown in Fig. 12 for a low density polyethylene melt (21). At higher shear rates the viscosity overshoots slightly and decreases, and at lower shear rates the location of the maximum, shifts to longer time and higher values. The envelope of the viscosity versus time curve corresponds to the curve given by shear rates low enough to be in the zero-shear rate viscosity region, i.e., the domain of linear viscoelasticity.

Step Strain or Stress Relaxation. The stress relaxation modulus in the linear viscoelastic regime for the polymers described are shown in Fig. 13. For the polymer solution (I) and low molecular weight melt (II), the modulus is low and relaxes quickly. For the high molecular weight polymer (III), the glassy plateau is observed at very short times, which corresponds to the plateau at high frequencies observed in the dynamic oscillatory measurements. The rubbery plateau arising from entanglements is seen at intermediate times, and at long times the stresses relax completely. For the crosslinked rubber (IV), the rubbery plateau extends to infinite time, since the stresses cannot relax in the permanently crosslinked system.

Constant Stress Creep. When a constant stress is applied to a polymeric material, the strain versus time behavior shown in Fig. 14

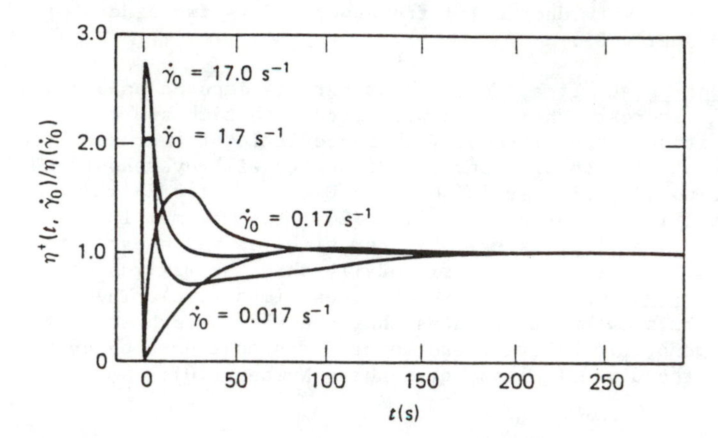

Figure 11. Stress shear growth function normalized by the equilibrium viscosity versus time. The data are for a 2.0% wt polyisobutylene solution in Primol. Increasing shear rates, $\dot{\gamma}_0$, applied at t = 0 lead to larger stress overshoot and the overshoot occurs at shorter times.

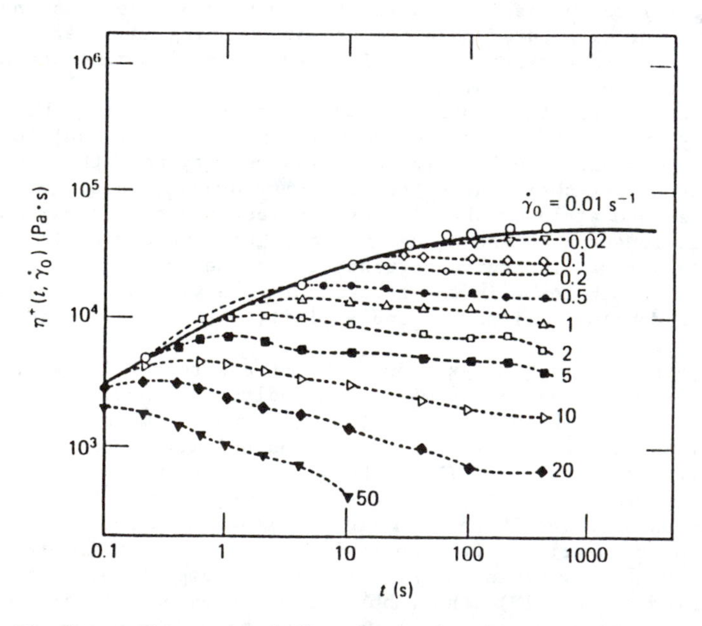

Figure 12. Stress shear growth function versus time at various imposed shear rates as shown on the figure.

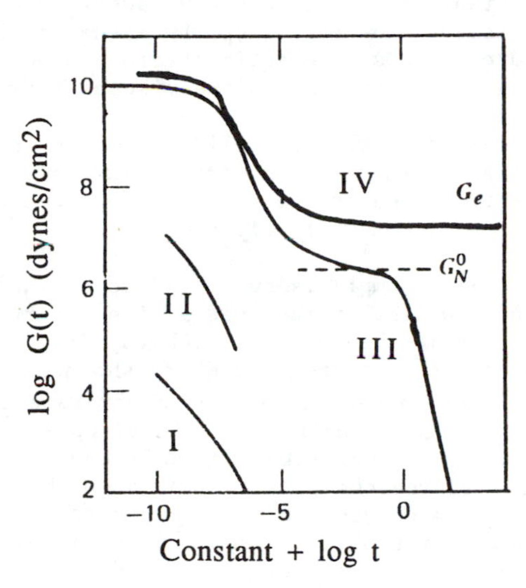

Figure 13. Stress relaxation modulus versus time for several classes of materials. Materials are identified in the text.

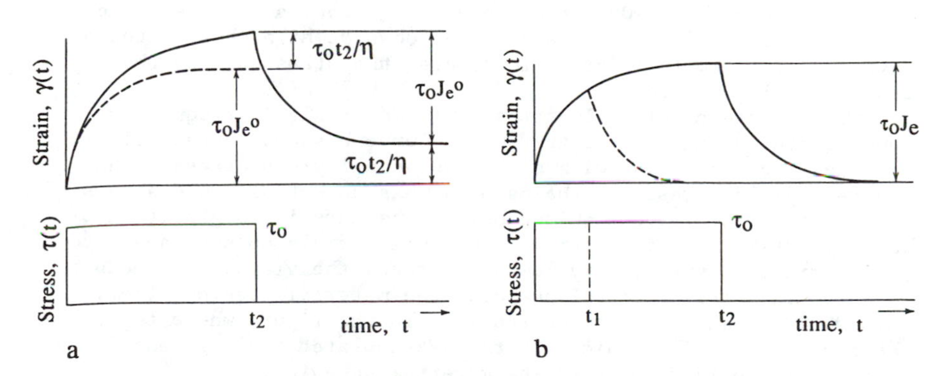

Figure 14. Shear creep and creep recovery for a stress τ_0 applied at time t = 0 and removed at t = t_2. The material in (a) is a viscoelastic fluid, and the material in (b) is a viscoelastic solid.

is observed. Initially, the sample will elastically deform and viscoelastically flow. From the ultimate slope of the strain versus time curve (i.e., shear rate) and imposed stress, the viscosity can be calculated. The elastically stored energy can be recovered when the stress is removed at time t and the material will recoil backwards. The recoverable strain is given by subtracting the strain associated from viscous flow ($\tau_{xy}\ t/\eta_0$) from the total strain, as is shown in the figure. The same elastic strain can be calculated from either the initial deformation minus the viscous flow or the final recovery.

If the material is an elastic solid (as shown in Fig. 14b), then the final strain will reach an equilibrium value corresponding to the compliance times the stress, $J\tau_{xy}$. Upon removal of the stress, the strain will be completely recovered.

Thixotropic Loop. If a sample is subjected to an increasing shear rate from zero to some final value over a time interval from t=0 to t_1 and the shear rate is then brought back to zero, several possible stress responses can be observed. These are shown in Fig. 15. If the fluid is Newtonian, the stress versus shear rate line will be straight with a slope that equals the fluid viscosity (Curve A). If the fluid is non-Newtonian but without significant memory effects over the time scale of the shear rate ramp, then the stress versus shear rate data will be curved but the increasing and decreasing ramps will coincide (Curve B). The shear rate divided by the stress at any point defines the viscosity.

If the sample has kinetic processes occurring that have time constants similar to the time scales of the ramps, the results will then be those shown as curve C. The stress will be higher on the increasing portion of the ramp and will be lower on the decreasing portion of the ramp. On the decreasing portion, structures, such as aggregated particles, have been broken down at high shear rates and have not had time to reform. Therefore, the stress during the decreasing leg is reduced.

The existence of yield stresses in materials is frequently investigated using shear rate ramps. Curve D shows characteristic behavior for a material with either an actual yield stress or an apparent yield stress. If the material has an actual yield stress, then a stress of τ_0 is required before the material begins to flow; this stress is the yield stress. Often materials do not have true yield stresses, but at very low shear rates the viscosities become so large that for practical purposes the material does not flow. This would be shown by the dotted line in the figure where the material is actually fluid, but the extrapolated yield stress gives a useful approximation as to the material behavior.

Models of Rheological Behavior

Generalized Newtonian Fluid. In this section we try to briefly review some simple but useful rheological models. Rheological models, especially for viscoelastic fluids, is a complete field by itself (5, 10) and so we skim only the surface. The simplest rheological equation of state is the Newtonian fluid where only one material function -- the viscosity -- is needed to characterize the

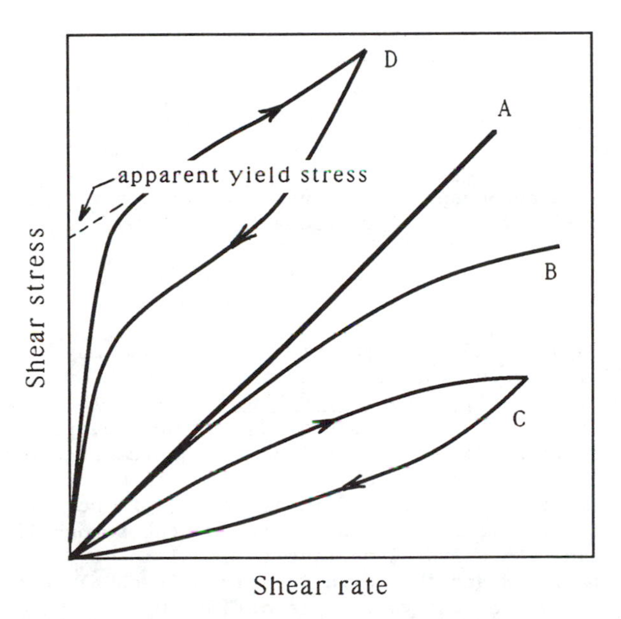

Figure 15. Thixotropic loop experiment in which the shear rate is increased from zero to some final value and then decreased to zero. Directions of arrows indicate the increasing and decreasing shear rate portions of the loop. The responses are for (A) Newtonian fluid, (B) pseudo-plastic shear thinning fluid, (C) thixotropic fluid, and (D) thixotropic material with an apparent yield stress.

fluid in any flow. The constitutive equation for a Newtonian fluid is,

$$\underset{\approx}{\tau} = -\eta \underset{\approx}{\dot{\gamma}} \tag{28}$$

where $\underset{\approx}{\dot{\gamma}}$ is the rate of deformation tensor defined by:

$$\underset{\approx}{\dot{\gamma}} = \nabla \underline{v} + (\nabla \underline{v})^{\dagger} \tag{29}$$

where the superscript ($\dagger$) denotes the transpose of the tensor. In component form this tensor becomes,

$$\dot{\gamma}_{ij} = \frac{\partial v_j}{\partial x_i} + \frac{\partial v_i}{\partial x_j} \tag{30}$$

where x_i are the components of the position vector $\underset{\sim}{x}$. For the shear flow given in Fig. 1, the shear stress is given by,

$$\tau_{xy} = -\eta \dot{\gamma}_{yx} \quad . \tag{31}$$

For non-Newtonian fluids the Newtonian fluid model may be generalized by allowing the viscosity to be a function of shear rate leading to the Generalized Newtonian Fluid model. The viscosity is made a function of the square root of the second invariant of the rate of deformation tensor. [Continuum mechanics arguments show that the viscosity can depend only on certain combinations of the rate of deformation tensor called invariants. Of the three independent combinations that can be formed, the second invariant is chosen because the first invariant is equal to zero for an incompressible fluid and the third invariant is equal to zero in a shear flow.] The second invariant is defined by

$$II = \Sigma_i \Sigma_j \dot{\gamma}_{ij} \dot{\gamma}_{ji} \quad . \tag{32}$$

Instead of II directly, the viscosity is made a function of the square root of 1/2 II, which is the magnitude of the rate of deformation tensor and is given by:

$$\dot{\gamma} = \sqrt{\frac{1}{2} \Sigma_i \Sigma_j \dot{\gamma}_{ij} \dot{\gamma}_{ji}} = \sqrt{\frac{1}{2} II} \quad . \tag{33}$$

For shear flow, the magnitude of the rate of deformation tensor is just equal to the velocity gradient or shear rate.

The Generalized Newtonian Fluid model works well for modeling steady shear flows but not for transient or elongational flows of polymeric fluids. One of the most popular Generalized Newtonian Fluid models is the power-law model. This is a two parameter model that describes the viscosity versus shear rate curve on a log-log plot as a straight line. The viscosity in terms of the shear rate and the magnitude of the shear stress (τ_{xy}) is

$$\eta = K\dot{\gamma}_{xy}^{(n-1)} = -\tau_{xy}/\dot{\gamma}_{xy} \tag{34}$$

where the power law parameters are n and K. This model is attractive because it can be used conveniently in the anlaytical solution of flow problems, it describes the high shear rate viscosity behavior of polymer solutions under process conditions, and it requires only two parameters. However, it cannot describe the constant low shear rate viscosity observed at low flow rates as shown in Fig. 3. Therefore, it fails for slow flows where the characteristic shear rates are in the Newtonian or transition regime. Other three-parameter models must be chosen to model these flows, and several models are shown in Table II in terms of the shear rate and the magnitude of the shear stress.

Models with Yield Stresses. Very concentrated filled melts, emulsions, foams, dispersions, and gels display yield stresses. A true yield stress implies the material will not flow until some minimum stress is achieved. In many sytems there may be a very slow flow at low shear stresses; however, over the time scale of practical importance the material can be treated as if it does not yield. This can be called an "apparent" yield stress.

Models describing the shear stress vs. shear rate for materials with yield stress have been reviewed (10) and are presented in Table III in terms of shear rate and the magnitude of the shear stress.

Models with Elastic Effects. Polymer rheologists have constructed an impressive array of models to represent viscoelastic fluids. A survey of these models can be found in Table 9.4-1 of reference 10 or reference 5. Most of these models were developed by researchers involved in polymer melt processing where elastic effects often are quite large.

To describe the flow of a visco-elastic fluid, models must be constructed that display both viscous and elastic behavior (10 (Ch 5), 5). The simplest model is the Maxwell model that represents the superposition of the equations for a Newtonian fluid and a Hookean solid, which for simplicity we write for a shear flow:

$$\tau_{yx} + \frac{\mu}{G}\frac{d}{dt}\tau_{yx} = \mu\dot{\gamma}_{yx} \tag{37}$$

In Eq. 37, μ is the viscosity and G is the elastic modulus. The ratio μ/G is a relaxation time, λ. Alternately, Eq. 37 can be written in an integral form

$$\tau_{yx} = \int_{-\infty}^{t} \left[\frac{\mu}{\lambda} \epsilon^{-(t-t')/\lambda} \right] \dot{\gamma}_{yx}(t')dt' \tag{38}$$

Table II. Generalized Newtonian Fluid Models for Shear-Thinning Fluids

Model	Parameters	Equation
Power Law	n -- power-law index	$\eta = K\,\dot{\gamma}_{xy}^{(n-1)}$
	K -- consistency index	
Ellis	η_0 -- zero-shear vixcosity	$\frac{\eta_0}{\eta} = 1 + \left[\frac{\tau_{xy}}{\tau_{1/2}}\right]^{\alpha-1}$
	$\tau_{1/2}$ --shear stress where $\eta = \eta_0/2$	
	$(\alpha$-1)--slope of log $[\eta_0/\eta-1]$ vs. log $[\tau_{xy}/\tau_{1/2}]$	
Carreau	η_0 -- zero-shear viscosity	$\frac{\eta - \eta_\infty}{\eta_0 - \eta_\infty} = [1 + (\lambda\dot{\gamma}_{xy}^2)]^{\frac{n-1}{2}}$
	η_∞ -- infinite-shear viscosity	
	λ -- time constant	
	n -- power-law index	

Table III. Fluid Models with Yield Stresses

Model	Parameters	Equation
Bingham	τ_0 -- yield stress η_p -- plastic viscosity	$\tau_{xy} = \tau_0 + \eta_p \dot{\gamma}_{xy}$ for $\tau_{xy} > \tau_0$ $\dot{\gamma}_{xy} = 0$ for $\tau_{xy} \leq \tau_0$
Casson	τ_0 -- yield stress η_p -- plastic viscosity	$\sqrt{\tau_{yx}} = \sqrt{\tau_0} + \sqrt{\eta_p} \sqrt{\dot{\gamma}_{xy}}$ for $\tau > \tau_0$ $\dot{\gamma}_{xy} = 0$ for $\tau_{xy} \leq \tau_0$
Herschel-Buckley	τ_0 -- yield stress K -- consistency index n --- shear rate exponent	$\tau_{xy} = \tau_0 + K \dot{\gamma}_{xy}^n$ for $\tau_{xy} > \tau_0$ $\dot{\gamma}_{xy} = 0$ for $\tau_{xy} \leq \tau_0$

where the the quantity in brackets is the relaxation modulus, G(t) for the Maxwell fluid. It can also be shown that this can be written in terms of <u>strain</u>, γ_{yx}, rather than <u>rate of strain</u>, $\dot{\gamma}_{yx}$,

$$\tau_{yx} = \int_{-\infty}^{t} \left[\frac{\mu}{\lambda^2} \epsilon^{-(t-t')/\lambda} \right] \gamma_{yx} \ (t')dt' \qquad (39)$$

where the term in brackets is the memory function, m, for the Maxwell fluid.

This model with a single time constant, and a single viscosity parameter (or modulus parameter) can qualitatively reproduce the creep behavior shown in Fig. 14a; however, it cannot quantitatively reproduce linear viscoelastic behavior for real polymers because real materials have a range of relaxation times and modulus parameters. The Maxwell model can be generalized by replacing the relaxation modulus by a sum of contributions:

$$\tau_{yx} = - \int_{\infty}^{t} \sum_{i} \left[\frac{\mu_i}{G_i} \epsilon^{-(t-t')/\lambda_i} \right] \dot{\gamma}_{yx}(t')dt' \qquad (40)$$

An example of the fitting of moduli data for a low density polyethylene melt with the sum of eight contributions are shown in Figs. 16a,b,c. Figure 16a shows the predicted G′ and G" values of the individual contributions and their resultant sum. And Fig. 16c shows a comparison of the experimental data with the fitted moduli (22). Excellent agreement can be seen. However, even with multiple relaxation modes the Maxwell model still predicts a viscosity that is independent of shear rate, in contrast to the data in Fig. 3. It can not predict stress overshoot as seen in the data of Fig. 11.

Attempts to introduce non-linearities into the constitutive equation have followed several tacks.

1. <u>Non-Linear Measures of Strain</u>: Modification of the Maxwell model (Eq. 39) to account for non-linearities inherent in large amplitude deformations (such as steady shear flow) begins with the replacement of the strain measure in Eq. 39, which is valid for the small amplitude strain experiments inherent in linear viscoelasticity, with strain measures that translate and rotate with the fluid elements (10 (Ch1), 5 (Ch7)). This substitution alone is still not adequate to predict flow behavior quantitatively.

A powerful set of theories arises from a picture of molecular motion that involves polymer chains being prevented from moving transversely because of steric interactions with their neighbors and, therefore, relaxing by motion occuring along the direction of their backbone contour. This constrained motion is termed "reptation" by analogy with the motion of snake through a grid of obstacles. The original formulation of the models in somewhat different forms are due to Doi and Edwards, deGennes, and Curtiss and Bird (10,15). The detailed analysis of this model of molecular

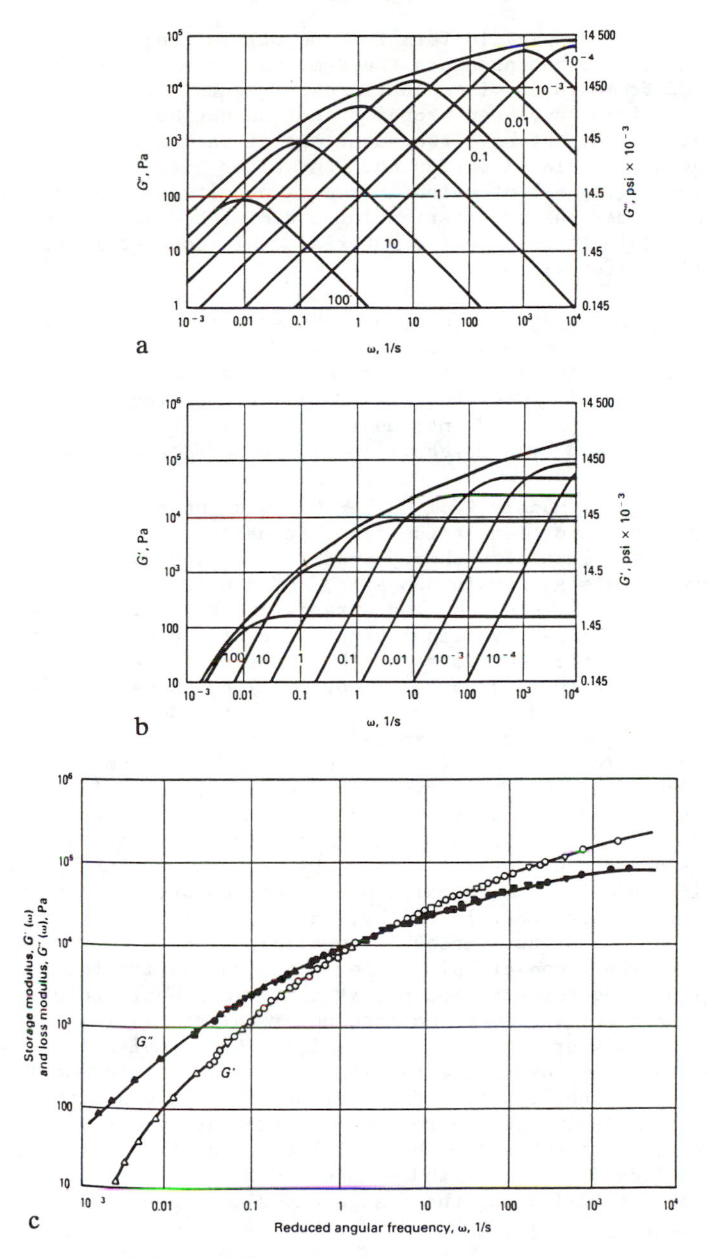

Figure 16. Fitting the modulus of a low-density polyethylene melt with a generalized Maxwell model with 10 time constants, λ_i, and ten modulus constants G_i. (a) and (b): Individual G′ and G′′ contributions from the terms comprising the Maxwell elements. (c): Comparison of fitted model to experimental data. (Reproduced with permission from reference 22. Copyright 1978 Steinkopff.)

motion leads to a specific form for the strain function that should appear in Eq. 39, but predicts the same memory function as determined from linear viscoelastic experiments. Other non-linear strain functions have been proposed that do not have a rigorous molecular origin, but have the advantage of mathematical simplicity. These include models of Gordon and Schowalter, Segalman and Johnson, and Larson (10,5) all of which introduce non-linearities associated with the polymer chains experiencing a diminished strain compared to the bulk strain imposed on the material: this concept is termed "non-affine deformation".

2. Non-Linear Stress Functions. White and Metzner introduced a non-linear coefficient that depended on the strain invariant (ref 10 Ch 5). This coefficient multiplies the derivative of stress appearing in Eq. 37. Giesekus (10 (Ch6)), pre-dating the molecular reptation ideas, proposed introducing a non-isotropic form for the stress experienced by a molecule in a deformed state.

3. Molecular Models from Rubber Network Theory. Another class of constitutive equations arise from extensions of rubber elasticity theory. Rubber elasticity theory predicts the modulus, or more generally the stress-strain behavior, of a crosslinked network in terms of the architecture of the strands and crosslinks. These theories are modified by "liquifying" some of the network strands and writing kinetic equations for the rates of formation and destruction of network strands. Theories along these lines were proposed first by Lodge and Yamamoto, and with additional refinements to better match experimental data by Phan-Thien and Tanner, and Acierno et al. (15). Also arising from the theory of rubber elasticity, but this time from continuum mechanics arguments about the thermodynamics of deformations, Leonov has proposed a model that fits material functions reasonably well (5).

This broad brush overview is highly schematic, in that we have not really gone into the details of any one constutive theory. The reason for this is two-fold. First, the complexity of the various models and the continuum mechanics preliminaries that are required are outside the scope of this review which has as its focus rheological measurements. Second, at this time the incroporation of complex constitutive equations into numerical simulations of complicated polymer flows is not possible. Numerical simulations most generally employ only Generalized Newtonian Fluid models for the rheology. Incorporation of constitutive equations with elastic effects is at the very forefront of research at this time. As the field evolves, undoubtedly, there will be a merging of complex numerical flow simulations and complex rheological constitutive equations that will allow the design and simulation of more complex flow processes.

References

1. Schoff, C. K. In *Encyclopedia of Polymer Science and Engineering,* 2nd ed.; Mark, H. F. et al., Eds.; Wiley: New York, 1988; Vol. 14, pp 454–451.
2. Walters, K. *Rheometry;* Wiley: New York, 1980.
3. Whorlow, R. W. *Rheological Techniques;* Wiley: New York, 1980.

4. Tanner, R. I. *Engineering Rheology;* Oxford: New York, 1985.

5. Larson, R. G. *Constitutive Equations for Polymer Melts and Solutions;* Butterworths: Boston, 1988.

6. Ferry, J. D. *Viscoelastic Properties of Polymers,* 3rd ed.; Wiley: New York, 1980.

7. Clark, D.; Miller, W. *Polym. Preprints* **1983,** *24,* 87.

8. Prud'homme R. K.; Uhl, J. T.; Poinsatte, J. P.; Halverson, F. *Soc. Petrol. Eng. J.* **1983,** *October,* 804.

9. Uhl, J. T.; "Rheological Studies of Water-Soluble Polymer Solutions with Interacting Solutes," Ph.D. Thesis, Department of Chemical Engineering, Princeton University, 1983.

10. Bird, R. B.; Armstrong, R. C.; Hassager, O. *Dynamics of Polymeric Liquids,* 3rd ed.; Wiley: New York, 1987; Vol. 1.

11. Han, C. D. et al. *Polym. Eng. Sci.* **1981,** *21,* 196–204.

12. Han, C. D. *Multiphase Flow in Polymer Processing;* Academic Press: New York, 1981.

13. Tanaka, H.; White, J. L. *Polym. Eng. Sci.* **1980,** *20,* 949–956.

14. Wildemuth, C. R.; Williams, M. C. *Rheologica Acta* **1985,** *24,* 75–91.

15. Laun, H. M. *Colloid Polym. Sci.* **1984,** *262,* 257–269.

16. Suetsugu, Y.; White, J. L. *J. Appl. Polym. Sci.* **1983,** *28,* 1481–1501.

17. Khan, S. A.; Prud'homme, R. K. *Rev. Chem. Eng.* **1987,** *4,* 205–269.

18. Brandrupt, J.; Emmergut, E. H., Eds. *Polymer Handbook,* 2nd Ed.; Wiley: New York, 1975.

19. Meissner, J. *Chem. Eng. Commun.* **1985,** *33,* 159–180.

20. Huppler, J. D.; Macdonald, I. F.; Ashare, E.; Spriggs, T. W.; Bird, R. B. *Trans. Soc. Rheology* **1987,** *11,* 181–204.

21. Wagner, M. H.; Meissner, J. *Macromol. Chem.* **1980,** *181,* 1533–1550.

22. Laun, H. M. *Rheologica Acta* **1978,** *17,* 1–15.

23. Meissner, J. *Kunststoffe* **1971,** *61,* 576–582.

24. Berry, G. C.; Fox, T. G. *Adv. Polym. Sci.* **1968,** *5,* 261–357.

25. Munstedt, H. *J. Rheol.* **1980,** *24,* 847–867.

RECEIVED March 11, 1991

Chapter 3

Extensional Viscometry of Polymer Solutions

Gerald G. Fuller and Cheryl A. Cathey

Department of Chemical Engineering, Stanford University, Stanford, CA 94305–5025

This paper reviews recent progress in understanding of the extensional flow behavior of polymer solutions. Classification of flows with extensional character according to flow strength is discussed along with a summary the current state of molecular modeling in this area. Different experimental approaches to measurement of the extensional viscosity and macromolecular conformation are discussed. Applications of extensional flow measurements to macromolecular characterization, chain scission and fluid dynamics are presented.

Polymer solutions subject to extensional flows can produce viscosities that are dramatically different compared with those obtained using simple shearing flow. High molecular weight, flexible chains, present in only minute quantities, can produce extensional viscosities that are predicted to be strain thickening and several orders of magnitude higher than the corresponding shear viscosities. These effects are the basis of numerous flow phenomena that impact processes such as jet stability [1], flow through porous media [2] and turbulent drag reduction [3]. Furthermore, many technologically important processes such as fiber spinning, high speed coating, and oil spill removal, involve flows with substantial extensional components, making a knowledge of extensional rheometry essential to modeling the flow behavior.

It is useful to classify flow fields according to the relative amounts of extension and rotation that they possess. In general, a velocity field, $\mathbf{u}(\mathbf{r})$, can be expanded about a position, $\mathbf{R}$, as

$$\mathbf{u}(\mathbf{r}) = \mathbf{u}(\mathbf{R}) + \nabla\mathbf{u} \cdot (\mathbf{r} - \mathbf{R}) + \frac{1}{2}(\mathbf{r} - \mathbf{R}) \cdot (\nabla\nabla\mathbf{u}) \cdot (\mathbf{r} - \mathbf{R}) + \ldots \quad (1)$$

The uniform flow term, $\mathbf{u}(\mathbf{R})$, will not affect the microstructure of the fluid and on the scale of a constituent macromolecule, only the velocity gradient term, $\nabla\mathbf{u}$, need normally be considered. For the purposes of flow field classification, this term is divided into its symmetric and antisymmetric parts according to,

0097–6156/91/0462–0048$06.00/0

$$\nabla \mathbf{u} = \frac{1}{2}(\nabla \mathbf{u} + \nabla \mathbf{u}^{+}) = \mathbf{E} + \mathbf{W} \tag{2}$$

The rate of strain tensor, **E**, characterizes extensional deformations and the vorticity tensor, **W**, produces rigid body rotations.

A simple shear flow has a velocity gradient

$$\nabla \mathbf{u} = \dot{\gamma}\begin{pmatrix} 0\,1\,0 \\ 0\,0\,0 \\ 0\,0\,0 \end{pmatrix} = \frac{\dot{\gamma}}{2}\begin{pmatrix} 0\,1\,0 \\ 1\,0\,0 \\ 0\,0\,0 \end{pmatrix} + \frac{\dot{\gamma}}{2}\begin{pmatrix} 0 & 1\,0 \\ -1 & 0\,0 \\ 0 & 0\,0 \end{pmatrix} \tag{3}$$

and therefore has an equal amount of extension and vorticity.

Linear stability analysis using simple linear spring models to represent flexible polymer chains is used to determine the tendency of a flow field to produce significant distortion [4]. The result of such a procedure is that simple shear represents the demarcation between flows that are strong and flows that are weak. In a weak flow, flexible chains remain in a coiled configuration regardless of the magnitude of the flow. In a strong flow, a coil–to–stretch transition is predicted when the condition

$$\lambda\tau > \frac{1}{2} \tag{4}$$

is satisfied. Here λ is the largest positive eigenvalue of the velocity gradient tensor and τ is the relaxation time of the linear spring dumbbell model. In this paper, extensional flows are taken to be any flow where the magnitude of the rate of strain tensor is larger than the magnitude of the vorticity tensor.

As in the case of the shear viscosity, the extensional viscosity is defined as the coefficient relating the stress tensor to the rate of strain tensor. For an ideal, uniaxial extensional flow,

$$\mathbf{u} = \dot{\gamma}\begin{pmatrix} 1 & 0 & 0 \\ 0 & -1/2 & 0 \\ 0 & 0 & -1/2 \end{pmatrix} \cdot \mathbf{r}, \tag{5}$$

the extensional viscosity is simply

$$\bar{\eta} = \frac{\tau_{xx} - \tau_{yy}}{\dot{\gamma}} \tag{6}$$

Here τ_{ij} is the stress tensor, $\dot{\gamma}$ is the velocity gradient and **r** is the position vector. For a Newtonian liquid, the extensional viscosity is given by Trouton's rule to be simply three times the shear viscosity.

This paper reviews current experimental methods in extensional rheometry. Mechanical techniques employing spinning and opposed jets geometries are discussed along with the use of optical methods to probe polymer deformation and orientation processes. The use of extensional rheometry to characterize chain rigidity, molecular weight distributions, and chain entanglements is presented. Theoretical models and the results of molecular simulations of dilute polymer solutions subject to strong, extensional flows are also reviewed.

MOLECULAR MODEL PREDICTIONS

Flexible Chain Models

Dilute solutions. Simple dumbbell models have been used to predict extensional viscosities [5]. In its simplest form, the model represents a polymer chain by a single vector, **R**, that spans two points of hydrodynamic friction. These points are tethered by an entropic spring, which resists deformation of the chain. If a linear spring is used, the model predicts that the extensional viscosity will increase from its Trouton rule value in weak flows and will diverge to infinity at the condition given by eq. (4). This divergence can be removed by using a spring force that itself diverges when $|\mathbf{R}|$ approaches the contour length of the chain. It is important to note that the convection–diffusion equation describing the probability distribution function of possible dumbbell conformations can be solved exactly for the case of steady, purely extensional flows for arbitrary choices of the functional form of the spring function.

The dumbbell model can be embellished to include a variety of nonlinear effects. These include electrostatic interactions, for the case of polyelectrolytes [6], and hydrodynamic interactions. Hydrodynamic interactions have been introduced in two ways. The simplest approach incorporates hydrodynamic interactions between the two ends of the dumbbell [5]. Including this effect only slightly modifies the quantitative predictions of the model. A second approach is to allow the hydrodynamic friction factor to be conformation dependent [7, 8]. Ordinarily, the form used is $\zeta \approx \zeta_0 \sqrt{N}(R/L)$, where L is the contour length of the polymer, N is the number of statistical subgroups and is proportional to the molecular weight, and ζ_0 is the friction factor of the coiled chain. This form is motivated by the fact that the friction factor of an object is roughly proportional to its largest length scale. The addition of this dependence can lead to the prediction of multiple steady states for the dumbbell's conformation if the inclusion of this nonlinear term is treated by preaveraging its value. This prediction, however, has been objected to by Bird and coworkers [9] who have demonstrated that multiple solutions are inherently not possible with a simple dumbbell model. Recent calculations using a multi–bead and spring model by Larson *et al.*, however, suggest that this hysterisis may exist for real chains [10].

Recent simulations using models with greater internal structure, such as multi–bead–rod models [11] and multi–bead and spring models [12] suggest that the distortion process of real chains in extensional flows may not proceed in a manner that a simple dumbbell model can ever capture. These simulations predict that flexible chains deform according to the three stage scenario described in Figure 1. This particular sequence is for a strong flow condition where the ultimate steady configuration should have the chain almost fully extended.

At small strains, the molecule is predicted to both orient and deform from its initial coiled state. At higher strains the segments become highly oriented but the end–to–end distance is only slightly larger than the coiled state.The final process is one in which the highly oriented segments negotiate very tight "hairpin" turns in order to fully extend the chain. A local probe of segmental orientation, such as birefringence, would be expected to saturate at lower strains than would properties such as the extensional viscosity that are sensitive to the overall deformation.

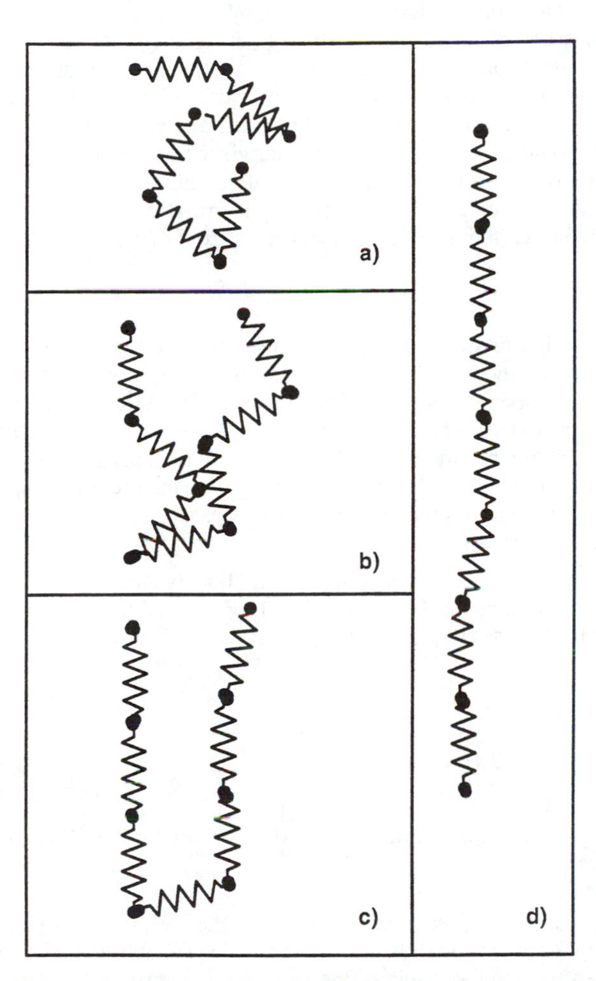

Figure 1. Sequence of the orientation/deformation process of a flexible chain modeled as a series of beads and springs. a, random coil; b, low strain deformation/orientation process; c, orientation saturated; and d, fully extended chain.

Semi–dilute and entangled solutions. Two types of microstructural models have been used for interacting chains at higher concentration. These are the dumbbell model of Deaguiar and Bird [13], having an anisotropic friction factor that depends on chain orientation, and the phenomenological transient network model of Yamamoto [14] and Lodge [15].

The transient network model has been developed in order to account for chain entanglements. These interactions are modeled by allowing the system to consist of a variable number of elastic, stress bearing segments. The population of segments is governed by destruction and creation functions that, in general, depend on the configuration of the segments. For a specific choice of such functions (constant rate of destruction and a Gaussian creation function), the model is equivalent to the linear elastic dumbbell model and predicts a singular viscosity. By choosing a destruction rate that increases with the distortion of the segments, this singularity can be removed, leading to an extensional viscosity that initially strain thickens, and then strain thins as a larger number of elastic segments are removed from the system [16].

Rigid Rod Models

Dilute solutions. The model most frequently used for dilute solutions of rigid chains is the rigid dumbbell model, consisting of two beads of hydrodynamic friction separated by a fixed distance [5]. For steady, purely extensional flows, this model predicts a strain thickening extensional viscosity that plateaus to a strain rate independent value. The strain thickening proceeds without the coil–to–stretch transition predicted using the elastic dumbbell. Inclusion of hydrodynamic interaction along the rod does not modify the qualitative predictions, but is necessary to obtain the correct high strain rate limiting viscosity.

Hydrodynamic interactions between neighboring chains that can occur at semi–dilute concentrations has been considered by Batchelor [17], and Acrivos and Shaqfeh [18], for the case of perfectly aligned rods. Both treatments predict $\bar{\eta}$ to be of order

$$\frac{\bar{\eta} - 3\mu}{3\mu} \approx \frac{nL^3}{\ln(\frac{h}{R})} \tag{7}$$

where n is the number concentration of the rods, L is their length, R is their radius, and h is their average separation.

This equation is frequently used to estimate the high strain rate limiting extensional viscosity for both rigid rods and fully extended flexible chains. From equation (7), one can readily see the origin of the very large extensional viscosity enhancements available with dilute quantities of high molecular weight polymers.

Semi–dilute, sterically interacting chains

The only model that has been proposed for the purpose of describing rigid rods in semi–dilute solution is due to Doi and Edwards [19]. In this model the solution consists of rods constrained to move within tubes formed by their neighbors. Although a rod can freely diffuse along its axis, lateral motions are limited to angular displacements prescribed by the dimensions of the tube. These dimensions depend on the number concentration and length of the rods. The impingements caused by neighboring chains lead to large enhancements in the shear and extensional viscosities. Alignment of the rods by hydrodynamic forces cause the viscosities to decrease with increasing velocity gradients. Extensional viscosity measurements, therefore, provide a qualitative distinction between semi–dilute rigid and flexible chains. The former show strain thinning whereas the latter are strain thickening. Both types of chains, however, generally produce thinning shear viscosities.

EXTENSIONAL FLOW INSTRUMENTATION

In rheometry, it is a desirable to consider flows with simple histories of the rate of strain tensor. With simple shear flow, rotational flow devices (Couette and cone–and–plate cells, for example) are capable of creating flows where molecules travel in closed streamlines with uniform histories of deformation. Extensional flows, however, cannot be created with closed streamlines and experiments are inherently transient in nature. In practice it is normally more meaningful to report extensional viscosities not only as a function of the velocity gradient, but also as a function of the average total strain produced by the measurement. In other words,

$$\bar{\eta} = \bar{\eta}(\dot{\gamma}, <\dot{\gamma} t_{res}>) \tag{8}$$

where $<\dot{\gamma} t_{res}>$ is the average of the product of the velocity gradient times the residence time of molecules involved with the measurement. High viscosity materials such as polymer melts can be handled with testing equipment that is similar in many ways to tensile instruments used for solids. The review by Meissner [20] summarizes many of the devices used to accomplish extensional viscosity measurements for melts. Low viscosity polymer solutions are very difficult to study in extension and the field is far less well developed for this class of materials.

Many devices have been developed over the years to approximate extensional flows. These devices are classified here as either "flow–through" devices or "stagnation–point" devices. Flow–through geometries would include the spinning experiment [21, 22], the tubeless siphon [23], and contraction flows [24, 25]. Stagnation point flows are generated by a variety of instruments including the four roll mill [26, 8] and opposed jets [27, 28]. With these flow cells, both mechanical measurements (pressure drops, tensile forces) and optical measurements (velocimetry, birefringence) have been performed.

APPLICATIONS OF EXTENSIONAL RHEOMETRY

Molecular Characterization

Simple molecular representations of flexible chains, such as the dumbbell model, predict a rapid transition from a coiled state to a fully stretched transition when the strength of the velocity gradient approaches or exceeds the inverse of the principal relaxation time of the chains. This transition can be readily detected using flow birefringence measurements [26, 27] with the result that the critical velocity gradient scales with molecular weight raised to the power of 1.5, regardless of the quality of the solvent. This outcome is the basis of a procedure to measure molecular weight distributions suggested by Keller and coworkers [29].

The inherent transient nature of the flow means that the experimental observables will depend on strain as well as strain rate. In stagnation point flows where there is a distribution of residence times, only streamlines carrying molecules subjected to sufficiently large total strains will respond significantly. A spatially resolved measurement, such as flow birefringence, will reflect this distribution by producing highly localized patterns of chain deformation. Figure 2 is a retardation trace from a sample of polystyrene (molecular weight 7.7×10^6) dissolved in dioctyl phthalate at a concentration of 300 ppm. This sample was subject to a stagnation point flow with a velocity gradient of 400 /s using an opposed jets apparatus with 1 mm nozzles. The curve indicates that the majority of chains oriented by the flow reside near the center, outgoing streamline. As discussed later, this localized pattern is typical of high molecular weight, flexible chains and is very different from the pattern produced by rigid, rodlike macromolecules. It must be emphasized that flow birefringence is a measure of local segmental orientation and cannot be used to directly infer the degree of overall chain extension.

The total deformed length of the chains will strongly affect mechanical measurements of effective extensional viscosities since this property is roughly proportional to the cube of the largest length scale of additives to a solvent. Figure 3 is a plot of both the effective extensional viscosity and the birefringence of a sample of polystyrene (molecular weight 3.8×10^6) dissolved in dioctyl phthalate at a concentration of 300 ppm. These measurements were obtained using an opposing jets instrument described in reference 28. When interpreting these results, it must be kept in mind that as the velocity gradient is increased, the residence time is decreasing. The birefringence was observed to rise monotonically to an asymptote independent of strain rate, indicating that the degree of local segmental orientation ultimately saturates. The effective extensional viscosity, and therefore the degree of overall deformation, appears to go through a maximum. These results bear some qualitative agreement with the recent molecular predictions using many bead–rod and bead–spring models by Marrucci et al. [11] and Weist and Bird [12] described earlier. Those calculations predict that chain deformation first undergoes a rapid alignment of segments by the flow. This is followed by a much slower unraveling of the chain, which must negotiate very tight, hair pin curves in order to elongate.

The response of rigid, rodlike chains is expected to be very different in extensional flow compared with flexible chains. This is in contrast to simple shear flow where solutions of either flexible or rodlike chains, although different in detail, show similar, shear thinning behavior. As presented above, flexible chains have strain thickening extensional viscosities. Rodlike chains, in the semi–dilute concentration regime, are predicted to have strain thinning extensional viscosities [19]. This is because the dominant mechanism for enhancement of the viscosity for semi–dilute rodlike chains are chain entanglements that will be reduced by alignment by the flow. With flexible chains, on the other hand, the primary effect will be the increase in the friction factor of the chains as they extend. Figure 4 is a plot of the effective extensional viscosity of semi–dilute solutions of collagen protein in a 90/10 glycerin/water mixture. These data clearly show the strain thinning behavior. Extensional viscosity measurements, therefore, can be used as sensitive tests of chain structure and rigidity.

The time–dependent behavior of chains will also depend on their rigidity. Rigid chains require far fewer total strain units compared with flexible chains in order to approach steady state behavior. For this reason, retardation patterns from flow birefringence experiments in stagnation point flows will be expected to be far less localized than the patterns generated using solutions of flexible chain polymers. Figure 5

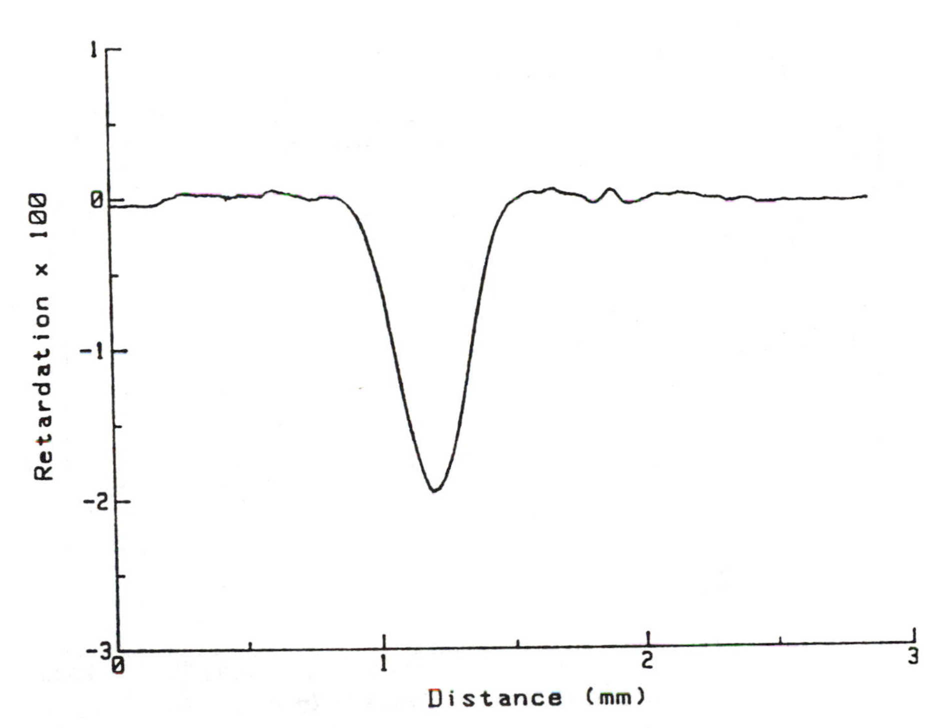

Figure 2. The retardation profile for 300 ppm 7.7x106 molecular weight polystyrene in dioctyl phthalate as a function of the distance the optical axis is translated between two 1 mm, opposing nozzles. These data were taken at a strain rate of 400 /s.

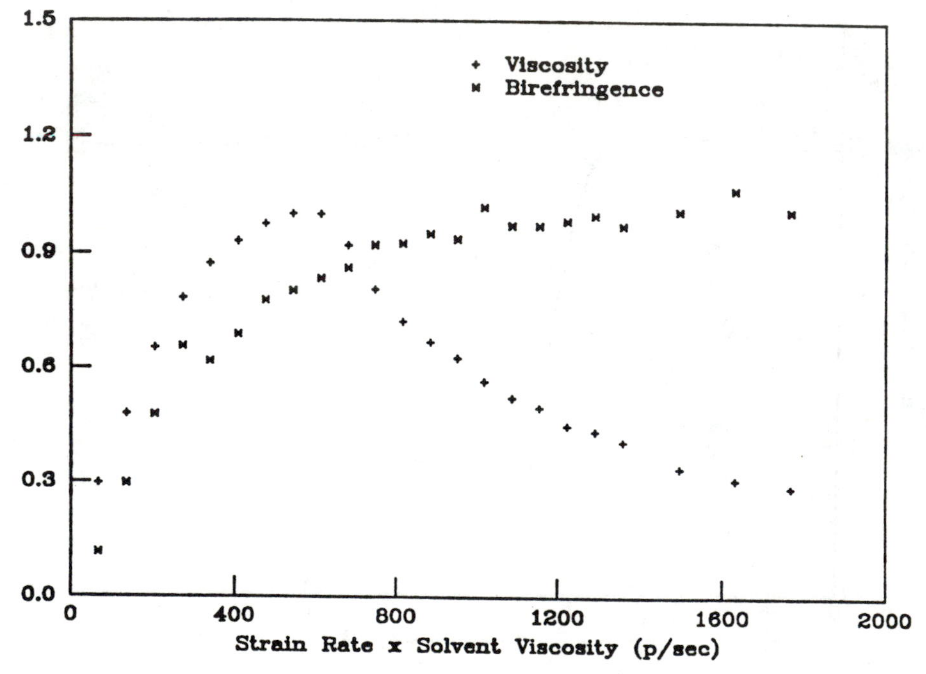

Figure 3. The normalized birefringence and viscosity vs. the strain rate times the shear viscosity of the solvent for 300 ppm 3.8x106 molecular weight polystyrene in dioctyl phthalate.

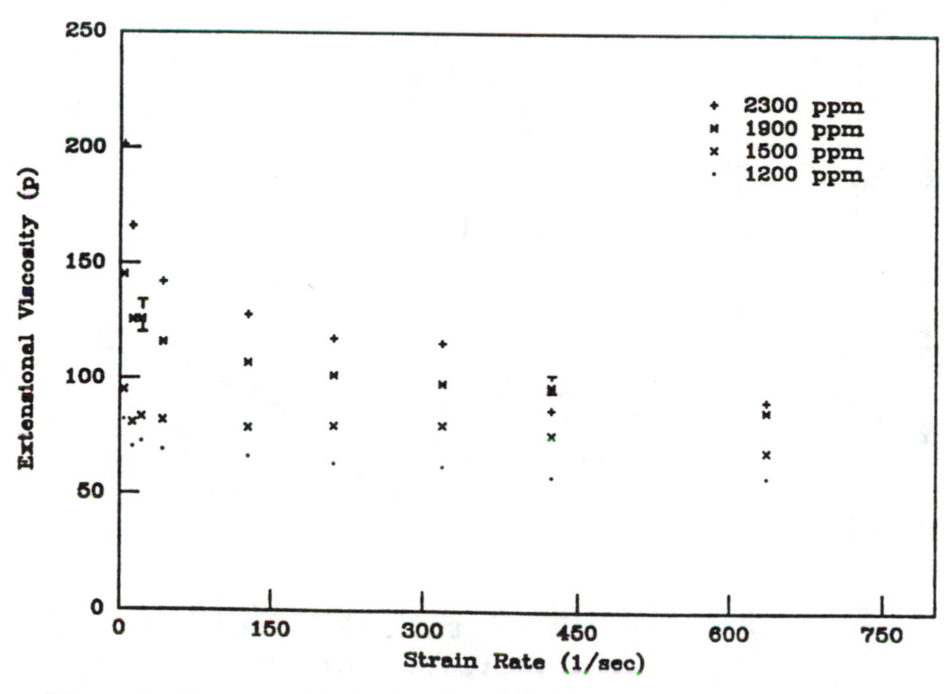

Figure 4. The extensional viscosity as a function of strain rate for semidilute solutions of collagen in 90/10 glycerin/water.

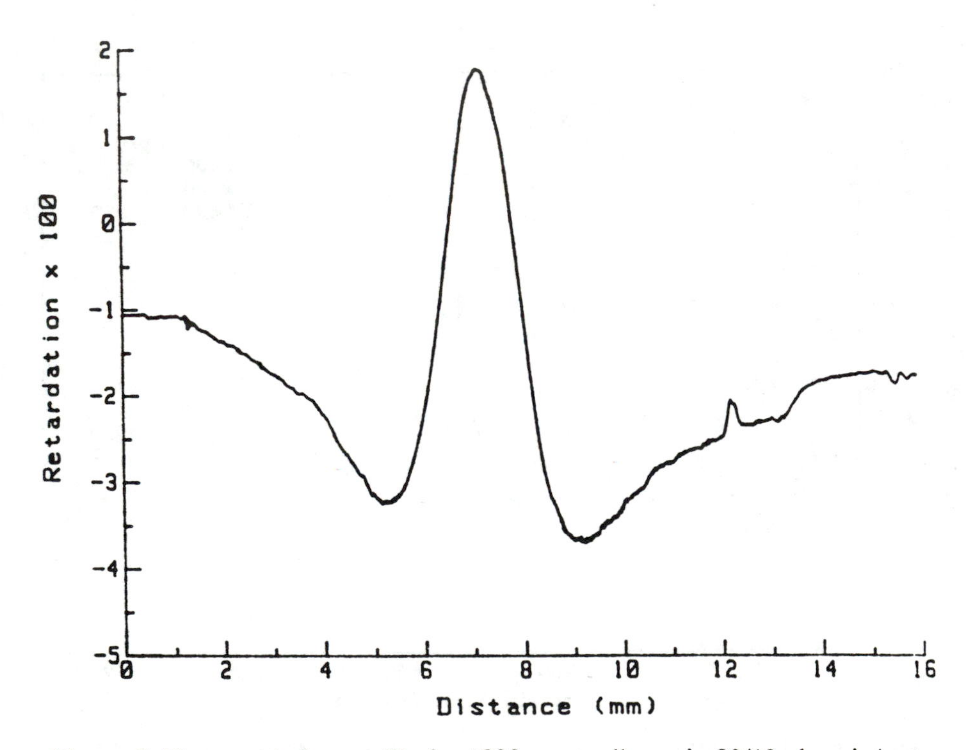

Figure 5. The retardation profile for 1200 ppm collagen in 90/10 glycerin/water as a function of the distance the optical axis is translated between the nozzles. These data were taken at a strain rate of 50 /s.

is a retardation sweep of a solution of 1200 ppm collagen in 90/10 glycerin/water subject to an effective strain rate of 50 /s using an opposed jets device. Unlike the pattern shown in Figure 1 for a flexible chain, the pattern in Figure 4 is very broad and exceeds the dimensions of the nozzles themselves (1 mm). The negative regions of birefringence are a result of chain orientation perpendicular to the centerline spanning the nozzles as the chains are first brought into the region between the nozzles. As the rodlike collagen chains are withdrawn into the nozzles, the orientation of the chains will be rotated by 90 degrees, producing a birefringence of opposite sign.

Chain Scission

Mechanical degradation of polymer chains limits the utility of additives for such applications as drag reduction and viscosity enhancement. Extensional flows, and especially stagnation point flows, are the most efficient means of producing hydrodynamic scission. Furthermore, applications such as drag reduction and viscosity enhancement in lubricants, often contain extensional viscosity components. Understanding the scission process and the forces required to break molecular bonds is also an important means of gaining insight into the ultimate mechanical properties of polymeric materials.

Originally pioneered by Merrill and coworkers using contraction flows [30], and later developed using stagnation point flows by Keller and coworkers [31], chain scission studies have been able to elucidate the deformation process of flexible chains in transient, extensional flows. Among the important findings have been the observation that dilute solutions of flexible chains normally break in the center, implying that the middle portion of the chains is the most highly stretched. The scaling of the critical strain rate, $\dot{\epsilon}$, to induce scission with molecular weight can also offer insight into the process of chain extension. Odell and Keller [16] have found that $\dot{\epsilon} \approx M^{-2}$. This exponent would imply that complete extension of the molecules is occurring. Conversely, experiments using contraction flows by Nguyen and Kausch [32] produced an exponent of 1.1 which infers that the chains are mostly locally deformed.

Rheology/Flow Interactions

The dramatic changes in extensional viscosity imparted by small amounts of high molecular weight polymer have been observed to significantly alter the basic nature of extensional flows. The modification of flow has been largely ignored in many extensional flow studies and measurements of rheology and structure combined with velocity field determinations have only recently been carried out. In stagnation point flows alterations in the flow field are characterized by a marked reduction in velocity and in the velocity gradient along the outgoing streamline emanating from the stagnation point (Leal [33]). In spinning flows, very large shearing gradients, which are normally assumed to be negligible, can appear at the air/solution interface (Matthys, [34]). These modifications of the flow often require direct, spatially resolved measurements of velocities and velocity gradients as opposed to images of streamlines. The insensitivity of streamline patterns to polymer induced velocity field changes is most likely due to the fact that streamlines are largely controlled by the symmetry of the geometries inducing the flows. Because the flow field is often dramatically altered in many extensional viscometers used for low viscosity materials, these instruments can only be expected to provide effective measures, or indexes of extensional viscosities.

REFERENCES

1. Hoyt, J. W., J. Rheol., **24** (1980) 685.
2. Interthal, JW. and R. Haas, *Flow Trans. Porous Media,* Proc. of Euromech, (1981) 157.
3. Virk, P. S., AIChE J., *21* (1975) 625.
4. Olbricht, W. L., J. M. Rallison, and L. G. Leal, J. N. N. F. M., *10* (1982) 291.
5. Bird, R. B., O. Hassager, R. C. Armstrong, and C. F. Curtiss, *Dynamics of Polymeric Liquids, 2nd Ed., Volume 2.*, John Wiley and Sons, NY, 1977.
6. Dunlap, P. N. and L. G. Leal, Rheol. Acta, *23* (1984) 238.
7. de Gennes, P. G., J. Chem. Phys., *60* (1974) 5030.
8. G.G. Fuller and L.G. Leal, Rheol. Acta *19* (1980) 580–600.
9. Fan, X.–J., R. B. Bird, and M. Renardy, J. NonNewt. Fluid Mech., *18* (1985) 255.
10. R. G. Larson, J. Magda, and M. Mackay, J. Chem. Phys. *89* (1988) 2504.
11. Acierno, D., G. Titomanlio, and G. Marrucci, J. Polym. Sci. Polym. Phys. Ed., *12* (1974) 2177.
12. Weist, J. M., R. B. Bird, and L. E. Wedgewood, *On Coil–Stretch Transitions in Dilute Polymer Solutions,* Rheology Research Center Report, University of Wisconsin–Madison, 1988.
13. Bird, R. B. and Deaguiar, J. NonNewt. Fluid Mech., *13* (1983) 149.
14. Yamamoto, M., J. Phys. Soc. Japan, *11* (1956) 413.
15. Lodge, A. S., Trans. Faraday Soc., *52* (1956) 120.
16. Fuller, G. G. and L. G. Leal, J. Polym. Sci. Polym. Phys. Ed., *19* (1981) 531.
17. Batchelor, G. K., J. Fluid Mech., *46* (1971) 813.
18. Acrivos, A. and E. S. G. Shaqfeh, Phys. Fluids, *31* (1988) 1841.
19. Kuzuu, N. Y. and M. Doi, Polym. J., *12* (1980) 883.
20. Meissner, J., Ann. Rev. Fluid Mech., *17* (1985) 45–64.
21. Talbott, W. H., Goddard, J. D. 1979. Rheol. Acta., *18* (1979) 505.
22. Khagram, M., Gupta, R. K. and Sridhar, T., J. Rheol., *29* (1985) 191.
23. Peng, S. T. J. and R. F. Landel, in *Rheology,* vol. 2, pp. 385, G. Astarita, G. Marrucci, and L. Nicolais, ed., 1980.
24. James, D. F., B. D. McLean, and J. H. Saringer, J. Rheol. *31* (1987) 453.
25. Wunderlich, A. M. and D. F. James, Rheol. Acta, *26* (1987) 522.
26. Pope, D. P. and A. Keller, Coll. Polym. Sci., *255* (1977) 633.
27. Mackley, M. R. and A. Keller, Phil. Trans. Royal Soc. Lond., *278* (1975) 29.
28. G.G. Fuller, C. Cathey, B. Hubbard, and B.E. Zebrowski, J. Rheology *31* (1987) 235–249.
29. Odell, J. A., Keller, A., and Rabin, Y., J. Chem. Phys., *88* (1988) 4022.
30. Merrill, E. W., Proceedings of the ACS Meeting in Chicago (Sept. 1985).
31. Odell, J. A. and A. Keller, J. Polym. Sci., Polym. Phys. Ed., *24* (1986) 1889.
32. Nguyen, T. Q. and H.–H. Kausch, J. Non–Newt. Fluid Mech., in press.
33. Leal, L. G., AIP Conf. Proc., *137* (1985) 5.
34. Matthys, E. F., J. Rheol., *32* (1988) 773.

RECEIVED March 1, 1991

GELS AND LATICES

Chapter 4

Preparation, Characterization, and Rheological Behavior of Water-Swellable Polymer Networks

H. Nottelmann and W.-M. Kulicke[1]

Institut für Technische und Makromolekulare Chemie, Universität Hamburg, Bundesstrasse 45, 2000 Hamburg 13, Germany

The swelling and rheological behavior has been investigated on highly absorbent water-swellable polymers prepared from anionic and cationic monomers with tetra- and hexafunctional cross-linking agents. Because of their technical application, the absorption capacity of gel blocks and their reversibility of swelling in salt solutions with multivalent cations have been studied. A method based on very few experimental data was introduced, which allows to determine the parameters necessary for predicting the degree of swelling at any desired salt concentration. In addition, rheological investigations on gel blocks and particle solutions provide information about regions where inhomogeneities occur and data for determining the absorption capacities and their elasticity values.

Water-swellable polymer networks, referred to generically as hydrogels, represent highly elastic, solvent-containing bodies with very specific properties and may be either synthetic or biopolymeric in nature. As has been stated in diverse publications (*1–8*), there is an increasing number of applications for covalently cross-linked hydrogels in the field of engineering and medicine. Their range of properties spans from a slight degree of swelling, such as that necessary in soft contact lenses, to the cyclically reversible absorption of more than 1000 times the weight of their solid content, which enables them to be used as additives for incontinence aids or as moisture regulators in areas with prolonged dry periods. They exhibit a high retention capacity under load, as well as resistance to biological and chemical attack over various time periods.

These properties, which may in part seem contradictory, can be influenced by varying a number of factors, namely, the chemical nature of the polymer

[1]Corresponding author

0097–6156/91/0462–0062$07.50/0

components, the proportion of ionic and nonionic groups, and the distribution of sequence length and mesh width, that is, network density. The result of this is that a particular hydrogel with specified properties can be allocated to the appropriate field of application.

To determine or predict the property profile, it is necessary to have an exact knowledge of the formation and structure of the hydrogels (*9–12*). The conditions required for this are laid down at the time of synthesis. As far as the published data on gel systems are concerned, the degree of comparability often has shortcomings. There would thus appear to be a need for the introduction of a reference state for the synthesis. The synthesis of uniform gel blocks was therefore carried out according to a preparation concept to ensure simple, direct, and partially quantitative comparability between differing gel systems.

As an extension of previously published results for synthetic, semisynthetic, and biopolymer gels based on polyacrylamide and starch gels (*13, 14*), synthetic ionic hydrogels based on 2-acrylamido-2-methylpropanesulfonic acid and *N,N*-dimethylaminoethylacrylate with triallylamine or bisacrylamide as cross-linking agents were synthesized according to the abovementioned concept.

Because the industrial use of hydrogels largely calls for gels in the form of dried particles, or solutions and suspensions of these particles, whose macroscopic material properties depend on their surface configuration and size, some dry gel particles of different particle sizes were synthesized on the basis of 2-acrylamido-2-methylpropanesulfonic acid and triallylamine.

The investigations deal with technically relevant, phenomenological aspects, such as the stability in salt solutions of different concentrations with charges ranging from 1 to 3, the equilibrium degrees of swelling within these solutions, the reversibility of swelling, and the rheological characterization with respect to the synthesis parameters.

The results are to be compared with the property profiles of gels from analogous syntheses, which have been obtained with starch as a basis. Correlating the recorded data with molecular parameters should enable qualitative and quantitative relationships to be derived, which can then be compared with network theories, such as the Flory–Rehner relationship, as modified by the Donnan theory (*15*).

Experimental Details

The total mass of components in each preparation batch, including solvent, was set at 10 g. The total substrate concentration of the network components is given in percentage by weight and represents a dimension that is technically very easy to manipulate. Because the mesh width in a network system is dependent upon the ratio of cross-linker concentration to the concentration of monomer or, where a polymer is used, to the number of repeating units, the proportion of cross-linking agent is given in mole percentage.

The gels will henceforth be identified by two numbers separated by an oblique stroke (% w/w/mol-%). The first number deals with the total substrate content, and the second with the associated cross-linking agent concentration.

All gels were synthesized by a radical-initiated, cross-linking copolymerization in aqueous solution. The redox initiator system ammonium peroxodisulfate/TEMED was used for the radical copolymerization. This has often been used in already published studies (*16–21*). The initiator content, irrespective of the total monomer content, was the same for each preparation batch: 30 mg of ammonium peroxodisulfate and 50 μL (38.8 mg) of TEMED per 10 g. The following gel systems have been prepared as gel blocks according to the above concept.

Anionic Gels.

PAAm-AAc-BisAAm	Poly-(acrylamide-*co*-acrylate-*co*-bisacrylamide)
PAAm-AAc-TriAA	Poly-(acrylamide-*co*-acrylate-*co*-triallylamine)
PAMPS-BisAAm	Poly-(2-acrylamido-2-methylpropane sulfonic acid-*co*-bisacrylamide)
PAMPS-TriAA	Poly-(2-acrylamido-2-methylpropane sulfonic acid-*co*-triallylamine)

Production of Dried PAMPS-TriAA Gel Particles. After synthesis, PAMPS-TriAA gel blocks were placed in a plastic bag and pounded to a uniform mass and, after 24 h, dried in an oven at 50 °C. The dried gel particles were pulverized in a mortar and sized through sieves of differing widths. The fractions 100–300 μm, 300–500 μm, and 500–720 μm were used for rheological investigations.

Cationic Gels.

PDAEA-TriAA	Poly-(*N,N*-dimethylaminoethylacrylate-*co*-triallylamine)

Swelling Measurements on Gel Blocks. A mould (Figure 1) was fabricated for the swelling measurements that enables the gels to be produced as uniform cylinders of a defined geometry and removed without being destroyed. Equilibrium swelling measurements were carried out in NaCl, $MgCl_2$, and $Al(NO_3)_3$ solutions with concentrations ranging from 1×10^{-4} to 1.0 mol/L and in distilled water. In addition, the gel cylinders obtained were weighed and swollen in 250 ml of their respective solution for at least a week. The gels that had attained equilibrium were then removed from the solution with as little damage as possible and swabbed dry with absorbent paper before being weighed.

Rheological Investigations on Gel Blocks. Gels may be regarded as entropy-elastic bodies and can be reversibly deformed over a wide range. At sufficiently small deformation, the Hookian Law applies so that the quotient from the resulting stress (σ_{12}) and the imposed deformation (γ) is constant. The constant in this so-called linear viscoelastic region is given by the shear modulus (G) (*22*). The investigations on the gel blocks were carried out by nondestructive mechanical oscillation measurements. The complex shear modulus (G*), calculated from the resulting torque, is composed of the storage or elasticity modulus (G′) and the loss modulus (G″), although the latter was negligibly small in every case.

As can be seen from Figure 2, gelation takes several hours to reach completion (constant value of the storage modulus). In the radical-initiated copolymerization of synthetic hydrogels, gelation takes approximately 2–3 h, whereas the base-induced cross-linking of starch hydrogels takes 10–12 h because of the slower ionic mechanism. After completion, frequency-dependent measurements were then carried out in the linear viscoelastic region to determine the storage modulus in the plateau region (G'_p). This value can be linked to the number of elastically effective chains (ν_e) via the theory of rubber elasticity (*15*).

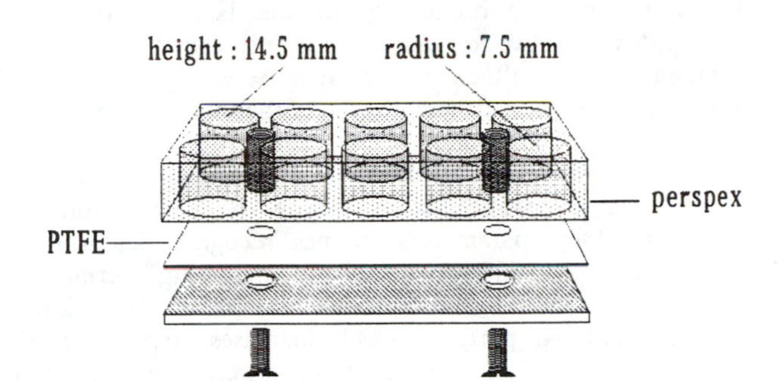

Figure 1. Mould for the preparation of cylindrical gel blocks.

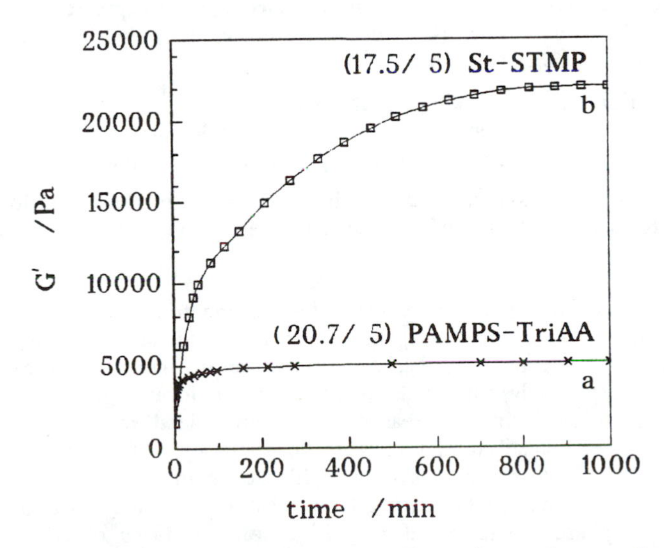

Figure 2. Time sweep measurement: Time dependence of gelation for a) radical-initiated copolymerization of AMPS and triallylamine b) base-induced cross-linking of starch with sodium trimetaphosphate.

Results and Discussion

All the investigations were carried out on uniform gel masses, so-called gel blocks. This is necessary to allow the results to be evaluated quantitatively, with a high degree of accuracy and reproducibility. The measurements of the material properties are thus restricted exclusively to the behavior of the macroscopic structure of the polymer networks and are not influenced by the particle size or surface phenomena. These influences can then be changed deliberately, with the aim of achieving qualitative and quantitative functions by comparison with the results obtained from the gel blocks.

The preparation of uniform gel blocks allows critical parameters to be recognized even during the preparation state; during the preparation of gel suspensions and dispersions, these parameters are not recognizable, or if so, only to a limited extent. This is particularly true of the increasing thermodynamic incompatibility during synthesis, which manifests itself as phase separation and inhomogeneity, and of the soluble portion, which increases strongly in the region of critical concentrations. The following limitations were therefore imposed on the gels to be investigated.

- All the components are to have as good a solubility as possible.
- The total polymer concentration (C) must be considerably greater than the critical overlap concentration (C*) of a correspondingly synthesized linear polymer (*21, 23, 24, 50;* C*-Theorem).
- The concentration of cross-linking agent must be chosen so that it lies in a range that supplies homogeneous, geometrically stable, and manipulable gel blocks that, nevertheless, must still have high degrees of swelling.
- Gelation should occur as rapidly as possible to prevent phase separation due to progressing incompatibility and a change in weight and volume due to evaporation.

The alkaline cross-linking molecules bisacrylamide and triallylamine were used for the gelation of the synthetic hydrogels. These molecules are characterized by their low solubility in neutral and basic solutions (approximately 150 mg/10 g and 300 mg/10 g, respectively) and, consequently, tend to form aggregates or give rise to phase separation. For this reason, they were added in strongly acidic solution, thus achieving a substantial increase in solubility. Because the redox initiator chosen, TEMED/persulfate, is only highly active in a narrow pH range (pH 6.5 to 8), gelation has to be initiated by a subsequent neutralization. To prevent a possible phase separation at very high concentrations of cross-linker, an unusually high concentration of initiator was employed, which resulted in gelation within 5 to 30 s in every case. When starch, which is insoluble in water, is used as the basis for covalently cross-linked, homogeneous gel blocks, it must first be pasted. This can be carried out by heating it to between 90 and 140 °C, or by sharply increasing the basicity of the solution.

The highly viscous solution that results from these processes no longer permits homogeneous addition of the cross-linker. If the cross-linker is slightly volatile (e.g., epichlorohydrin), gelation can only be achieved by pasting in basic solution at low temperatures. As follows from Figure 3 in the light of the pasting of a 10% w/w starch solution, a dramatic rise in viscosity does occur until immediately after a certain minimum concentration of NaOH has been attained

(*25, 26*). If the basicity is increased further, the viscosity decreases dramatically. A further fact emerging from the figure is that the pasting process requires a certain time to reach completion, during which the viscosity decreases further (Figure 3, curves a and b).

The region of minimum hydrolysis concentration is of particular interest for the formation of covalently cross-linked gels, because the viscosity here is sufficiently high to obtain geometrically stable, highly elastic gel blocks. In addition, the time-dependent reduction in viscosity appears to be at its lowest in this region. If the cross-linker is added shortly before reaching the minimum pasting concentration either in a homogeneous form (e.g., trimetaphosphate) or at least in a uniform suspension (e.g., epichlorohydrin), the addition of a few drops of NaOH solution is sufficient to bring about an immediate and dramatic rise in viscosity, which should prevent phase separation as far as possible. The submicroscopic crystalline regions can then be broken down by subsequent heating.

Knowledge of the mechanism and the reaction conditions can provide a rough guide to the expected network structure. For synthetic hydrogels, a network with a large number of free chain ends may be inferred from the high concentration of initiator. A correspondingly high percentage of cross-linker molecules is thus required to combine the multitude of chains that have arisen. The number of side chains and free chain ends increases considerably if cross-linking is not fully effective. In radical copolymerization, the reactivity is very high and is associated with a sharp rise in viscosity, thus making a relatively low effectiveness of cross-linking very probable. This structural assumption is supported by the smallest cross-linker concentration that could be employed experimentally and yet could still produce geometrically stable and manipulable gel blocks. This concentration lay between 2.5 and 10 mol-% for the gels prepared. Figure 4 (left) shows diagrammatically how such a network may be conceived.

As is represented schematically in Figure 4 (right), a structural conception of the network construction can also be derived for starch gels from the reaction conditions. If it is taken into account that the starch molecules possess a certain surface configuration in the native state and that they only start to swell and dissolve slowly when subjected to high temperatures or strongly alkaline conditions, then the covalent cross-linking that occurs immediately after the minimum alkalinity has been exceeded will take place largely in the surface region of the starch spherules. Cross-linking within the unpasted regions will therefore only occur slowly with increasing dilution and takes place to a considerably lesser degree. The cross-links that have previously occurred on the surface can impose structural limitations before the pasting has completely terminated. As a result, regions of greater and lesser polymer density may occur. A further indication of this is given by the marked turbidity immediately after synthesis, which slowly decreases with thermal treatment.

Theory of Swelling. The potential difference that arises when a polymer is diluted by a solvent is compensated for by concentration equalization and the change in the hydrodynamic volume of the polymer. The potential difference can be described by the change in the free energy of mixing alone, as deduced by Flory and Huggins (*27, 28*). On the other hand, network structures can only expand to a limited extent because of their network parameters. During this process, they produce elastic stress within the polymer chains that counteract the dilution forces. As a consequence of the classic swelling theory of Flory and Stockmeyer (*15, 29*), swelling only takes place until the free energy of mixing,

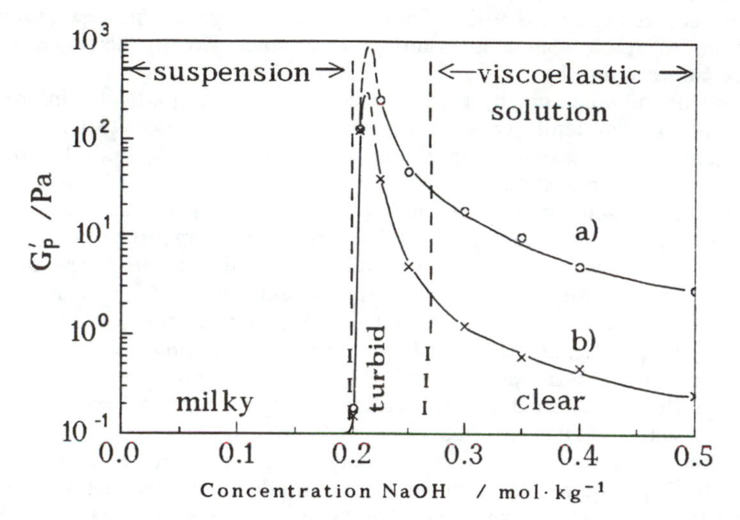

Figure 3. The swelling and solution behaviour during the basic pasting process of native starch represented by the plateau modulus as a function of the alkali concentration and its time dependence.
a) 10 minutes after preparation b) 48 hours later

Figure 4. Diagrams of network structures for (left) the radical-initiated copolymerization of bifunctional and tetrafunctional monomers and (right) the base-induced cross-linkage of native starch.

ΔF_{mix}, is equal to the free elastic energy, ΔF_{el}. The dynamic equilibrium state of swelling is thus described by:

$$\Delta F_{mix} = \Delta F_{el} \qquad (1)$$

A relationship has been derived from these conditions by Flory and Rehner (*30*), which is represented below in its most original and most general form for a network with tetrafunctional cross-linking points:

$$v_1 (\nu_e/V_o) (\upsilon_{2m}^{1/3} - \upsilon_{2m}) = -[\ln (1 - \upsilon_{2m}) + \upsilon_{2m} + \chi_1 \upsilon_{2m}^2] \qquad (2)$$

where v_1 is the molar volume of solvent, ν_e/V_o is the elastically effective number of chains in a real network according to the volume of the undeformed polymer network, υ_{2m} is the volume fraction of polymer in swollen equilibrium with pure solvent, and χ_1 is the Flory–Huggins polymer/solvent interaction parameter divided by kT.

The volume fraction of polymer is assumed to be equal to the reciprocal of the degree of swelling ($\upsilon_{2m} \sim 1/q$) and therefore refers to the solvent-free volume of polymer and not to the preparation state, which contains solvent. The swelling equation was derived on the basis of polymers cross-linked in the absence of solvent.

In addition, this equation was derived under the condition that the cross-linking points have a fixed position within the network and are thus displaced uniformly as the dimensions change (affine network). The theory was not, however, able to explain some effects in real networks and has therefore been superseded by a considerably more complex theory (Flory and Erman; *31, 32*), which permits free mobility of the entanglement points (phantom network). However, the fact that this theory has a great number of parameters that are not easy to determine would appear to render it extremely unsuitable for practical purposes. Studies by Queslel and Mark (*33, 34*) provide excellent synopses and explanations of these problems.

The swelling equations must be extended if the swelling behavior of hydrogels with strongly interactive functional groups is to be described. The swelling behavior of these kinds of gels is strongly influenced by the number and strength of the salt ions and polymer ions, as well as the dielectricity of the medium, both within and outside the gel. By inserting an additional term into the swelling equation, which takes into account the ionic exchange and the resulting osmotic pressure on the basis of the Donnan theory, Flory was able to establish a relationship that is valid for ionic networks. According to this, the swelling equilibrium is attained when the additional forces ΔF_{ion}, resulting from the osmotic pressure of the ions, are in equilibrium with the mixing forces and the elastic retention forces. The following then applies (*15*):

$$\Delta F_{ion} + \Delta F_{mix} + \Delta F_{el} = 0 \qquad (3)$$

Here it is assumed that there is an exchange of mobile ions that ensures a charge neutrality so that no membrane potential builds up between the gel and the external solution. According to this, the following generally applies:

$$\upsilon_{2m}(i/v_u) - \nu(c_s^* - c_s) = (1/v_1)\,[\ln(1 - \upsilon_{2m} + \upsilon_{2m} + \chi\upsilon_{2m}^2] + (\nu_e/V_o)(\upsilon_{2m}^{1/3} - \upsilon_{2m}/2) \quad (4)$$

where (i/v_u) is the concentration of fixed charges referred to the unswollen network; $\nu = \nu_+ + \nu_-$ are stoichiometric factors of the given salt; and c_s and c_s^* are the molar salt concentrations within a polyelectrolyte and in the surrounding medium, respectively. A generally valid transformation of the equation according to υ_{2m} is not possible so that a simplification was only carried out for the extreme cases $c_s^* \ll c_s$ (in pure water) and $c_s^* \gg c_s$ (in highly concentrated salt solution):

$$c_s^* \ll c_s\text{:} \quad q^{2/3} \sim (i/v_u)/(\nu_e/V_o) \quad (5)$$

$$c_s^* > c_s\text{:} \quad q^{5/3} \sim [(i/v_u)^2 (1/4\ I^*) - (0.5 - \chi_1)/v_1)/(\nu_e/V_o) \quad (6)$$

I^* $(1/2\ \nu\ c_s^*\ z_+z_-)$ is the ionic strength of the counter ions in the external solution.

Results of the Swelling Investigations. To prevent any misunderstanding, it should be pointed out once again that all the swelling investigations were carried out on gel blocks that had been polymerized in solution. The degree of swelling is therefore given by the quotient of the swollen volume (V) and the volume in the preparation state (V_o). This is not identical to the degree of swelling (q), which is given by the reciprocal of the volume fraction of polymer and is only valid for investigations on gels synthesized in the absence of solvent.

Such a differentiation is necessary because the preparation state is distinguished by a minimum of elastic stress in the network. Both decreasing (mesh compression) and increasing (mesh elongation) gel volume should result in a rise in the network stress (*35*).

Figures 5a–5d show swelling investigations on anionic 14.52% w/w PAMPS-TriAA gels with varying initial concentrations of cross-linker. The degree of swelling is represented as a function of the ionic strength of cationic counter ions having different charges. In pure water and in solutions with a very low counter ionic strength, the degree of swelling (Q) decreases as the initial concentration of cross-linker increases (reduction of mesh width). With the increase of counter ionic strengths, the intermolecular and intramolecular repulsion forces of the network ions will be noticeably compensated (ΔF_{ion} becomes smaller). Consequently, there is a continual decrease in the degree of swelling, as can be seen from the examples of gel behavior in solutions of the counter ions Na^+, Mg^{2+}, and Al^{3+}. The higher the charge on the counter ion is, the greater influence it has ($Al^{3+} > Mg^{2+} > Na^+$).

Considering, first of all, the region of very high ionic strength in Figure 5d, it can be seen that the degree of swelling strives to attain the same limiting value, irrespective of the type of counter ion. The polymer ions are completely screened from one another in this region ($\Delta F_{ion} \sim 0$). The swelling behavior should thus match that of a correspondingly uncharged polymer network having the same topology, and it should therefore be possible to describe it by $\Delta F_{mix} = \Delta F_{el}$ (*36*).

In the regions of moderate counter ionic strength, effects occur in the presence of multivalent counter ions that can no longer be explained by charge

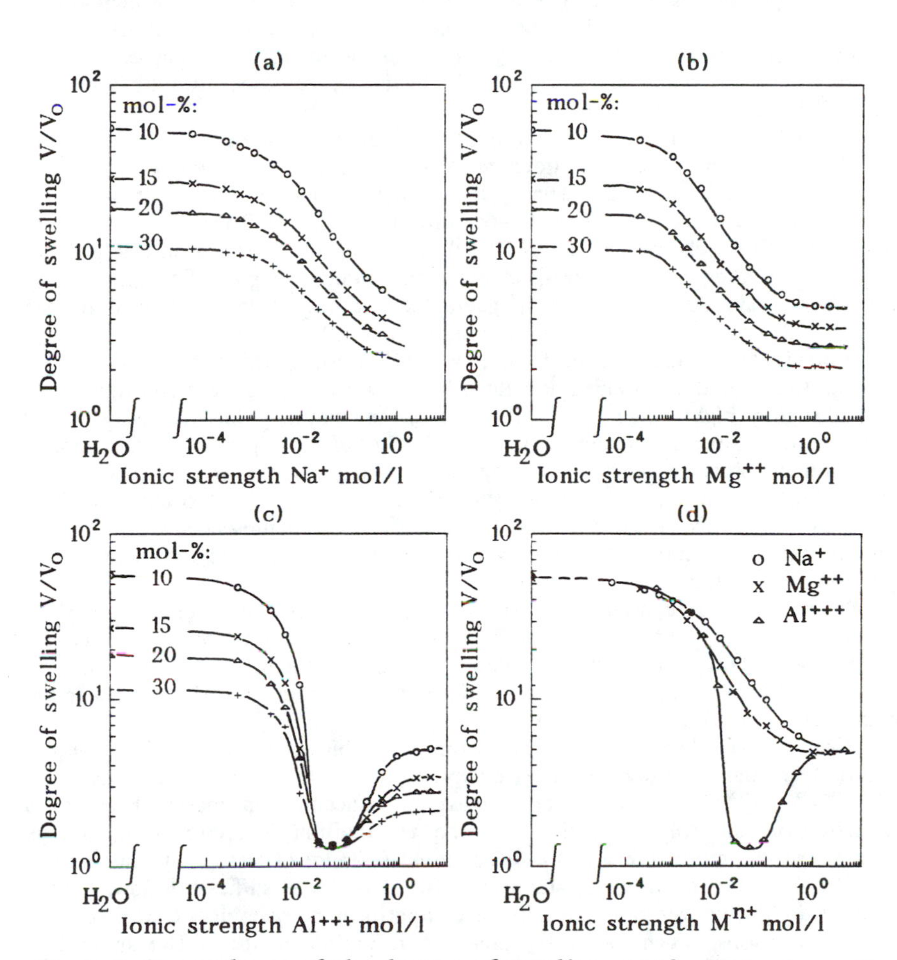

Figure 5 . Dependence of the degree of swelling on the ionic strength of the cationic counter ions (a) Na^+ (b) Mg^{++} (c) Al^{+++} of 14.51 % w/w PAMPS-TriAA gels having cross-linker concentrations of 5, 10, 15 and 20 mol-% TriAA. (d) Comparison of the swelling behaviour of 14.51 % w/w PAMPS gels with 10 mol-% TriAA in Na^+, Mg^{++} and Al^{+++}.

screening of the polymer ions alone (*37*). As Figure 5d shows from the swelling behavior towards Al^{3+} solutions, the swelling curve passes through a minimum. This behavior has already been observed for polyacrylamide-*co*-acrylate-*co*-bisacrylamide gels in Mg^{2+} solutions. If the degree of swelling falls below that of the preparation state (10°), the gel blocks becomes turbid and increasingly deformed. As observed in the swelling of polyacrylamide-*co*-acrylate-*co*-bisacrylamide gels in Al^{3+} solutions, this can result in gel collapse. Similar results have been observed by Ricka and Tanaka during their swelling investigations with Cu(II) salt solutions (*38*).

To obtain a clear characterization of this counter ionic effect, the gels that had been swollen in salt solutions were then examined to see whether they would regain their maximum degree of swelling in pure water. This reversible behavior is illustrated in Figures 6a–6c, with the 17.62/10 PAMPS-TriAA gel as examples. The gels swollen in Na^{+} solutions regain their maximum possible degree of swelling in water irrespective of the ionic strength. For gels treated with Mg^{2+} and Al^{3+} this is only guaranteed if the ionic strength does not exceed $1–2 \times 10^{-2}$ mol/L (concentrations of 5×10^{-3} mol/L).

Above Mg^{2+} ionic strengths of 1×10^{-1} mol/L ($[Mg^{2+}] > 5 \times 10^{-2}$ mol/L), the degree of swelling merely achieves an approximately constant value that is slightly higher than half of the possible maximum. Gels treated with Al^{3} with an ionic strength greater than 1×10^{-1} mol/L ($[Al^{3}] > 2 \times 10^{-2}$ mol/L) can even be observed to shrink during the swelling in pure water. The degree of swelling is constant and assumes the minimum value of the original swelling curve in Al^{3}. This behavior of the anionic gels in the presence of multivalent cations can be attributed to the formation of complexlike junctions (ionotropic gelation) and is illustrated in Figure 7.

With increasing concentration, the number of cations that are connected to more than one polymer segment increases, leading to ionotropic junctions. Beyond a certain concentration, which depends on the charge density of the polyelectrolyte gel, the multivalent cations are coordinated with a maximum number of segments.

How the gel then behaves depends on the stability of the resulting complex bonds. If the stability of the complex is relatively high, for example, $[(R{-}COO)_{x}Al]^{(x\,-\,3)+}$, a further increase in concentration merely leads to a reduction in the solvent quality. As the Θ-condition is approached, the gel becomes visibly more turbid and a discontinual collapse of the gel system occurs.

On the other hand, if the stability of the complex is sufficiently low, further increase in the concentration results in competition between the cations due to the polymer ions, which lowers the mean coordination number. The amount of coordinated or ionotropic junctions decreases and the degree of swelling rises again. In the extreme case of high concentrations, the mean coordination number is smaller than 1 and the swelling behavior of the gels can be described by the ionic strength, irrespective of the charge number of the cations.

As follows from the reversibility studies on concentrated solutions of complex-forming cations, a certain proportion of the multivalent cations remains irreversibly bound in the gel, which then assumes an almost constant degree of swelling.

Quantitative Evaluation of Swelling Investigations. If Flory's swelling equations 5 and 6, which were derived for the two extremes, in pure water and at high salt concentrations, are resolved according to the reciprocal of the degree of swelling

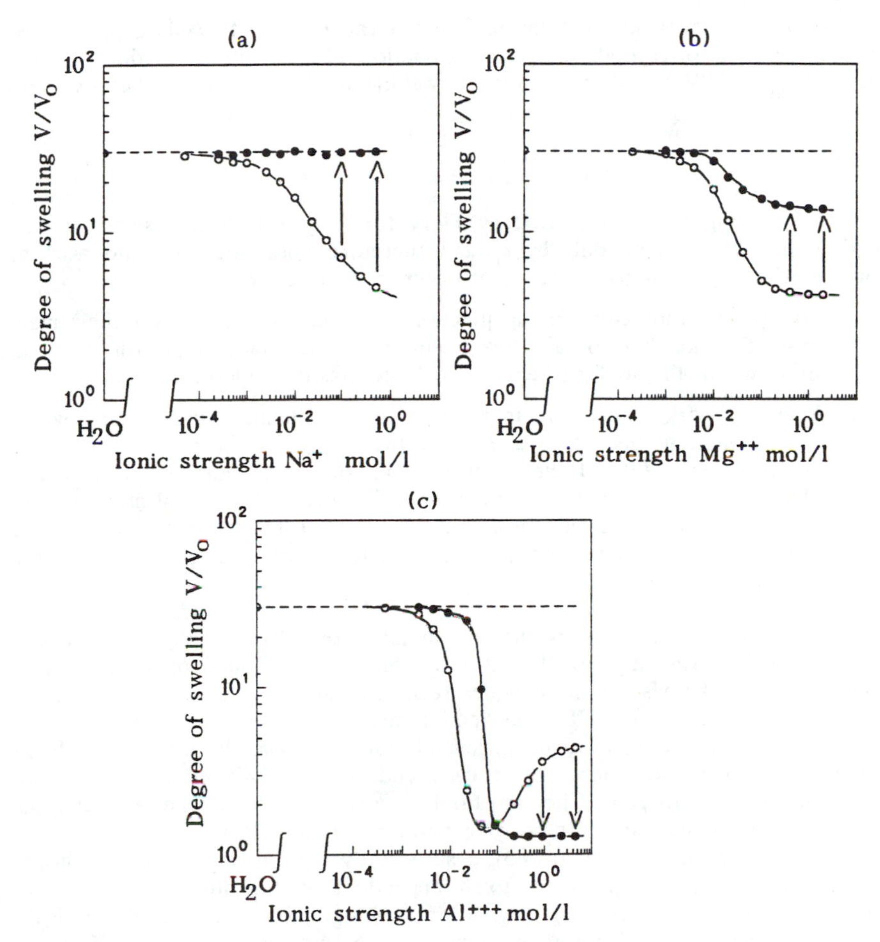

Figure 6. Reversible behaviour of 17.62 % w/w PAMPS-TriAA gels in H_2O, which had previorsly been swollen in Na^+, Mg^{++} and Al^{+++} solutions.

(1/q), the following ratio results:

$$1/q \sim (\nu_e/V_o)^{\beta} \qquad (7)$$

where $\beta = 3/5$ at high salt concentrations and $\beta = 3/2$ in pure water. Consequently, the following should apply to the intermediate concentration ranges: $3/5 < \beta < 3/2$.

Equation 8 gives the relationship between the degree of swelling (q) defined by Flory, which corresponds to an approximation of the volume fraction of polymer ($1/\upsilon_{2m} = V/V_p$), and the degree of swelling relating to the synthesis volume ($Q = V/V_o$):

$$q = Q\, V_o/V_p \qquad (8)$$

The swelling equation derived by Flory for polyelectrolytes assumes a network structure described only by ν_e and therefore does not take into account some of the important parameters that occur in real gel systems:

1. Gel systems with cross-linking junctions that do not react with their maximum functionality contain free chain ends and branches, which exert an influence on the swelling behavior and the elastic properties.
2. A real network system may, in addition to the chemical links, have physical entanglement points, which also affect the swelling behavior and the elastic properties (*39, 40*). Their proportion in the total number of elastically effective chains depends on the chemical nature of the polymer network components, the cross-linking density, the total substrate concentration, the number of free chain ends, and the physical conditions pertaining to the preparation state.

What emerges from this is that the number of effective, physical entanglement points in covalently cross-linked gels cannot be calculated or, if so, only to a limited extent because of the complex relations (*24, 41*).

As follows from Figure 8, a distinction has to be made between two types of elastically effective, physical entanglement points. Physically interlinked chains (Figure 8a), which are connected at each end by elastically effective junctions, are permanent or trapped. They are firmly incorporated in the network and can thus only be disentangled by destroying covalent bonds. However, if there is at least one free chain end (Figure 8b), a sufficiently high stress can lead to disentanglement. Their quantity is not solely dependent on the preparation state but also diminishes with increasing degree of swelling. The number of elastically effective chains (ν_e) is therefore made up of the number of both chemical and physical links :

$$\nu_e = \nu_{e(ch)} + \nu_{e(ph)} \qquad (9)$$

The degree of swelling of gels with the same theoretical network degree of polymerization ($M_c/M_o = n_{tot}/\nu_o$) should, in the ideal, increase proportionally with the total substrate content. However, a counteractive effect is observed for most real gel systems. The number of physical entanglement points in real gels generally increases with the number of network chains. This influence alone, however, is unlikely to suffice as an explanation of this effect. The effectiveness

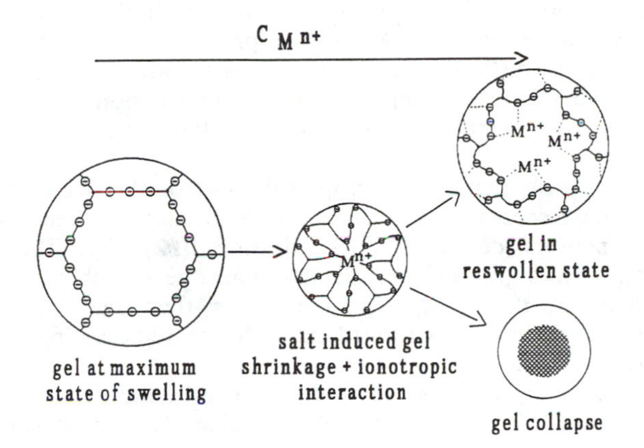

Figure 7. : Influence of multivalent cations on the swelling behaviour of anionic polyelectrolyte hydrogels.

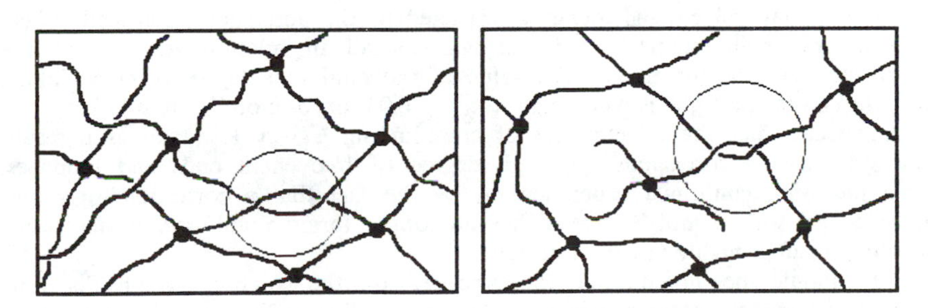

Figure 8. Types of elastically effective, physical entanglement points a) permanent or trapped b) temporary

of the chemical cross-linking also increases with total substrate content at the same level of theoretical network degree of polymerization. In analogy to equation 7, and taking equation 8 into account, the following ratio then applies between the reciprocal degree of swelling (1/Q) and the reciprocal of the mean network degree of polymerization (ν_e/n_{tot}):

$$1/Q \sim (\nu_e/n_{tot})^{\beta} = (\nu_{e(ch)}/n_{tot} + \nu_{e(ph)},n_{tot})^{\beta} \qquad (10)$$

Because ν_e/n_{tot} cannot be determined in theory and also depends on the swelling state of the gel, it has proved to be practically useful to plot the reciprocal degree of swelling against the reciprocal of the theoretical network degree of polymerization. The latter is associated with the mole fraction of cross-linker (degree of cross-linking: X_c) in the preparation state via the functionality of the cross-linking molecules (f). The following is valid: $\nu_o/n_{tot} = [(f/2)X_c]_o$.

Figure 9 is a linear plot of the reciprocal value of the degree of swelling as a function of the respective Na^+ concentration against the reciprocal value of the theoretical network degree of polymerization $[(f/2)X_c]_o$. The lines linking the experimental values give an approximately linear relationship.

The gradient of the straight lines rises as the ionic strength of Na^+ increases, with the gradient in saturated salt solution representing an upper limit and that in pure water a lower limit.

If no physical entanglements occurred in the gels, the curves would have to intersect at the point $V_o/V = 0$ (infinitely swellable, i.e., not cross-linked), which represents the minimum critical degree of cross-linking in the synthesis mixture. If the line through the experimental measurements in pure water is extrapolated to $V_o/V = 0$, a degree of cross-linking is obtained that approximates very closely that of the minimum critical degree of cross-linking $(X_c^*)_o$.

At the high degree of swelling in pure water, it can be assumed that the number of physical entanglements is restricted to the permanent, elastically effective entanglement points and is negligibly small in relation to that of the covalently effective junctions. The value of the minimum degree of cross-linking for the given example is very high ($X_c \sim 0.05$ or 5 mol-% cross-linker) and demonstrates that the effectiveness of cross-linking is very low, and as a result, the gels have a correspondingly high degree of free chain ends and branches. This has been confirmed experimentally by the fact that a correspondingly synthesized gel with 5 mol-% cross-linker no longer forms a uniform, geometrically stable gel but a highly viscous solution.

It should be noted that the linearity of the swelling graphs is only guaranteed to any extent within a certain range of cross-linking. The concentration of cross-linker employed must be sufficiently larger than the minimum critical concentration of cross-linker. Otherwise the influence of the physical entanglements and the increasing sol content can no longer be ignored. If cross-linking agents are used with a tendency towards phase separation at higher concentration and that consequently have macroscopic inhomogeneities, deviations from the linear course of the graph will occur. Within the given range of cross-linker concentration, the swelling behavior can be approximated by a linear relationship:

$$(1/Q)_i = a_i[(f/2)X_c]_o + b_i \qquad (11)$$

Figure 10 shows the swelling curves for various hydrogel systems that were swollen in pure water. Some of the examples have been taken from publications and transformed correspondingly. As can be seen from these examples, the linearity is not just accidentally valid for one network system, but can also be regarded as a very good approximation within the limits given.

As is illustrated later by rheological investigations, the deviations in the behavior of the starch gels can be attributed to an increase in inhomogeneities. The functionality of the cross-linking points in the starch gels was set at a value of $f = 4$ for both epichlorohydrin and the trimetaphosphate conversion. The starch–epichlorohydrin gels have small degrees of swelling so that it can be assumed that the proportion of elastically effective entanglement points in the gel is not negligible. Because of their phosphate groups, the St-STMP gels have a certain proportion of ionic character, which explains their considerably higher degrees of swelling in comparison with St-ECH gels. The gradients of the straight lines are directly proportional to the effectiveness of the chemical cross-linking. If $1/Q$ is plotted against $[(f/2)X_c]_o$, the different gel systems can be compared with one another. According to this, the PMAAc-DVB gels have the highest effectiveness of cross-linking. The minimum concentration of cross-linking agent required to produce a gel is approximately 0.4 mol-% of divinylbenzene. The comparison of the PAMPS-BisAAm and PAMPS-TriAA gels proves conclusively that bisacrylamide represents the considerably more suitable cross-linking reagent. The effectiveness of cross-linking with bisacrylamide is some 5 times higher than that with triallylamine.

The effectiveness of cross-linking in PDAEA-TriAA gels is extremely low because of the limited compatibility of the alkaline monomer DAEA with triallylamine in solution. Consequently, these two components have a limited reactivity with one another.

Apart from the ionic strength, the swelling equation derived by Flory contains the decisive χ-parameter, which is not at all easy to determine experimentally. As shown before, the swelling equation $1/q \sim (\nu_e/V_o)^\beta$, with $0.6 < \beta < 1.5$ as the ionic strength variable, can be replaced by a linear equation with $\beta = 1$ and an additive term (equation 11).

If the straight lines in Figure 9, which are dependent upon the ionic strength, are extrapolated beyond the negative region, they are all found to intersect within another very small region from which a mean value can be determined. This mean is assumed to be the intersection of the straight lines in the coordinates $[(f/2)X_c]_p/(V_o/V)_p$. Via this intersection, it is possible to substitute one of the two individual constants a_i or b_i.

With the relationship $b_i = (V_o/V)_p - a_i[(f/2)X_c]_p$ and equation 11, the following extended form can be written:

$$(1/Q) = (V_o/V_i) = a_i\{[(f/2)X_c]_o - [(f/2)X_c)_p]\} + (V_o/V)_p \qquad (12)$$

The coordinates of the intersection can be determined with sufficient accuracy from relatively few readings in concentrated salt solution and in pure water. The only remaining unknown term is the gradient of the individual straight lines a_i, which must be a function of the ionic strength and the charge density of the polymer ions. For the 17.63% w/w PAMPS-TriAA gels, the coordinates of the intersection were determined as $[(f/2)X_c]_p = -0.247$ and $(V_o/V)_p = -0.053$. In Figure 11, the values of the gradient are plotted against the ionic strength of the counter ions in the external solution.

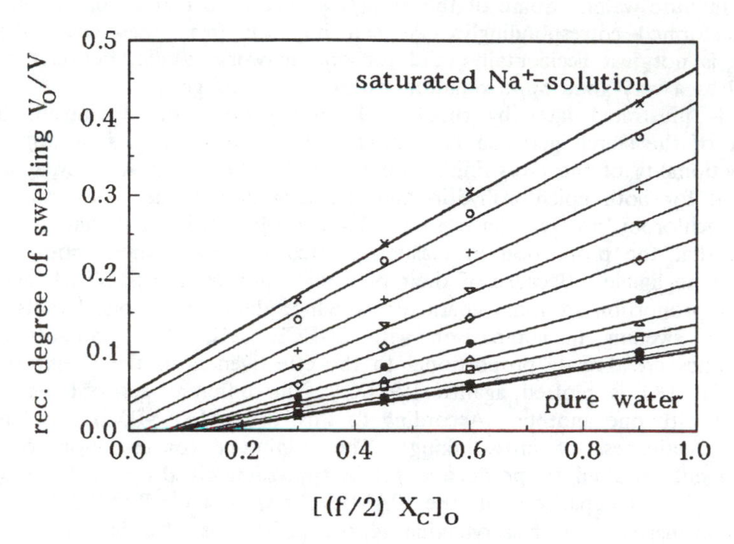

Figure 9. Relationship between the reciprocal degree of swelling in Na^+ solutions and the reciprocal value of the theoretical network degree of polymerization, exemplified by the 14.51% w/w PAMPS-TriAA gels.

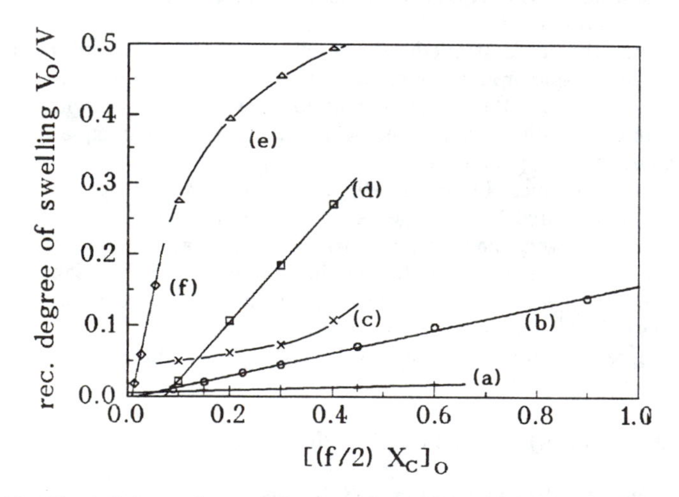

Figure 10. Plot of the reciprocal degree of swelling against the reciprocal value of the theoretical network degree of polymerization $(f/2X_c)_o$ for a, 25.06% w/w PDAEA-TriAA; b, 20.73% w/w PAMPS-TriAA; c, 10% w/w starch-sodium-trimetaphosphate (St-STMP); d, 10.36% w/w PAMPS-BisAAm; e, 10% w/w starch-epichlorohydrin (St-ECH); f, polymethacrylic acid-co-divinylbenzene (PMAAc-DVB). Data taken from Flory (15).

The limits of the gradients in pure water and in concentrated salt solution are labeled a_H and a_s, respectively. The curve may be described by an equation that has similarities to Langmuir's adsorption isotherms: $1/Y = (K/S)1/X + 1/S$. Taking $1/Y$ to be equal to $1/(a_i - a_H)$, a consideration of the limiting values must give $S = a_s - a_H$. Thus the following relationship is obtained for X as the ionic strength (I_{Na^+}) of the external Na^+ ions:

$$1/(a_i - a_H) = K/(a_s - a_H)\ 1/(I_{Na^+}) + 1/(a_s - a_H) \quad (13)$$

In Figure 12, $1/(a_i - a_H)$ is plotted against the reciprocal of the ionic strength of Na^+. The value a_s, which can only be determined inaccurately from Figure 11, can be obtained from the intercept of the ordinates, by using the limiting value for pure water (a_H). Good experimental results can be determined for the latter.

According to this, a value of ~0.513 is obtained for the limiting value of the gradient (a_s) in solution of high ionic strength and a value of ~1/19 for the constant K. This value correlates directly with the network density. For uncharged polymer networks, $K = 0$ so that the same gradient a_s is obtained irrespective of the ionic strength. This confirms the assumption put forward at the start of the chapter that, in solution of very high salt concentrations, an ionic gel behaves very differently from the corresponding uncharged gel.

By using equations 11 and 12, very little effort and very few measurements are required to determine the constants specific to any one system a_s, a_H, K, and the coordinates at the intersection $[(f/2)X_c]_p/(V_o/V)_p$ and the critical minimum concentration of cross-linking agent required in the synthesis.

Results of Rheological Studies. The rheological characterization of the gels was carried out by means of dynamically mechanical oscillation measurements in the disc/plate system of a rheometer at room temperature. The mean molecular weight between entanglement points (M_c) is proportional to the quotient of the total substrate concentration and the number of elastically effective network chains (ν_e). Rheological characterization of network structures by measurement of the plateau modulus (G'_p are specified by the number of elastically effective entanglement points via the theory of rubber elasticity. The following relationship is valid for a perfect network (*40, 42, 43*):

$$G'_p = [1 - (2/f)]\nu_e RT \ \langle r^2\rangle/\langle r_o^2\rangle \quad (14)$$

where $\langle r^2\rangle/\langle r_o^2\rangle$ is defined as the ratio of the mean square end-to-end distance of a segment in the swollen state to the mean square end-to-end distance that would be assumed for comparable segments if there was no constraint due to cross-links. For unswollen gels, the ratio is approximately equal to unity at constant temperature.

Investigations on Gel Blocks. Due to their extremely high degree of swelling, it was not possible to produce thin slices from the gel blocks without destroying them. Investigations on gel blocks were therefore carried out exclusively in the preparation state, which means that a comparison with the swelling investigations is only possible to a limited extent. It is worth noting that the number of elastically effective network chains of gel blocks includes a high proportion of physical entanglement points.

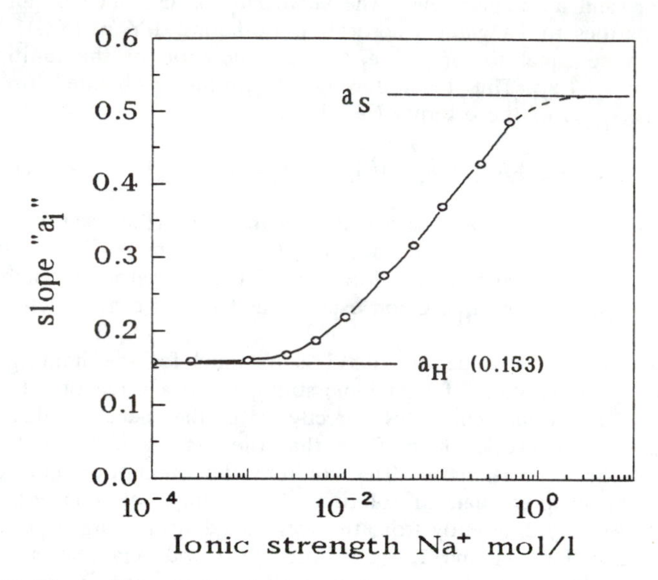

Figure 11. Plot of the individual gradients "a_i" of 17.62 % w/w PAMPS-TriAA gels as a function of the ionic strength of Na^+ ions.

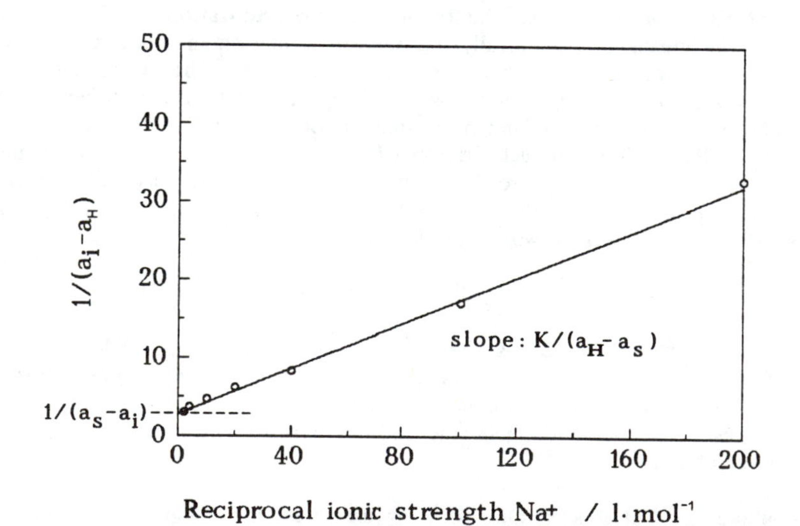

Figure 12. Determination of the limiting value for the gradient (a_S) in concentrated salt solution, the constant "K" from the intercept of the ordinates and the gradient in accordance with equation 12.

For this reason, a quantitative evaluation using equation 14 is subjected to a very high degree of error, because this equation was derived for ideal and perfect network systems. As descriptions of Ferry and Graessley state, a correction can be made by introducing a probability term for the proportion of physical entanglement points. However, this only makes sense for model networks that have been produced by cross-linking polymers with known molecular weights (*44, 45*). Investigations on real gel systems have established that networks that are as homogeneous as possible can be described qualitatively very well (*9, 19, 21, 38*).

Figure 13 shows the linear relation between the elastic modulus and the reciprocal of the theoretical network degree of polymerization for PAMPS-TriAA gels in the preparation state. As emerges from Figure 14, measurements of the plateau modulus as a function of the concentration of cross-linking agent represents a good method of detecting inhomogeneities in real networks. The peaks of the PAAm-BisAAm and starch gels show a maximum behavior that point to incompatibility of the network components. In contrast to PAMPS-BisAAm gels, the PAAm-BisAAm gels were cross-linked without neutralization. Because of the poor solubility in neutral and alkaline solutions, phase separation of the cross-linking agent occurs if the saturation point is exceeded. Regions of aggregation are generated in the cross-linker molecules that react like individual cross-linking sites, thus reducing the effective number of junctions. As was the case for the swelling measurements, the gradient in the linear region can be regarded as a measure of the efficiency of cross-linking.

Investigations on Solutions of Gel Particles. Technical applications of hydrogels involve their use in form of solutions of gel particles, suspensions, or dispersions. The following, therefore, aims to demonstrate some of the ways in which they can be rheologically characterized, by using solutions of gel particles with defined particle size.

Figure 15 represents the plateau moduli of aqueous particle solutions for different initial concentrations of cross-linking agent. The initial total substrate content used in all syntheses was 14.51% w/w. The curves show a rapid change of the slope at a critical gel concentration with respect to the cross-linker concentration. The values at the inflections of the curves, represented by the broken lines, describe the maximum water absorption capacity.

Concentration-dependent rheological measurements therefore permit both the equilibrium degree of swelling and the accompanying elasticity value to be determined. If the readings are taken with the solution being subjected to a defined load, then it is possible to ascertain the corresponding retention capacity of the gels.

Corresponding investigations for gel particle solutions of the same size have been prepared with different initial substrate contents. The results are shown in Figure 16. As emerges from a comparison with Figure 15, the gradients through the intersections of the maximum degree of swelling have the same value of ~1.7. The plateau modulus of the fully swollen PAMPS-TriAA gel particles is therefore a function of the particle concentration and can be described by the following relationship:

$$G'_p \sim C_{GEL}^{1.7} \qquad (15)$$

Investigations by Hild, Munch and Zrinyi, and Horkay (*47–49*) on polystyrene–divinylbenzene and polyvinylacetate gels in thermodynamically good

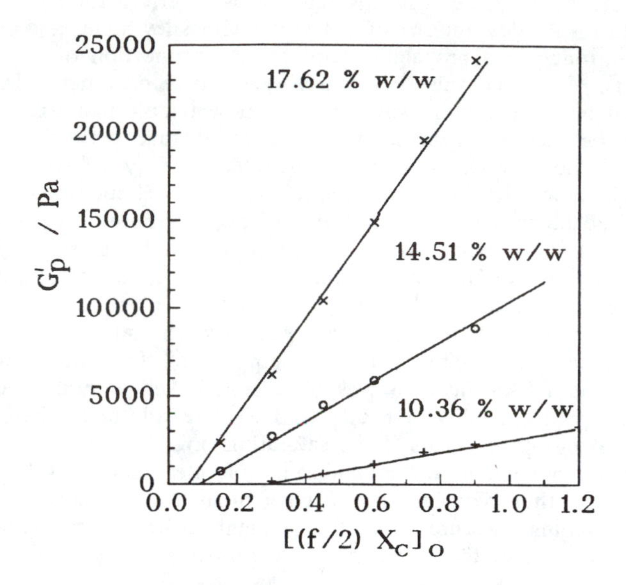

Figure 13. Plateau modulus as a function of the reciprocal of the theoretical network degree of polymerization $[(f/2)\ X_c]_o$ for PAMPS-TriAA gels with different initial total substrate contents.

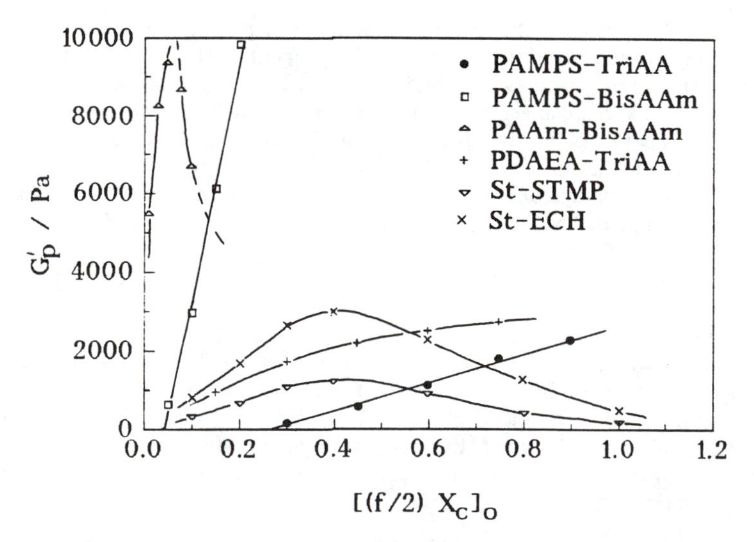

Figure 14. The plateau modulus as a function of the theoretical degree of cross-linking for synthetic and biopolymer gel blocks.

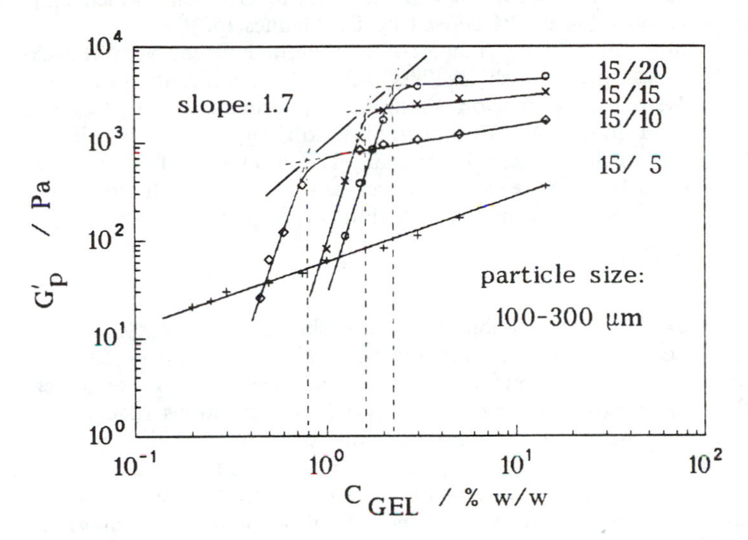

Figure 15. Plateau modulus as a function of the concentration for solutions of 14.51% w/w PAMPS-TriAA gel particles with various contents of cross-linking agent.

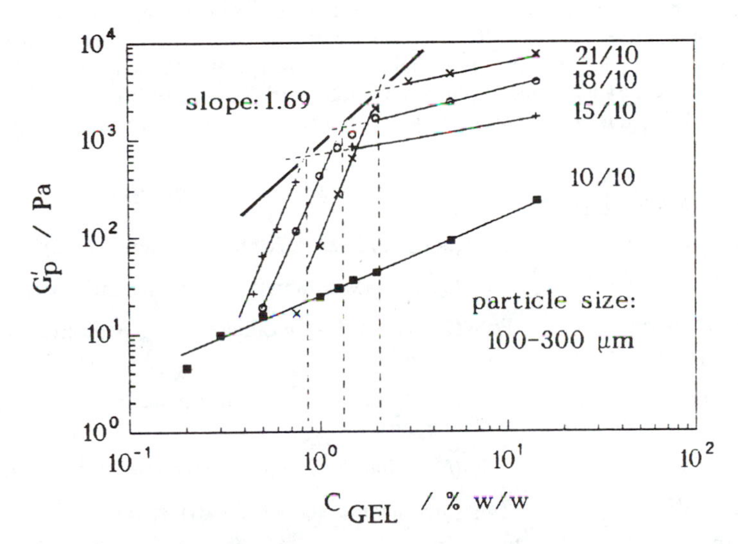

Figure 16. Plateau modulus as a function of the concentration for solutions of 14.51% w/w PAMPS-TriAA gel particles with different initial substrate contents.

solvents give values of between 2.0 and 2.6 for the exponent, which agree relatively well with the value of 9/4 derived by De Gennes (*50*).

These results can only be compared to a limited extent with the exponent for PAMPS-TriAA gels, which exhibits relatively significant deviations. The investigations were carried out on solutions of particles, which means particle size may have an influence. This possibility still requires further investigation before definitive conclusions can be drawn. The influence of ionic groups may also be another factor. However, this effect can only be clearly detected in fully swollen gel blocks, which are in fact very difficult to prepare.

Conclusions

By means of swelling investigations on highly absorbent hydrogels, it was shown that the degree of swelling in (monovalent) salt solutions can be determined quantitatively. A simplified method is used which avoids the χ parameter, which is difficult to determine experimentally. Two linear equations (equations 12 and 13) can be used with very few experimental data to provide the parameters necessary for predicting the degree of swelling at any desired salt concentration. This offers a very useful method, above all for the practical applications of gels with a strong degree of swelling, because it allows a direct comparison to be made of different gel systems. In conjunction with rheological measurements, regions where inhomogeneities occur can be determined for any gel system.

The methods used in industry for determining the absorption capacity of gel particle solutions and suspensions, which involve weighing the gel, may be subject to a high degree of error because of the incorporation of solvent molecules or adhesion to the surface. Rheological investigations of gel particle solutions provide data for determining not only the absorption capacity but also the accompanying elasticity values, which correlate with the retention capacity. For highly absorbent gels, this represents a rapid and highly accurate method of characterization.

Abbreviations and Symbols

PAAm-AAc-BisAAm	Poly-(acrylamide-*co*-acrylate-*co*-bisacrylamide)
PAAm-AAc-TriAA	Poly-(acrylamide-*co*-acrylate-*co*-triallylamine)
PAMPS-BisAAm	Poly-(2-acrylamido-2-methylpropane sulfonic acid-*co*-bisacrylamide)
PAMPS-TriAA	Poly-(2-acrylamido-2-methylpropane sulfonic acid-*co*-triallylamine)
PDAEA-TriAA	Poly-(*N,N*-dimethylaminoethylacrylate-*co*-triallylamine)
PMAAc-DVB	Poly-(methacrylic acid-*co*-divinylbenzene)
St-STMP	Starch–sodium trimetaphosphate
St-ECH	Starch–epichlorohydrin
TEMED	*N,N,N′,N′*-tetramethylethylenediamine
a_i, b_i	Individual constants
a_H, a_s	Limit of the gradients in pure water and in concentrated salt solution, respectively

c_s, c_s^*	Molar salt concentration within a gel and in the surrounding medium, respectively
f	Functionality of cross-linking unit
G^*, G', G''	Complex shear modulus, storage modulus, and loss modulus, respectively
G'_p	Storage modulus in the plateau region
I^*	Ionic strength of the counter ions in the external solution
(i/v_u)	Concentration of fixed charges referred to the unswollen network
M_o	Mean average molecular weight of the monomer or repeating units
M_c	Mean molecular weight per chain of a perfect network
n_{tot}	Total number of monomer or repeating units
q	Degree of swelling according to Flory ($\sim 1/\upsilon_{2m}$)
Q	Degree of swelling (V/V_o) referred to the preparation volume
v_1	Molar volume of solvent
V, V_o	Volume of the gel in the equilibrium state of swelling and the preparation state, respectively
X_c	Mole fraction of cross-linking agent for a perfect network
X_c^*	Minimum critical mole fraction of cross-linking agent for a perfect network
z_+, z_-	Valencies of cation and anion, respectively
ΔF_{el}	Free energy change for elastic deformation
ΔF_{ion}	Free energy change from osmotic pressure of the ions
ΔF_{mix}	Free energy change on mixing
ν_o	Number of chains of a perfect network
ν_e	Total number of elastically effective chains in a real network
$\nu_{e(ch)}$, $\nu_{e(ph)}$	Number of elastically effective chains generated by chemical and physical links, respectively.
υ_{2m}	Volume fraction of polymer in swollen equilibrium with pure solvent
χ_1	Flory–Huggins polymer/solvent interaction parameter divided by kT.

References

1. Brand, K. A.; Goldman, S. A.; Inglin, T. A. U.S. Patent 4 654 039, 1987.
2. Masduda, F. *CEER, Chem. Econ. Eng. Rev.* **1983,** *15,* 19.
3. Stuckenbrock, K.-H.; Werner, G. *Melliand Textilber.* **1985,** *65,* 173.
4. Park, K. *Biomaterials* **1988,** *9,* 435.
5. Ng, Chiong O.; Tighe, B. J. *J. Biomed. Mater. Res.* **1976,** *8,* 118.
6. Kickhoefen, B.; Wokalek, H.; Schell, D.; Ruh, H. *Biomaterials* **1986,** *7,* 67.
7. Refojo, M. F.; Leong, F. L. *J. Biomed. Mater. Res.* **1981,** *15,* 497.
8. Kossmehl, G.; Klaus, N.; Schäfer, H. *Angew. Makromol. Chem.* **1984,** *124,* 241.
9. Dusek, K.; Prins, W. *Adv. Polym. Sci.* **1969,** *6,* 1.
10. Clark, A.H.; Ross-Murphy, B. *Adv. Polym. Sci.* **1987,** *83,* 57.
11. Fanta, W. *Adv. Polym. Sci.* **1965,** *4,* 157.
12. Funke, W. In *Polymer Yearbook*; Ellias, H. G.; Pethrick, H. A. Eds.; Harwood Academic Press: New York, 1989; p 102.
13. Kulicke, W.-M.; Nottelmann, H. In *Polymers in Aqueous Media*; Glass, J. E. Ed.; Advances in Chemistry 223; American Chemical Society: Washington DC, 1989; p 15.
14. Kulicke, W.-M., Aggour, Y. A.; Nottelmann, H.; Elsabee, M. Z. *Starch/Stärke* **1989,** *41,* 140.
15. Flory, P. J. *Principles of Polymer Chemistry*; Cornell University Press: Ithaca, London, 1953.
16. Ilavsky, M.; Hrouz, J.; Havlicek, I. *Polymer* **1985,** *26,* 1514.
17. Oppermann, W.; Rose, S.; Rehage, G. *Br. Polym. J.* **1985,** *17,* 175.
18. Katayama, S.; Hirokawa, Y.; Tanaka, T. *Macromolecules* **1984,** *17,* 2641.
19. Oppermann, W. *Angew. Makromol. Chem.* **1984,** *124,* 229.
20. Janas, V. F.; Rodrigues, F.; Cohen, C. *Macromolecules* **1980,** *13,* 977.
21. Nossal, R. *Macromolecules* **1985,** *18,* 49.
22. Gerth, C. *Ullmann's Enzyklopedie der technischen Chemie, Bd. 5,* Verlag Chemie: Weinheim, 1980.
23. Pearson, D. S.; Graessley, W. W. *Macromolecules* **1978,** *11,* 528.
24. Graessley, W. W. *Adv. Polym. Sci.* **1974,** *16,* 100.
25. Leach, H. W.; Schoch, T. J.; Chessman, E. F. *Starch/Stärke* **1961,** *13,* 200.
26. Maywald, E. C.; Leach, H. W. Schoch, T. G. *Starch/Stärke* **1968,** *20,* 189.
27. Flory, P. J. *J. Chem. Phys.* **1942,** *10,* 51.
28. Huggins, M. L. *J. Phys. Chem.* **1942,** *46,* 51.

29. Stockmayer, W. H. *J. Chem. Phys.* **1943,** *11,* 45.

30. Flory, P. J.; Rehner, J. *J. Chem. Phys.* **1943,** *11,* 512, 521.

31. Flory, P. J. *Macromolecules* **1982,** *15,* 99.

32. Flory, P. J.; Erman, B. *Macromolecules* **1982,** *15,* 800.

33. Queslel, J. P.; Mark, J. E. *Adv. Polym. Sci.* **1984,** *71,* 229.

34. Queslel, J. P.; Mark, J. E. *Adv. Polym. Sci.* **1984,** *65,* 135.

35. Gnanou, Y.; Hild, G.; Rempp, P. *Macromolecules* **1987,** *20,* 1662.

36. Ohmine, I.; Tanaka, T. *J. Chem. Phys.* **1982,** *77,* 5725.

37. Ricka, J.; Tanaka, T. *Macromolecules* **1984,** *17,* 2916.

38. Ricka, J.; Tanaka, T. *Macromolecules* **1985,** *18,* 83.

39. Langley, N. R. *Macromolecules* **1968,** *1,* 348.

40. Ferry, J. D. *Viscoelastic Properties of Polymers, 3rd ed.* Wiley: New York, 1980; p 408.

41. Heinrich, G.; Straube, E.; Helmis, G. *Acta Polymerica* **1980,** *31,* 275.

42. Langley, N. R.; Polmanteer, K. E. *J. Polym. Sci.; Polym. Phys. Ed.* **1974,** *12,* 925.

43. Heinrich, G.; Straub, E.; Helmis, G. *Adv. Polym. Sci.* **1988,** *85,* 33.

44. Mark, J. E. *Adv. Polym. Sci.* **1982,** *44,* 1.

45. Herz, J. E.; Rempp, P.; Borchard, W. *Adv. Polym. Sci.* **1978,** *26,* 105.

46. Hild, G.; Okasha, R.; Macret, M.; Gnanou, Y. *Makromol. Chem.* **1986,** *187,* 2271.

47. Munch, P. J.; Candau, J.; Herz, G.; Hild, J. *J. Phys. Lett.* **1977,** *38,* 971.

48. Zrinyi, M.; Horkay, F. *J. Polym. Sci.; Polym. Lett. Ed.* **1982,** *20,* 815.

49. Zrinyi, M.; Horkay, F. *Macromolecules* **1984,** *17,* 2805.

50. De Gennes, P.-G. *Scaling Concepts in Polymer Physics*; Cornell University Press: Ithaca and London, 1979; p 128.

RECEIVED August 22, 1990

Chapter 5

Chemical Grafting of Poly(ethylene oxide) onto Polystyrene Latex

Synthesis, Characterization, and Rheology

H. J. Ploehn[1] and J. W. Goodwin

Department of Physical Chemistry, University of Bristol, Bristol, England

A novel method utilizing a water-soluble carbodiimide was used to chemically graft 100,000 and 615,000 g/mol PEO onto PS latex particles with mean diameters of 375 and 347 nm. Electrophoretic mobility measurements show that the PEO is truly grafted, not just adsorbed. Capillary viscometry yields a layer thickness of 23 nm for grafted 615,000 g/mol PEO. Steady shear and oscillatory rheological measurements indicate a transition from primarily liquid-like behavior to solid-like behavior at an effective volume fraction between 0.528 and 0.550. The rheological behavior is interpreted in terms of long-range repulsive particle interactions due to the grafted polymer layers. These measurements thus provide a useful probe of the relationship between polymeric interactions, suspension microstructure, and macroscopic material properties.

Economic and environmental considerations have led to renewed interest in polymer latices which employ water as the suspending medium. It is well known that interparticle electrostatic forces play an important role in controlling the stability and rheological properties of aqueous suspensions. Although the particle electrostatics can be adjusted so that the latex has desirable properties and behavior, in many situations the electrostatics impose serious limitations on the

[1]Current address: Department of Chemical Engineering, Texas A&M University, College Station, TX 77843–3122

0097–6156/91/0462–0088$06.00/0

formulation of the suspension and its performance in practical situations.

Dissolved polymer also modifies particle interactions through any of several mechanisms (1). For instance, polymer in solution can be excluded from the interstitial region between two particles; the resultant osmotic attraction between the particles may lead to depletion flocculation. Physical adsorption of homopolymers, block and random copolymers, polyelectrolytes, and proteins onto particles occurs in many systems. At low adsorbed amounts, polymer molecules may adsorb on two or more particles simultaneously, producing attraction between the particles. The polymer-mediated interaction also has a repulsive component that increases with adsorbed amount due to the exclusion of segments from saturated surfaces and unfavorable polymer-polymer interactions in good solvents.

Strongly repulsive interactions which impart "steric" or polymeric stabilization (2) can be achieved by chemically grafting polymer to the particles. Relatively low molecular weight stabilizers (<10,000 g/mol) give high graft densities and produce steep but short-ranged repulsions (3). For higher molecular weights, the ratio of the grafted layer thickness to the particle radius increases and the interparticle repulsion becomes less steep and longer-ranged, i.e. "softer" (4). The details of the interactions of grafted polymer layers have a strong influence on the suspension's equilibrium properties and equilibrium behavior.

Although chemical grafting techniques have been widely applied in organic media, aqueous systems have not been studied as extensively (5). In this work, we investigate a novel method of homopolymer grafting in aqueous media. An extensive body of literature describes the grafting of biochemical ligands onto particle substrates for agglutination tests (6) and the use of chemical cross-linkers for analysis of protein structure and activity (7). These biochemical methods are applied in the chemical grafting of poly(ethylene oxide) (PEO) to colloidal polystyrene (PS) particles. Electrophoresis measurements show that the PEO is truly grafted, not just physically adsorbed. Preliminary rheological measurements indicate that the polymer-mediated interactions between particles are repulsive and long-ranged, i.e. "soft", as contrasted with the short-range repulsions found in "hard-sphere" suspensions (8).

Experimental Details

Materials. Polystyrene latices were synthesized using standard emulsion polymerization techniques (9). Following the recipe of Goodwin *et al.* (10), anionic latices were prepared using 4,4′-azobis-4-cyanopentanoic acid (Wako Chemicals GmbH) as the initiator; dissociated carboxyl groups thus stabilized the suspensions. Results for suspensions of 375 nm and 347 nm diameter particles, denoted as PS1 and PS2, are reported here.

Two poly(ethylene oxide) (PEO) samples were used as received: PEO1 (Aldrich), a commercial grade with $M_w \approx 1.0 \times 10^5$ g/mol and $M_w/M_n \approx 6.6$, and PEO2 (Polymer Labs), a polymer standard with $M_w \approx 6.15 \times 10^5$ g/mol and $M_w/M_n \approx 1.10$.

In order to graft PEO onto PS via a water-soluble carbodiimide (WSC), the hydroxyl end group of PEO was replaced by an amine group. First, the PEO was dissolved in pure toluene; depending on the molecular weight, concentrations of 5×10^{-3} to 5×10^{-2} g/cm^3 were convenient. Since toluene is a marginal solvent for PEO, slight warming was required. A tenfold molar excess of tosyl chloride (TsCl) (toluene-4-sulphonyl chloride, B.D.H.) and a small excess of pyridine (0.5 ml) were added. The pyridine scavenged chloride ions which were released when the tosyl chloride combined with the PEO hydroxyl end groups:

$$\begin{aligned} &\mathrm{Bu{-}(OCH_2CH_2)_n{-}OH + TsCl + Pyr} \\ &\Rightarrow \mathrm{Bu{-}(OCH_2CH_2)_n{-}OTs + Pyr\,HCl} \end{aligned} \qquad [1]$$

to produce tosylate leaving groups; insoluble pyridinium hydrochloride (PyHCl) precipitated from solution. Ammonia gas was bubbled through the warm (≈30°C) stirred solution for five hours to produce aminated PEO (APEO):

$$\begin{aligned} &\mathrm{Bu{-}(OCH_2CH_2)_n{-}OTs + NH_3} \\ &\Rightarrow \mathrm{Bu{-}(OCH_2CH_2)_n{-}NH_2 + TsOH} \end{aligned} \qquad [2]$$

Most of the toluene was then removed through evaporation under vacuum. The APEO was dissolved in dioxane and freeze-dried to remove all organic solvents; then, water was added so that the final solution was 0.5-5% APEO.

The grafting reaction utilized a water-soluble carbodiimide (WSC) which acted as a coupling agent. Two alternate routes are available. If the polymer to be grafted is sensitive to the WSC (as is the case for most biochemical ligands), or if the grafted latex must kept free of contaminants, then the grafting reaction can be preceded by an activation step:

$$\begin{aligned} &\mathrm{PS{-}COOH + HO{-}N_3C_6H_5 + R'{-}N{=}C{=}N{-}R} \\ &\rightarrow \mathrm{PS{-}COO{-}N_3C_6H_5 + R'{-}NH{-}CO{-}NH{-}R} \end{aligned} \qquad [3]$$

whereby the WSC 1-ethyl-3-(3-dimethylaminopropyl) carbodiimide (R'-N=C=N-R, Sigma) links 1-hydroxybenzotriazole ($HO{-}N_3C_6H_5$, Fluka) to the carboxyl groups on the latex (denoted PS) to form an active ester. The activated PS latex may be cleaned, if desired, by centrifugation/serum replacement or dialysis. Subsequent addition of APEO leads to polymer grafting. Alternately, WSC and APEO may be added directly to the carboxylated PS latex with

$$\begin{array}{c} \text{PS-COOH} + \text{NH}_2\text{-}(\text{CH}_2\text{CH}_2\text{O-})_n\text{Bu} + \text{R'-N=C=N-R} \\ \Downarrow \\ \text{PS-CO-NH-}(\text{CH}_2\text{CH}_2\text{O-})_n\text{Bu} + \text{R'-NH-CO-NH-R} \end{array} \qquad [4]$$

giving particles with grafted polymer.

The former procedure ([3] plus grafting) was used to graft APEO1 to PS1, while the latter ([4]) was used to graft APEO2 to PS2. In both cases, simple precautions were taken to minimize flocculation of the suspension. First, sufficient WSC was added to activate up to a maximum of 20% of the estimated number of surface carboxyl groups so that enough groups remained to provide electrostatic stabilization. The polymer dosage, estimated from the equilibrium adsorption data reported by Cohen Stuart *et al.* (11), was gauged to provide an adsorbed amount corresponding to the plateau value on the adsorption isotherm in order to maximize grafting and minimize bridging which would lead to flocculation. Cohen Stuart et al. (11) measured adsorbed amounts of PEO on PS latex as a function of molecular weight; for 6.6×10^5 g/mol PEO, they found an adsorbed amount of 1.42×10^{-7} g/cm^2 in equilibrium with a solution concentration of 2×10^{-4} g/cm^3. These values were used in conjunction with the known particle diameter and volume fraction of PS2 to fix the amount of APEO2 to be added to the latex. Finally, the volume fraction of the latex was kept low (0.01-0.015) and the latex was added to the diluted polymer solution with moderate stirring.

Analytical Methods. Particle sizes were measured by transmission electron microscopy (TEM) using a JEOL CX100 instrument with a diffraction grating replica for internal calibration.

Electrophoretic mobility was measured using a Pen Kem System 3000 Automated Electrophoresis Apparatus. Samples were diluted into filtered 10^{-3} M sodium acetate buffer with pH ≈ 6.3; the pH was adjusted to other values by dropwize addition of 0.1 M sulphamic acid or 0.1 M sodium hydroxide. Samples containing nonionic surfactant $C_{12}E_6$ (hexa-ethylene glycol mono n-dodecyl ether, Nikko Chemicals) were diluted into sodium acetate buffer that was 4×10^{-4} M in $C_{12}E_6$.

Steady shear and oscillatory rheological measurements utilized a Bohlin VOR rheometer fitted with either the double concentric cylinder or Couette (C14) measuring geometries. At constant shear rate, the shear stress and viscosity generally reached steady-state values within 10-20 s after the start of a measurement; these values were then averaged over 10-20 s to give the reported results. Shear stress and viscosity were reproducible for both increasing and decreasing shear rate sweeps. Variation of the strain amplitude for oscillatory measurements produced nonlinear storage and loss modulii for all latex volume fractions considered. The storage modulus was essentially linear for strains below 1% but the samples displayed strain softening at higher amplitudes; frequency sweeps were performed both in the linear and nonlinear strain ranges. The loss modulus was nonlinear for all strain amplitudes. Storage modulii were reproducible for both increasing and decreasing frequency sweeps, but the loss modulii occasionally showed hysteresis which correlated with drying of the sample.

Rheological measurements were performed on latex PS2 bearing grafted APEO2. The sample was treated with $C_{12}E_6$ surfactant in order to desorb non-grafted polymer, prevent bridging flocculation, and improve the ease of redispersion after centrifugation. The suspension was centrifuged and redispersed several times in a buffer of 10^{-3} M sodium acetate and 4×10^{-4} M $C_{12}E_6$. Volume fractions of each sample were determined though dry weight analysis after the final centrifugation and removal of the appropriate amount of supernatant.

Results and Discussion

Synthesis and Characterization. TEM showed that latex PS1 had a mean diameter of 375 nm with a coefficient of variation of 5%, while the corresponding values for latex PS2 were 347 nm and 2%. Conductometric titration of PS1 indicated a surface charge density of about 0.5 $\mu C/cm^2$.

Electrophoretic mobility measurements distinguished between bare and polymer-coated latices. Figure 1 shows the mobility of bare PS1, PS1 with adsorbed APEO1 (PS1-a-APEO1), and PS1 with grafted APEO1 (PS1-g-APEO1) as functions of pH. The mobilities of the coated particles were very similar and much smaller than that of the bare PS1. In simplistic terms, the polymer layer displaced the hydrodynamic slipping plane outward, thus reducing the zeta potential and mobility. The mobility decrease can be used to define an electrophoretic layer thickness δ_e (12) which approaches the hydrodynamic layer thickness δ_h when $1/\kappa$, the characteristic thickness of the double layer, is much larger than δ_h. For the data in Figure 1, $1/\kappa \approx 0.3$ nm; the fivefold drop in the mobility of PS1-g-APEO1 gives $\delta_e \approx 5$ nm within the approximations of Debye-Huckel theory. The data of Cohen Stuart *et al.* (11)

suggest $\delta_h \approx 25$ nm. This difference indicates that more sophisticated models (12) must be applied to rationalize the zeta potentials and mobilities for this system.

Nevertheless, the comparable mobilities of PS1-a-APEO1 and PS1-g-APEO1 imply that the adsorbed and grafted APEO layers had similar structures. These layers were probably rather diffuse with polymer configurations similar to random coils. In contrast, low molecular weight stabilizers grafted at high density tend to assume more stretched configurations (3,5,8,13) with no tendency to adsorb at other points along the chain. For high molecular weight grafted polymer, the hypothesis that adsorption behavior controls grafting characteristics is plausible but needs to be tested more thoroughly.

The results in Figure 1 lead to a more fundamental question: is the polymer really grafted, or merely adsorbed? A conclusive test should show that adsorbed polymer can be desorbed but grafted polymer remains near the particle surface. A nonionic surfactant, $C_{12}E_6$, was added to displace the APEO from the PS latex surface. Based on adsorption isotherms measured by Partridge (14), sufficient $C_{12}E_6$ was added to samples of PS1, PS1-a-APEO1, and PS1-g-APEO1 to ensure monolayer coverage of the particles in equilibrium with 4×10^{-4} M surfactant in solution. The electrophoretic mobilities of both polymer-coated and surfactant-treated particles are shown in Table I.

Table I. Electrophoretic Mobilities of PS1, PS1-a-APEO1, and PS1-g-APEO1 Before and After Treatment with Surfactant

Species	Electrophoretic Mobility /10^{-8} m^2/Vs Before Treatment	With Added Surfactant
PS1	-4.598	-3.477
PS1-a-APEO1	-1.816	-3.409
PS1-g-APEO1	-1.774	-2.069

As in Figure 1, the bare PS1 was much more mobile than PS1-a-APEO1 or PS1-g-APEO1, but the polymer-coated particles had similar mobilities. The mobility of surfactant-treated PS1 was about 20% less than that of bare PS1. The mobility of treated PS1-a-APEO1 rose to about the same value as treated PS1, implying that the $C_{12}E_6$ had displaced the adsorbed APEO1. However, the mobility of treated PS1-g-APEO1 remained low - only slightly more than that of the untreated PS1-g-APEO1. This result indicates that the APEO1 remained near the particle surfaces despite the adsorption of surfactant.

We regard this as strong evidence that the APEO1 is indeed chemically grafted.

Table II shows a similar comparison of the mobilities of PS2 particles, PS2 particles with adsorbed APEO2, and PS2 particles with grafted APEO2. Again, the mobilities of surfactant-treated PS2 and PS2-a-APEO2 were similar, presumably due to the desorption of APEO2. The mobility of PS2-g-APEO2 particles remained relatively low after exposure to $C_{12}E_6$. Since the grafted APEO2 was not able to move away from the particle surfaces, the local electrostatic environment, and hence the mobility, did not change as much as when the polymer completely desorbed.

Table II. Electrophoretic Mobilities of PS2, PS2-a-APEO2, and PS2-g-APEO2 Before and After Treatment with Surfactant

Species	Electrophoretic Mobility /10^{-8} m^2/Vs Before Treatment	With Added Surfactant
PS2	-2.046	-3.574
PS2-a-APEO2	-1.767	-3.376
PS2-g-APEO2	-1.513	-2.051

Finally, an attempt was made to adsorb and to graft APEO2 onto PS2 which had been pretreated with surfactant. In both cases, the mobility of the particles remained constant, indicating that the surfactant prevented adsorption as well as grafting of APEO2.

Rheology. Samples for rheological measurements were concentrated using an MSE-18 high speed centrifuge. Upon centrifugation, bare PS2 latex formed a very hard, compact sediment that could not be redispersed. Latex PS2-g-APEO2 produced a much less compact sediment that was easily dispersed by stirring. Visual observation through the Pen Kem apparatus and electrophoresis measurements indicated, however, that the suspension was not completely redispersed into single particles, even after considerable tumbling. Sonication was not used because of the danger of degrading the grafted APEO2. The samples were therefore treated with sufficient $C_{12}E_6$ nonionic surfactant to provide monolayer coverage in equilibrium with a solution concentration of $4x10^{-4}$ M. After centrifugation, the sediment readily redispersed into single particles; the average mobility equalled the initial value. Several cycles of centrifugation and redispersion into $4x10^{-4}$ M $C_{12}E_6$ with 10^{-3} M sodium acetate were used to remove any desorbed APEO2. While the $C_{12}E_6$ complicates interpretation of the rheological measurements in terms of polymer configurations, the surfactant simplifies the handling and reduces the chances of bridging in the concentrated suspensions.

Capillary viscometry provides one measure of the thickness of the grafted polymer layers. Using the relationship proposed by Saunders (15)

$$\frac{\phi}{\ln(\eta_r)} = \frac{1}{k_1 f} - \frac{\phi}{k_1 \phi_m} \qquad [5]$$

between $\eta_r = \eta/\eta_o$, the relative viscosity, and φ, the latex volume fraction, a plot of $\varphi/\ln(\eta_r)$ vs. φ yields an intercept $(k_1 f)^{-1}$. Here $k_1 = 2.5$ is the Einstein coefficient for rigid, noninteracting spheres, φ_m is a packing factor, and

$$f \equiv \frac{\phi_e}{\phi} = \left[1 + \frac{2\delta}{d}\right]^3 \qquad [6]$$

defines an effective volume fraction φ in terms of the particle diameter d and the layer thickness δ. A plot of $\varphi/\ln(\eta_r)$ vs. φ for PS2-g-APEO2 yielded a straight line; the best fit of the points gave $f = 1.451$ and $\delta \approx 23$ nm. Thus the particle radius/layer thickness ratio was about 7.6.

Figure 2 shows the viscosity of PS2-g-APEO2 as a function of shear stress and volume fraction ($\varphi_e = 1.451\ \varphi$ according to [6]). At low volume fractions, the suspension viscosity was a few times greater than that of water and displayed a minor amount of shear thinning. These curves showed limiting viscosities at both high and low shear stresses. With increasing volume fraction, shear thinning became much more pronounced. Between $\varphi = 0.364$ ($\varphi_e = 0.528$) and $\varphi = 0.379$ ($\varphi_e = 0.550$), the suspension behavior changed considerably: iridescence was observed, and the Newtonian limit at low shear stress disappeared, at least for stresses attainable with the VOR for these suspensions. The onset of an apparent yield stress characterized a transition from "liquid-like" to "solid-like" behavior. Shear thinning dominated at higher volume fractions; for $\varphi = 0.561$ ($\varphi = 0.814$), the greatest measured viscosity was seven orders of magnitude greater than that of water.

The transition from liquid-like to solid-like rheology can also be seen in the storage modulus, plotted as a function of frequency in Figure 3. For low volume fractions, the ability of the suspension to store energy varied with the frequency of the applied deformation. The suspension microstructure had sufficient time to relax and dissipate energy at low frequency. At high frequency, the characteristic time of the deformation was much shorter than that of microstructural rearrangement, and so the suspension stored energy in the interactions of the grafted polymer layers. As the volume fraction

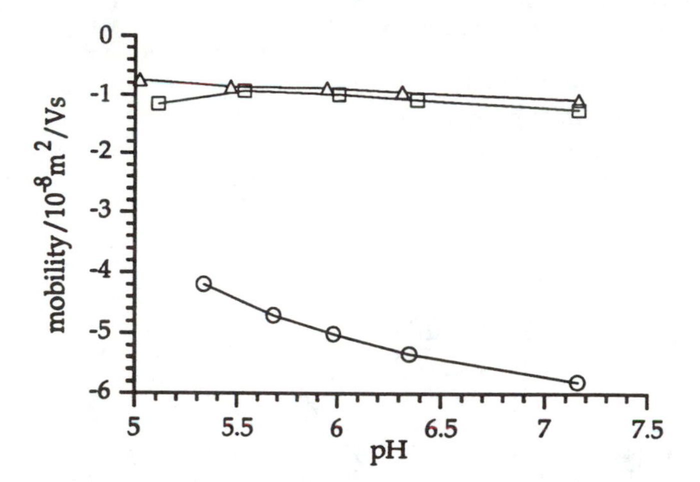

Figure 1. Electrophoretic mobility of PS1 (circles), PS1-a-APEO1 (squares), and PS1-g-APEO1 (triangles) as functions of buffer pH. For these data only, the buffer was 0.01 M sodium acetate.

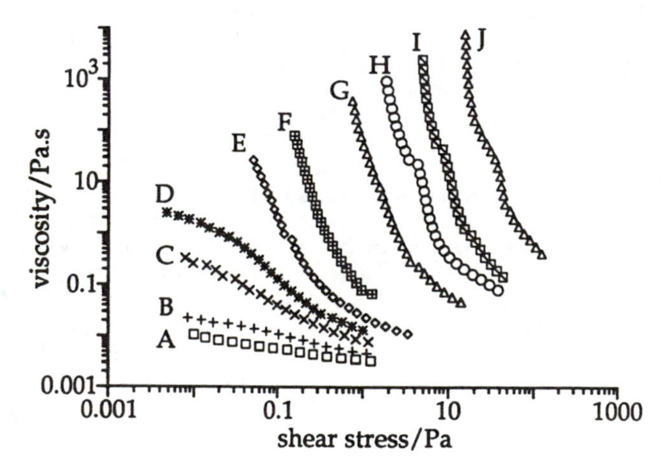

Figure 2. Viscosity as a function of shear stress and volume fraction for PS2-g-APEO2. Symbols, PS volume fractions, and effective volume fractions (in parentheses) are as follows: A = 0.262 (0.380); B = 0.294 (0.423); C = 0.340 (0.493); D = 0.364 (0.528); E = 0.379 (0.550); F = 0.400 (0.580); G = 0.472 (0.685); H = 0.514 (0.746); I = 0.520 (0.755); J = 0.561 (0.814).

increased, interactions between polymer layers became stronger and more numerous; hence the storage modulus increased. Meanwhile, the characteristic time for microstructural rearrangement grew longer as seen in the shift of the plateau value to lower frequency. As in the steady shear case, a transition was observed between φ_e = 0.528 and φ_e = 0.550. For volume fractions $\varphi \geq 0.379$, or $\varphi_e \geq 0.550$, the polymer layers overlapped so much that very little structural rearrangement occurred. Deformation at all applied frequencies led to storage and recovery of most of the supplied energy via polymer-mediated particle interactions.

Figure 4 highlights the variation of the storage and loss modulii as volume fraction increases. Between the effective volume fractions of $\varphi_e \approx 0.50$ and $\varphi_e \approx 0.60$, the storage modulus at 1 Hz increased by three orders of magnitude. The scatter of the points in the loss modulus curve were a consequence of sample drying. In spite of this inaccuracy, it is apparent that the storage modulus became greater than the loss modulus at some volume fraction in the transition range. Similar behavior was observed by Prestidge and Tadros (13) who found a transition at $\varphi_e \approx 0.62$. Their transition volume fraction is somewhat greater than that reported here due to the shorter range and greater steepness of the interparticle repulsions characteristic of the shorter grafted polymer in their system (13).

Conclusions

This preliminary study has demonstrated the efficacy of water-soluble carbodiimides in the chemical grafting of PEO onto PS latex particles. The synthesis follows several simple steps that are easy to apply in practice and yield grafted APEO on PS particles in aqueous media. Electrophoresis measurements give strong evidence that the APEO is truly grafted. Capillary viscometry and steady shear and oscillatory rheological measurements indicate that the polymer layers produce long range repulsive interactions which dominate the rheology of the suspension. Further work is in progress in each of these areas.

Several questions could be addressed in the synthesis stage. Foremost among these is the problem of controlling graft density. For low molecular weight polymers ($M_w < 10^4$) grafted in organic media, very high graft densities are routinely achieved (3,5,8,13). High molecular weight APEO, on the other hand, readily adsorbs on PS latex; thus graft density may be controlled by the preceding adsorption behavior. This hypothesis can be tested by (a) varying the adsorption conditions, (b) measuring adsorbed amounts via mass balances, (c)

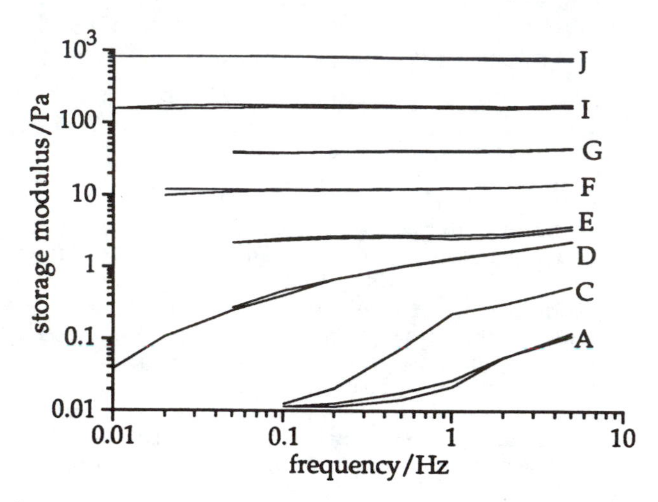

Figure 3. Storage modulus as a function of frequency and volume fraction for PS2-g-APEO2. Symbols as in Figure 2.

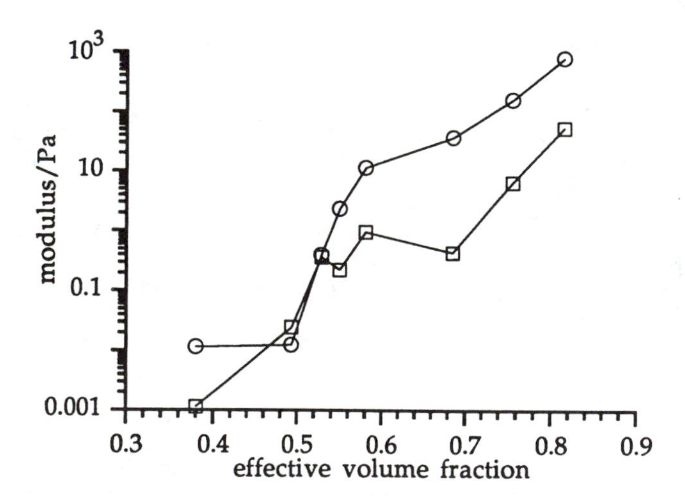

Figure 4. Storage and loss moduli (circles and squares) at 1 Hz as functions of effective volume fraction for PS2-g-APEO2.

desorbing any non-grafted polymer, and then (d) quantifying the graft density through a spectroscopic technique such as NMR.

Characterization of the grafted polymer layers would be complemented by dynamic light scattering (DLS) measurements of the layer thickness. Very few studies have compared the hydrodynamic layer thicknesses derived from capillary viscometry and DLS. DLS studies would also complement the electrophoresis measurements in probing the efficacy of the grafting procedure: changes in δ_h due to the desorption of adsorbed polymer can be readily seen, but δ_h for grafted layers should remain nearly constant.

Current efforts are primarily focussed on the rheology of PS-g-APEO suspensions. Variation of the particle radius/layer thickness ratio provides a means of varying the softness of the interparticle interactions. For high values of this ratio, the rheological behavior of the suspensions should be similar to that of the "effective Brownian hard sphere" systems studied by Choi and Krieger (3) and Frith and Mewis (4). The relative viscosity then depends only on a dimensionless shear stress and the particle volume fraction. Comparison of the scaled results for aqueous and organic media would be very interesting. With increasing polymer layer thickness, Brownian hard sphere scaling breaks down due to the importance of interparticle interactions (4). In this regime, rheological measurements are capable of probing the details of these interactions. Finally, for very thick layers and at high effective volume fractions, overlapping layers give strong interactions which prevent particle self-diffusion on experimental time scales and produce solid-like behavior.

Comparisons with the rheology of filled polymer systems and concentrated polymer solutions will be illuminating. In contrast with inert fillers, though, the polymer in the present system is strongly interacting: grafting induces conformational changes and promotes an inhomogeneous spatial distribution of polymer. The particles introduce extra entanglement points in the polymer network; the implications for the system's rheology are still to be assessed. Rheological studies of suspensions stabilized by grafted polymer can thus provide information not only on polymer-mediated particle interactions, but also on the structure and dynamics of more concentrated polymeric materials.

Acknowledgements

The authors would like to thank Wako Chemicals GmbH for supplying the initiators used in this study. We are also grateful to Dr. Keith Ryan and Mr. John Dimery for providing technical assistance.

Literature Cited

1. Ploehn, H. J.; Russel, W.B. *Adv. Chem. Eng.*, 1990, 15, 137.
2. Napper, D. H. Polymeric Stabilization of Colloidal Dispersions; Academic Press: New York; 1983.
3. Choi, G. N.; Krieger, I. M. *J. Colloid Interface Sci*. 1986, 113, 94; 1986, 113, 101.
4. Frith, W. J. Ph.D. thesis, Catholic University, Leuven, Belgium, 1987. Mewis, J.; Frith, W. J.; Strivens, T. A.; Russel, W. B. *AIChE J*. 1989, 35, 415.
5. Ryan, K. *Chem. Ind*. 6 June 1988.
6. Guilford, H. *Chem. Soc. Rev*. 1973, 2, 249.
7. Lundblad, R. L.; Noyes, C. M. Chemical Reagents for Protein Modification; CRC Press: Boca Raton, FL; Vol. 2.
8. de Kruif, C. G.; van Iersel, I. M. F.; Vrij, A.; Russel, W. B. *J. Chem. Phys*. 1985, 83, 4717.
9. Vanderhoff, J. W.; Van den Hul, H. J.; Tausk, R. J. M.; Overbeek, J. Th. G. In Clean Surfaces: Their Preparation and Characterization for Interfacial Studies; Goldfinger, G., Ed.; Marcel Dekker: New York, 1970; p 15.
10. Goodwin, J. W.; Ottewill, R. H.; Pelton, R.; Vianello, G.; Yates, D. E. *Brit. Polym. J*. 1978, 10, 173.
11. Cohen Stuart, M. A.; Waajen, F. H. F. H.; Cosgrove, T.; Vincent, B.; Crowley, T. L. *Macromolecules* 1984, 17, 1825.
12. Koopal, L. K.; Hlady, V.; Lyklema, J. *J. Colloid Interface Sci*. 1988, 121, 49.
13. Prestidge, C.; Tadros, Th. F. *J. Colloid Interface Sci*. 1988, 124, 660.
14. Partridge, S. J. Ph.D. thesis, University of Bristol, Bristol, England, 1987. Goodwin, J. W.; Hughes, R. W.; Partridge, S. J.; Zukoski, C. F. *J. Chem. Phys*. 1986, 85, 559.
15. Saunders, F. L. *J. Colloid Sci*. 1961, 16, 13.

RECEIVED September 17, 1990

Chapter 6

Electrosteric Stabilization of Oil-in-Water Emulsions by Hydrophobically Modified Poly(acrylic acid) Thickeners

R. Y. Lochhead[1]

Technical Center, BFGoodrich Company, Moore and Walker Roads, Avon Lake, OH 44012

Hydrophobic modification of poly(acrylic acid) thickeners yields products that are useful as primary emulsifiers for oil-in-water systems. The resulting emulsions are stable for years, but they break and coalesce almost instantly when a sufficient concentration of electrolyte is introduced into the aqueous phase. The effect of electrolytes on these electrosterically stabilized emulsions is complex. The polymer is anchored to the oil droplet by hydrophobic interaction. Such anchoring should theoretically be strengthened by the presence of water-structure-enhancing electrolytes. On the other hand, coulombic interaction between the electrolyte and the polyelectrolyte causes shrinkage of the overall polyelectrolyte configuration and this should theoretically make coalescence more likely.

The objective of this study was to separate these two effects; namely enhanced anchoring of the polymer and collapse of the polymer chain, by studying the relative effects of electrolytes selected from the classical Hofmeister series. Expansion of the polyion was controlled by varying the pH of the system.

Hydrophobically Modified Polyacrylic Acid

In a previous communication (*1*), a unique polymer primary emulsifier was described. This polymeric emulsifier was a hydrophobically modified poly(acrylic acid) thickener. Unlike conventional emulsifiers which are usually designated by the CTFA name "Carbomers", this polymer was free of the constraints of HLB and therefore it displayed the ability to form

[1]Current address: Department of Polymer Science, University of Southern Mississippi, Southern Station Box 10076, Hattiesburg, MS 39406–0076

0097–6156/91/0462–0101$06.00/0

stable emulsions with any oil (Figure 1). The emulsions were easily prepared at room temperature and were storage-stable for periods of years at ambient temperature.

These novel polymeric emulsifiers displayed another unique property which set them apart from other hydrophobically modified polymers (*2*), such as cationic derivatives of hydroxyethyl cellulose (*3*). The uniqueness lies in the fact that emulsions prepared with the hydrophobically modified poly(acrylic acid) coalesced spontaneously when applied to surfaces, such as human skin, that contained low-molecular- weight electrolytes. The oil that was released quickly formed a thin-film hydrophobic barrier on the substrate. This property renders these polymers useful in a number of industrially important areas.

A mechanism was proposed for the emulsion stabilization and the unique triggered release of oil (Figure 2). It was assumed that the polymeric microgels were anchored by hydrophobic interaction to the oil–water interface. This seemed reasonable, because polymers that were not hydrophobically modified gave unstable emulsions. The coalescence of the emulsion when it was exposed to electrolyte was explained by consideration of the Donnan equilibrium of counterions in polyelectrolytes. Addition of salt causes collapse of the polyelectrolyte microgels that are adsorbed at the oil–water interface. Subsequent destabilization of the emulsion could be due to:

1. Approach of the polyion chains to their unperturbed dimensions, when steric stabilization would be lost

2. Collapse of the stabilizing electrical double layer around the droplets.

3. A reduction in the surface coverage of the oil interface as a consequence of rapid shrinkage of the polyelectrolyte microgels

However, another possibility exists. It is possible that the oil droplets are merely trapped within hydrophobic domain inside a gel. This study was aimed at distinguishing whether the hydrophobically modified poly(acrylic acid) was acting as a primary emulsifier or merely as a gel matrix with hydrophobic domains.

Hydrophobic Interactions

Hydrophobic substances are defined as substances which are readily soluble in nonpolar solvents, but only sparingly soluble in water. This definition draws an important distinction from "lyophobic" substances, which have low solubility in all solvents, in general, as a consequence of strong intermolecular cohesion within the substance itself (*7*). When oil separates from water, it is tempting to ascribe this phenomenon to "like attracting like." This explanation is incorrect. The mutual attraction of non-polar groups plays a minor role in hydrophobic interaction. Hydrophobic

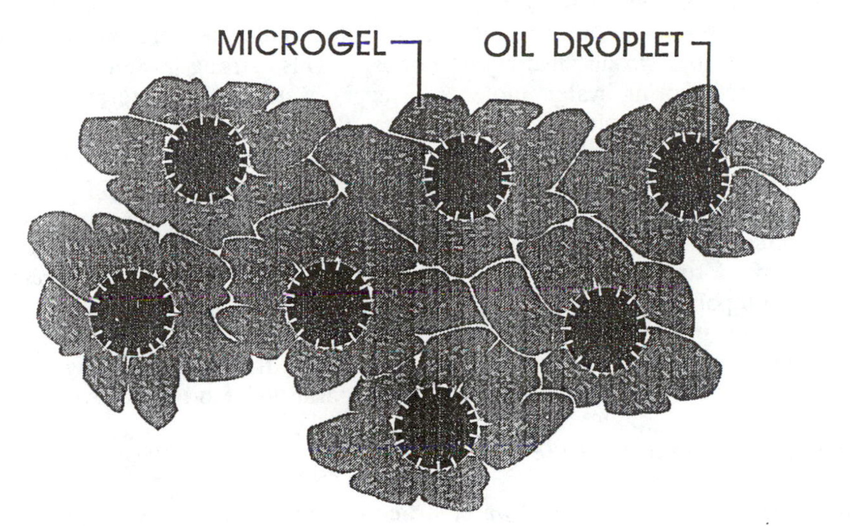

Figure 1. Schematic of emulsion stabilized with hydrophobically modified poly(acrylic acid).

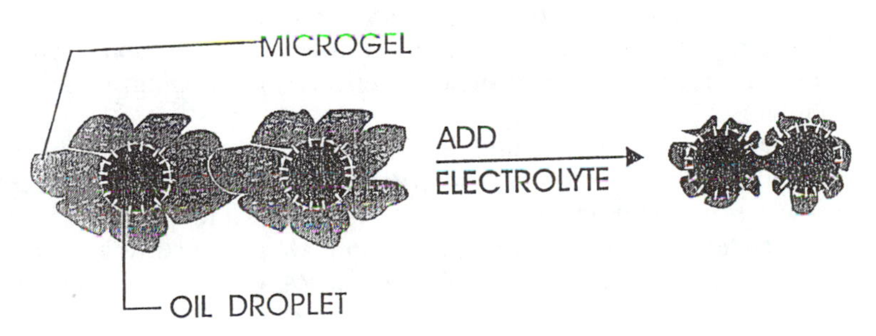

Figure 2. Triggered collapse of the microgels upon contact with electrolyte causes emulsion instability.

interaction actually arises from strong intermolecular forces between water molecules. In liquid water, the forces between water molecules are isotropically arranged, and when a solvent is dissolved in water, these forces must be distorted or disrupted. Strongly ionic or polar solutes can form strong bonds with the water molecules that more than compensate for the deformation of the water structure. Nonpolar substances are attracted to water molecules by weak dispersion forces such as van der Waals and dipole–induced-dipole interaction. The water molecules are attracted to each other by strong hydrogen bonds. This strong water structure between neighboring water molecules favors water–water interaction and eliminates the nonpolar molecules from the immediate vicinity. Thus, hydrophobic interaction arises from the water structure "pushing" the hydrophobic groups together to minimize the interfacial area (*8*).

In the immediate vicinity of the nonpolar solute molecules, the hydrogen bonds between the molecules are maintained but distorted. This leads to enhanced structuring of the water molecules in the immediate vicinity of the nonpolar substance, and consequently, there is a loss in entropy of these water molecules. This loss of entropy, rather than the nonpolar substance's bond energy, leads to an unfavorable free energy change associated with dissolution and results in the separation of oil from water.

Hydrophobic Interaction: the Effect of Salts

Salts can be classified as "water-structure makers" or "water-structure breakers" according to their position in the Hofmeister Series. Structure-making salts enhance water structure and thus enhance the hydrophobic effect. This is why sodium chloride is often added to organic compounds in "aqueous dispersion," in order to "salt them out."

Water-structure breakers, such as thiocyanate, "loosen" the water structure and reduce the hydrophobic effect, encouraging dissolution of hydrophobic solutes.

Since hydrophobic interaction is the postulated mechanism of anchoring the hydrophobically modified poly(acrylic acid) (HMPAA) molecules to the oil–water interface in an emulsion, it was of interest to examine the effect of added water-structure-making and water-structure-breaking salts on the stability of emulsions containing HMPAA as the primary emulsifier.

From this study, we hoped to be able to distinguish between the two possible mechanisms of emulsification using HMPAA, namely, true electrosteric stabilization or merely entrapment in hydrophobic domains within a hydrophilic gel.

Experimental Procedures

Materials. Hydrophobically modified poly(acrylic acid) (HMPAA-EX-231) has been described previously (*1*). This was a crosslinked poly(acrylic acid) that contained less than 1 mol % of a long-chain alkyl methacrylate.

Carbopol EX-230 is identical to HMPAA-EX-231, except for the fact that it was not hydrophobically modified. Water was USP-grade, organic-free, deionized water. Potassium hydroxide was 45% w/w solution certified grade from Fisher Scientific. Potassium chloride was certified ACS-grade from Fisher Scientific. Potassium thiocyanate was certified ACS-grade from Fisher Scientific. Mineral Oil was Drakeol 19 Mineral Oil-USP from Penreco Corp.

Mucilage Preparation. Each polymer was dispersed carefully in water to give the appropriate concentration. This dispersion of polymer in water was mixed for 20 min to ensure hydration of the polymer, then neutralized to exactly a pH of 7 by addition of the 45% solution of potassium hydroxide.

Viscosity and yield value were measured, and then each of the salts were added to separate mucilages in a series of additions. After each addition, viscosity and yield value were measured.

Emulsion Preparation. Unneutralized polymer dispersions were prepared as described above. Sufficient oil was added to this unneutralized dispersion to yield a final oil content of 10% by weight. The oil was added to the rapidly mixed aqueous polymer dispersion. The emulsion was neutralized to a pH of 7 with potassium hydroxide (45%) and an aliquot was taken for observation and storage. Each of the salts were added sequentially to appropriate emulsions. After each addition, the emulsion was mixed for 30 min, and an aliquot was removed for observation and storage.

All of the emulsions were stored at ambient temperature for 6 months before their stability was assessed. **Yield value** was determined using a B.P. Plastometer (*4*).

The analysis of yield value, σ_y, was derived from the work of Voet and Brand (*5*), who derived their equation from the work of Houwink (*6*). Houwink showed that

$$\sigma_y = \frac{12Wh}{\pi d^3}$$

where σ_y is the yield value in dynes cm^{-2}, W is the effective weight of the top plate in grams, h is the distance between the plates in centimeters, and d is the mean zone diameter in centimeters.

Assuming uniform spread, h can be expressed in terms of the volume of a cylinder ($h = 4\ V/d^2$), and the equation becomes

$$\sigma_y = \frac{VW}{\pi^2 d^5}$$

Replacing W by the downward force, P, yields

$$\sigma_y = \frac{48PV}{\pi^2 d^5}$$

From this equation, the yield force can easily be calculated from measurement of the mean zone diameter of a sample placed between the glass plates and allowed to reach equilibrium stress.

Results and Discussion

Tables 1–6 show clearly that emulsion prepared with conventional poly(acrylic acid) thickeners in the absence of conventional emulsifier, are incapable of preventing coalescence of the emulsion. This result, which has been shown previously (*1*), persists despite polymer concentration, oil loading or conferred viscosity or yield value on the continuous phase of the original emulsion.

Emulsions prepared using hydrophobically modified poly(acrylic acid) are stable even in many cases where the viscosity and yield value of the continuous phase is low. (Tables 7–9). In this series, emulsion instability is observed only at high salt concentrations or low polymer concentrations. Chloride is a water-structure maker and thiocyanate is a water-structure breaker. The effect of each of these salts on measured viscosity is, to all intents and purposes, identical, regardless of which anion is present. Thus, the potassium counterion dominates in determining the overall configuration of these polyelectrolyte molecules (Figures 3 and 4).

The effect of these ions on emulsion stability can be seen most clearly from phase diagrams which were constructed after storage of the compositions for 6 months at ambient temperature (Figures 5–7). The overall compositional extent of emulsion stability decreases with added salts in the order KCl > KOH > KSCN. The addition of potassium thiocyanate causes most rapid loss of emulsion stability.

It is significant that a large area of the KCl phase diagram represents creamed emulsion, whereas the area of creamed emulsion in the KSCN diagram is much smaller and is removed to lower salt levels. In these creamed emulsions, the droplets are stable, even though they are in "intimate" contact. The stability of these creamed emulsions must be due to the presence of a polymeric barrier between the droplets. As noted earlier, for such steric stabilization to be effective, it is essential that the polymers be anchored firmly to the interface. The polymers used in this investigation can be anchored to the oil–water interface only by

Table 1. Effect of pH on Mucilage Viscocity, Yield Value, and Emulsion Stability of an Emulsion Prepared with 2% Poly(acrylic acid) Thickener PAA EX-230 and with Potassium Hydroxide as Neutralizing Agent

pH	Brookfield viscosity (mPa.s) (2.5 rpm) (Model RVT)	Yield Value (Nm_{-2})	Emulsion Appearance (6 months)
3.1	7350	9.0	Coalesced
3.8	42,000	8.5	Coalesced
4.0	44,000	14.0	Coalesced
4.5	45,600	17.0	Coalesced
6.9	56,000	12.0	Coalesced
12.5	50,000	9.0	Coalesced
13.0	38,400	8.5	Coalesced
13.3	32,000	4.9	Coalesced
13.5	26,800	4.1	Coalesced
13.8	14,000	2.3	Coalesced

Table 2. Effect of pH on Mucilage Viscocity, Yield Value, and Emulsion Stability of an Emulsion Prepared with 1% Poly(acrylic acid) Thickener PAA EX-230 and with Potassium Hydroxide as Neutralizing Agent

1% w/w. PAA EX-230 pH	Brookfield viscosity (mPa s) (2.5 rpm. Model RVT)	Yield Value (Nm^{-2})	Emulsion Appearance (6 months)
2.9	13,528	3.2	Coalesced
3.6	28,000	4.9	Coalesced
3.9	29,600	6.5	Coalesced
4.2	29,600	6.0	Coalesced
4.7	30,000	5.6	Coalesced
6.6	32,000	5.2	Coalesced
11.9	31,300	3.0	Coalesced
12.7	20,000	2.4	Coalesced
13.3	12,400	0.93	Coalesced
13.2	8,400	0.6	Coalesced

Table 3. Effect of pH on Mucilage Viscocity, Yield Value, and Emulsion Stability of an Emulsion Prepared with 0.5% Poly(acrylic acid) Thickener PAA EX-230 and with Potassium Hydroxide as Neutralizing Agent

pH	Brookfield viscosity (mPa s) (2.5 rpm. Model RVT)	Yield Value (Nm^{-2})	Emulsion Appearance (6 months)
3.0	6400	1.6	Coalesced
4.0	19,200	1.65	Coalesced
4.4	19,200	3.25	Coalesced
4.7	19,600	3.0	Coalesced
5.0	20,000	2.7	Coalesced
10.0	19,200	3.25	Coalesced
12.0	14,000	2.3	Coalesced
12.5	9,000	1.15	Coalesced
13.1	2,800	0.44	Coalesced
13.2	1040	0.31	Coalesced

Table 4. Effect of pH on Mucilage Viscocity, Yield Value, and Emulsion Stability of an Emulsion Prepared with 0.25% Poly(acrylic acid) Thickener PAA EX-230 and with Potassium Hydroxide as Neutralizing Agent

pH	Brookfield viscosity (mPa s) (2.5 rpm. Model RVT)	Yield Value (Nm^{-2})	Emulsion Appearance (6 months)
3.15	3040	0.43	Coalesced
4.3	13000	0.93	Coalesced
4.6	13,200	2.3	Coalesced
4.8	13,400	3.0	Coalesced
6.1	14,600	1.95	Coalesced
11.5	10,400	1.95	Coalesced
11.8	7,560	1.7	Coalesced
12.0	5,600	0.87	Coalesced
12.8	560	0.67	Coalesced

Table 5. Effect of pH on Mucilage Viscocity, Yield Value, and Emulsion Stability of an Emulsion Prepared with 0.1% Poly(acrylic acid) Thickener PAA EX-230 and with Potassium Hydroxide as Neutralizing Agent

pH	Brookfield viscosity (mPa s) (2.5 rpm. Model RVT)	Yield Value (Nm^{-2})	Emulsion Appearance (6 months)
3.5	688		Coalesced
4.4	4190	<0.3	Coalesced
5.8	9000	<0.3	Coalesced
6.7	9200	<0.3	Coalesced
8.5	9600	<0.3	Coalesced
11.0	4400	<0.3	Coalesced
12.2	48	<0.3	Coalesced
12.5	28	<0.3	Coalesced

Table 6. Effect of pH on Mucilage Viscocity, Yield Value, and Emulsion Stability of an Emulsion Prepared with 0.05% Poly(acrylic acid) Thickener PAA EX-230 and with Potassium Hydroxide as Neutralizing Agent

pH	Brookfield viscosity (mPa s) (2.5 rpm. Model RVT)	Yield Value (Nm^{-2})	Emulsion Appearance (6 months)
3.7	112	<0.3	Coalesced
4.9	2475	<0.3	Coalesced
6.2	5280	<0.3	Coalesced
6.6	6400	<0.3	Coalesced
10.2	3200	<0.3	Coalesced
10.8	640	<0.3	Coalesced
11.2	160	<0.3	Coalesced
11.8	20	<0.3	Coalesced

Table 7. Effect of Addition of Salts to Mucilages and Emulsions Neutralized to pH 7 of an Emulsion Prepared with 1% w/w Polymeric Emulsifier HMPAA EX-231

Added Salt Molarity	Brookfield Viscosity (mPa.s) (2.5 rpm) (Model RVT)	Yield Value (Nm^{-2})	Emulsion Appearance (6 months)
Potassium Hydroxide			
0.0	64,800	12.21	Some Coalesced
0.4	20,600	6.93	Stable
0.75	8,200	4.3	Stable
1.0	2,400	<0.3	Creamed
<u>Potassium Thiocyanate</u>			
0.4	19,200	3.15	Stable
0.75	6,000	2.0	Coalesced
1.0	1600	<0.3	Coalesced
<u>Potassium Chloride</u>			
0.4	22,400	2.9	Stable
0.75	7,200	0.3	Stable
1.0	2,240	<0.3	Creamed

Table 8. Effect of Addition of Salts to Mucilages and Emulsions Neutralized to pH 7 of an Emulsion Prepared with 0.5% w/w Polymeric Emulsifier HMPAA EX-231

Added Salt Molarity	Brookfield Viscosity (mPa.s) (2.5 rpm) (Model RVT)	Yield Value (Nm^{-2})	Emulsion Appearance (6 months)
Potassium Hydroxide			
0.0	34,400	2.85	Stable
0.8	3,400	<0.3	Stable
0.16	1,200	<0.3	Stable
0.3	1,000	<0.3	Coalesced/Creamed
Potassium Thiocyanate			
0.8	3,200	3.2	Creamed
0.16	640	<0.3	Coalesced
0.3	240	<0.3	Coalesced
Potassium Chloride			
0.8	3,400	3.0	Stable
0.16	1,200	<0.3	Creamed
0.3	800	<0.3	Creamed

Table 9. Effect of Addition of Salts to Mucilages and Emulsions Neutralized to pH 7 of an Emulsion Prepared with 0.25% w/w Polymeric Emulsifier HMPAA EX-231

Added Salt Molarity	Brookfield Viscosity (mPa.s) (2.5 rpm) (Model RVT)	Yield Value (Nm^{-2})	Emulsion Appearance (6 months)
Potassium Hydroxide			
0.0	23,200	1.2	Stable
0.04	400	<0.5	Creamed/Coalesced
0.08	120	<0.5	Coalesced
0.15	80	<0.5	Coalesced
Potassium Thiocyanate			
0.04	320	<0.5	Coalesced
0.08	120	<0.5	Coalesced
Potassium Chloride			
0.04	400	<0.5	Creamed/Coalesced
0.08	120	<0.5	Coalesced

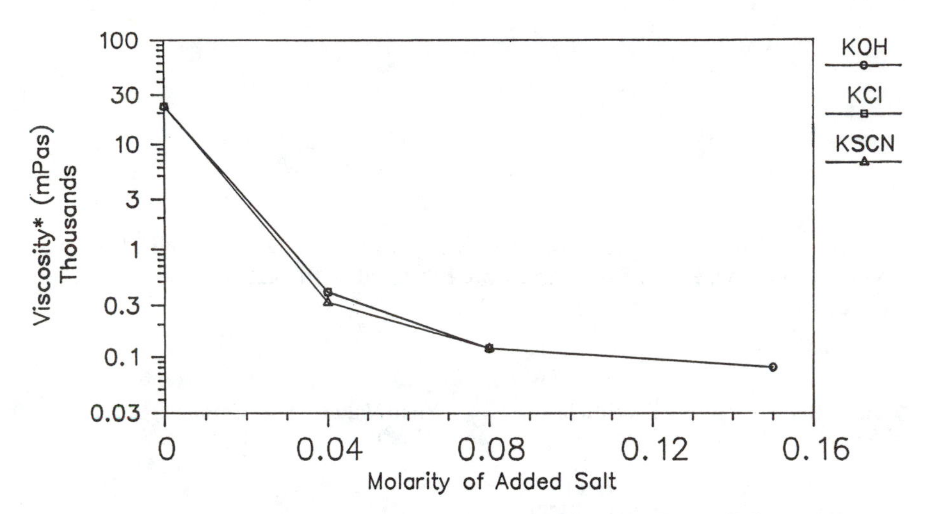

Figure 3. The effect of addition of salts on the Brookfield viscosity of mucilages and emulsions neutralized to pH 7 at a polymeric emulsifier EX-231 concentration of 0.25% w/w. All measurements were made at a spindle speed of 2.5 Hz.

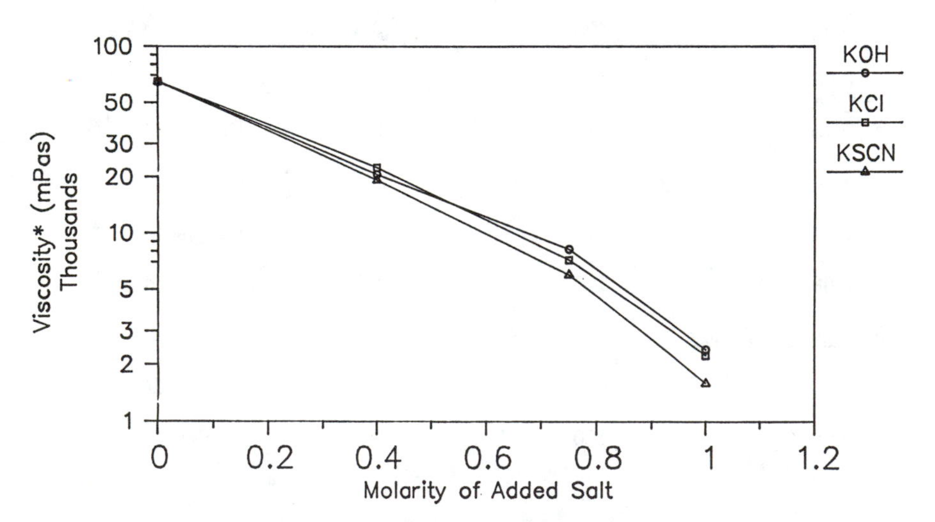

Figure 4. The effect of addition of salts on the Brookfield viscosity of mucilages and emulsions neutralized to pH 7 at a polymeric emulsifier EX-231 concentration of 1.0% w/w. All measurements were made at a spindle speed of 2.5 Hz.

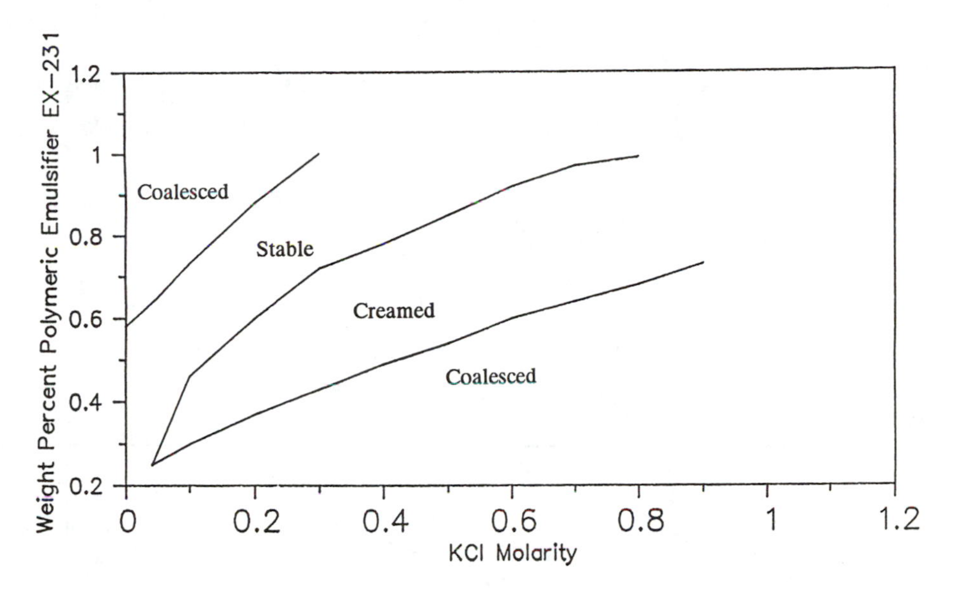

Figure 5. Phase diagram showing the effect of KCl concentration and polymeric emulsifier EX-231 concentration on the stability of 10 wt % mineral oil emulsions.

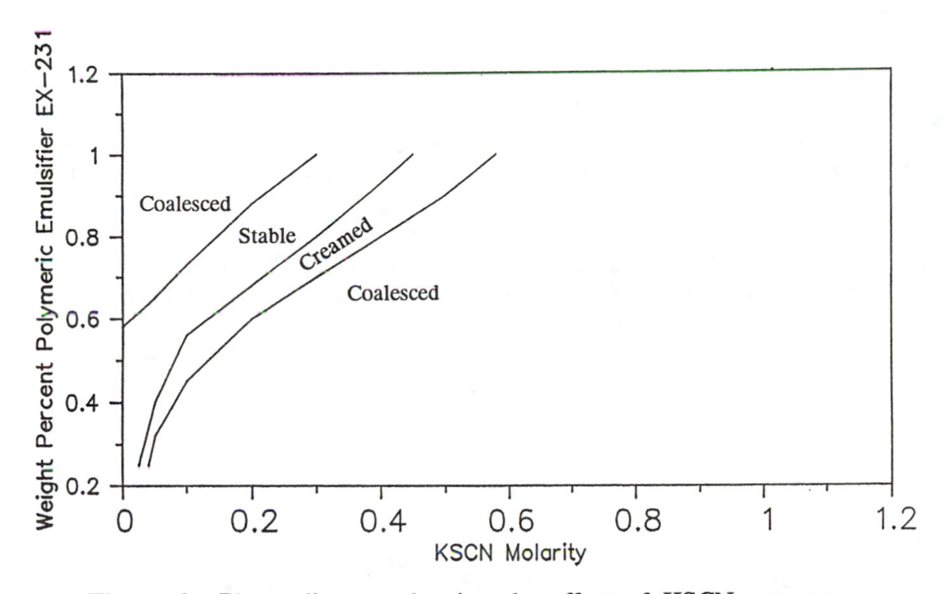

Figure 6. Phase diagram showing the effect of KSCN concentration and polymeric emulsifier EX-231 concentration on the stability of 10 wt % mineral oil emulsions.

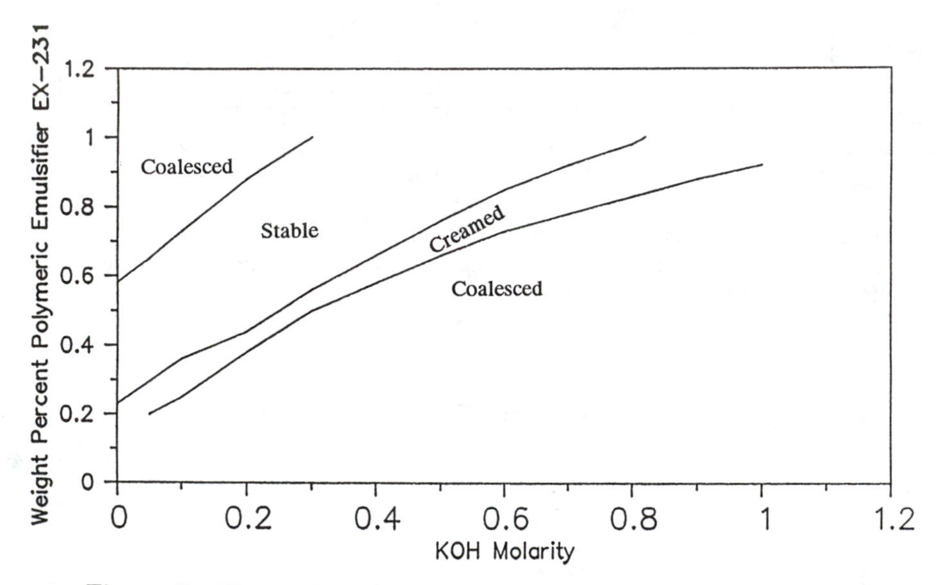

Figure 7. Phase diagram showing the effect of KOH concentration and polymeric emulsifier EX-231 concentration on the stability of 10 wt % mineral oil emulsions.

hydrophobic interaction. Thiocyanate is a water-structure-breaking ion and it would be expected to decrease the strength of the hydrophobic interaction, leading to disengagement of the polymer from the interface and subsequent emulsion instability. This is observed in the relatively large area of coalescence in the KSCN phase diagram.

The chloride ion, on the other hand, is a water-structure-making ion and as such it would be expected to enhance the strength of the hydrophobic interaction. In the systems studied here, the enhanced hydrophobic interaction might be expected to be manifested by a larger area of the phase diagram being occupied by stable or creamed emulsion as a consequence of stronger anchoring of the polymer at the oil–water interface.

In the presence of potassium chloride, the droplets in the creamed emulsions were stable against coalescence for a period greater than 6 months even when the viscosity of the continuous medium was close to that of water and the yield value was below detectable limits. In this case, the distance between the oil droplets would be determined by the polymeric "spaces" between the droplets (*9*). The fact that these creamed emulsions are stable to coalescence is a strong argument in favor of these hydrophobically modified poly(acrylic acids) being true emulsifiers, rather than gellants with hydrophobic domains.

A region of instability is found at high polymer concentrations. This has been observed and reported previously (*1*) and has been attributed to touching of the coils of adjacent polymer molecules in the semi-dilute region. Vincent (*10*) has shown that, in this concentration range, dimensional collapse of the polymer chains occurs and stabilization is lost.

Conclusions

1. Addition of micro-ion salts causes loss of viscosity of cross-linked poly(acrylic acid) rheology modifiers. This viscosity loss is dominated by the cations of the added electrolyte.

2. Conventional poly(acrylic acid) rheology modifiers cannot by themselves emulsify oil in water, but they can stabilize conventional emulsions against creaming.

3. Emulsions prepared using hydrophobically modified poly(acrylic acid) are generally stable, even in the absence of conventional emulsifier. The emulsions so formed spontaneously coalesce upon addition of sufficient salt to the system.

4. Addition of water-structure breakers, such as thiocyanate ion, to these emulsions, causes coalescence at lower molar concentrations than either potassium hydroxide or potassium chloride.

5. Added potassium chloride confers a larger compositional range of emulsions that cream, but are stable to coalescence, than comparable systems with added potassium hydroxide or potassium thiocyanate.

6. Hydrophobically modified poly(acrylic acid) rheology modifiers are true emulsifiers.

7. At higher polymer concentrations, a region of emulsion instability is evident.

References

1. Lochhead, R. Y.; Davidson, J. A.; Thomas, G. M. In *Polymers in Aqueous Media: Performance Through Association;* Glass, J. E., Ed.; ACS Advances in Chemistry Series 223; American Chemical Society: Washington, DC, 1989; Chapter 7.

2. Liang, S.-J.; Fitch, R. M. In *Polymer Adsorption and Dispersion Stability;* Goddard, E. D.; Vincent, B., Eds.; ACS Symposium Series 240; American Chemical Society: Washington, DC, 1984; Chapter 13.

3. Ananthapadmanabhan, K. P.; Leung, P. S.; Goddard, E. D. In *Polymer Association Structures;* El-Nokaly, M. A., Ed.; ACS Symposium Series 384; American Chemical Society: Washington, DC, 1989; Chapter 18.

4. Brown, D. H.; Topham, J. D.; Martin, G. P. *J. Pharm. Pharmacol.* **1984,** *36,* 20P.

5. Voet, A; Brand, J. S. *American Ink Maker* **1950,** *9,* 28–91.

6. Houwink, R. *Physikalische eigenshaften and feinbrau von atur-und Kinstharzen;* Akademische Verlagsgesellschaft MBH: Leipzig, 1934; pp 33–36.

7. Tanford, C. *The Hydrophobic Effect;* Wiley: 1973.

8. Franks, F., *Water;* Plenum: 1975; Volume 4, Chapter 1.

9. Cabane, C.; Wong, K.; Duplessix, R. In *Polymer Association Structures;* El-Nokaly, M. A., Ed.; ACS Symposium Series 384; American Chemical Society: Washington, DC, 1989; Chapter 19.

10. Vincent, B.; Whittington, S. In *Colloid and Surface Science;* Matijevic, E., Ed.; Plenum: 1982.

RECEIVED January 31, 1991

Chapter 7

Effects of Process Variables on the Emulsion and Solution Properties of Hydrophobically Modified Alkali-Swellable Emulsion Thickeners

Gregory D. Shay[1], Fran K. Kravitz, and Peter V. Brizgys

Administrative and Research Center, DeSoto, Inc., 1700 South Mount Prospect Road, Box 5030, Des Plaines, IL 60017

Emulsion polymerization process conditions and chemical modifiers often affect the physical characteristics of the latex polymers produced. They may also affect the aqueous solution properties of alkali-swellable or alkali-soluble latex emulsions which may be either non-associative (ASE), or associative by hydrophobe modification (HASE). Using a constant HASE terpolymer composition, a study was conducted to examine the effects of twelve independent reaction variables on latex stability during processing (reactor plating and grit formation), the final latex emulsion properties (pH, particle size, viscosity and THF solubility dried polymer), and the aqueous solution properties upon neutralization (viscosity, and clarity). Some direct relationships observed with increasing component concentration were an increase in latex viscosity as a function of copolymerizable surfactant and an increase in latex and solution viscosity as a function of conventional surfactant. Among the inverse relationships observed with increasing component concentration were a decrease in reactor grit and plating as a function of copolymerizable surfactant and a decrease in aqueous solution viscosity as a function of crosslinking monomer and chain transfer agent. The most surprising result observed which occurred in several variable sets was a relationship between aqueous solution thickening efficiency and a common dependent time variable (start of continuous monomer addition relative to the observed exotherm peak during seed formation). Maximum solution viscosity was obtained when the timing of these events coincided.

Discussed in a recent review (1) were the classification of AST's (alkali-swellable or alkali-soluble thickeners) by chemical composition and polymerization process, and the commercial importance of AST's produced by emulsion polymerization (ASE's and HASE's). Also reviewed were investigations into the effects of monomer composition on the swelling, dissolution and thickening behavior of ASE's. Among the compositional variables examined were the type or amount of carboxylic acid monomer present (2-6) and the resulting acid distributions within the emulsion particles produced (3-10). Complementary

[1]Current address: Union Carbide Corporation, 410 Gregson Drive, Cary, NC 27511

0097-6156/91/0462-0121$06.25/0

studies on the type of acid and its effect on acid distribution for emulsion polymers in general were also reported (10-14). Some of these and other important works examined the effects of variations in the type or amount of the non-carboxylic comonomers present in ASE copolymers (2,3,6,8,9,15-20).

Some limited data were also reported on the effects of compositional variations within HASE polymers. Since most HASE thickeners are terpolymers (ASE's that have been modified with a third hydrophobically terminated associative monomer to provide an additional thickening mechanism), prior data on the effects of carboxylic acid and non-associative comonomer in these polymers were generally reaffirmed. However, these studies also provided additional insight into the important effects of the associative monomer architecture on dissolution and thickening behavior (1,16,21,22).

Little information has been reported, however, on the effects of polymerization processing variables on the emulsion polymer physical properties or on the swelling, dissolution or thickening behavior of ASE or HASE polymers. The situation is particularly complex in the case of the HASE polymers because of multiple thickening mechanisms involving hydrodynamic thickening due to high molecular weight or swelling of crosslinked particles (2,19,23) and associative thickening via hydrophobe interaction (24,25) or ion-dipole interactions (26,27). In some of the work that has been reported, Nishida and co-workers (8,9) quantitatively determined the effects of batch vs semi-continuous emulsion polymerization on the thickening response of methacrylic acid containing ASE's. Morgan and Jensen examined the effects emulsion polymerization process variables on the homogeneity of low molecular weight styrene/acrylic acid ASE's (28,29).

Experimental Details

Polymer Composition. A single HASE terpolymer composition was selected to examine several independent polymerization processing variables. The terpolymer was composed by weight of 30% ethyl acrylate (EA), 35% methacrylic acid (MAA) and 35% of a urethane-functional associative macromonomer (UMM) containing 50 moles of ethylene oxide and a nonylphenol terminal hydrophobe. The chemical description and general procedures used to prepare the macromonomer and HASE terpolymer have been previously described (16,30,32), and commercial thickener products baring the POLYPHOBE (DeSoto Inc.) trade name encompass this technology. Except for reagent grade sodium chloride, all components for polymer syntheses were commercial grade and were used as received without further purification. Symbols and suppliers of the raw materials are detailed in List I.

Apparatus. The polymerization apparatus for preparation of the HASE terpolymer consisted of a 3L reaction flask equipped with stirrer, thermometer, nitrogen sparge inlet, and monomer and initiator solution delivery pumps. The flask was immersed in a water bath which was maintained at constant temperature throughout the polymerization process. The the water bath and reactor temperature were monitored with a calibrated dual trace recorder to observe monomer pre-charge exotherm (seed formation) and polymerization temperature during continuous monomer addition.

List I - Materials: Symbol, Trade Name and Source

EA	Ethyl Acrylate	Union Carbide
MAA	Methacrylic Acid	Rohm & Haas
UMM	Urethane Macromonomer	DeSoto Inc.
TMPTA	Trimethylolpropane Triacrylate	Sartomer
NaSS	Sodium Styrene Sulfonate	Toyo Soda U.S.A.
NaDOSS	Sodium Dioctyl Sulfosuccinate	American Cyanamide
NaPS	Sodium Persulfate	FMC Corporation
NaCl	Sodium Chloride	Baker
TDDM	t-Dodecyl Mercaptan	Pennwalt

Trade Names: (a) SR-351, (b) Spinomar NaSS, (c) Aerosol OT-75

Polymerization Procedure. After setting the agitator speed, deionized water was added (water pre-charge) to the reactor, and a nitrogen sparge was started to remove dissolved oxygen. The water bath was heated to a preselected equilibrium temperature and maintained at that temperature for the duration of the polymerization process. A monomer mix was prepared in a separate stirred vessel consisting of the EA, MAA, 90% aqueous (UMM), 83.5% of the total NaDOSS surfactant, and 2% deionized water based on total active monomer. An initiator solution was prepared in a third vessel consisting of 70% of the total NaPS in deionized water. The water content of this solution varied depending on specified water pre-charge but was 16.5% for most formulations based on total active monomer. Upon reaching reactor temperature equilibrium, the reactor was sparged with nitrogen for an additional 30 min, and the sparge was then switched to a nitrogen blanket. The remaining NaDOSS surfactant (surfactant pre-charge) was added to the reactor followed by a percentage of the monomer mix (monomer pre-charge). After reactor temperature equilibrium was again established (usually about 2.5 degrees C below the water bath temperature), the remaining initiator (30% of total) was added to the reactor (initiator pre-charge) and this point was recorded as reaction time 0. The elapsed time to monomer pre-charge exotherm peak was recorded, and constant rate additions of the remaining monomer mix and initiator solution were started simultaneously at a preselected time with the initiator solution ending 15 min after the monomer mix. The water bath was held at constant temperature for an additional hr, and the reactor was then cooled to ambient temperature after a minor water adjustment for 30% calculated weight fraction solids.

When the following components were present in a formulation, NaCl (neutral salt) was added to the reactor just prior to the NaDOSS pre-charge, NaSS (copolymerizable surfactant) was added to the reactor immediately after the NaDOSS pre-charge, and the TDDM (chain transfer agent) and TMPTA (polyfunctional monomer) were incorporated into the monomer mix before the monomer pre-charge. When water pre-charge was varied, total water content was held constant by adjusting the amount of water in the continuous initiator solution.

Aqueous Solutions. Aqueous solutions of the HASE polymers were prepared by diluting some of the latex to 1.0% total weight solids concentration in deionized water followed by neutralization to pH 9.0 with standard 28% aqueous ammonia. The dilutions were based on the actual weight solids obtained using procedure [g] in List II. A laboratory mixer was utilized during neutralization to facilitate inversion from latex to solution which normally occurred at pH 6.0 to 6.5. Solution viscosity was relatively constant for each neutralized polymer in the 8 to 10 pH range.

Independent Variables. Twelve independent variables were studied, and one common HASE polymer formulation (X) was included in all of the variable sets as a control (See Table I). Control formulation (X) was prepared five times with data averaged to improve statistical significance (See Table II). With the exception of the independent variable being studied, each polymerization run was identical to the control (X) formulation.

Table I - Independent Reaction Variables: Codes, Units, and Control HASE Polymer (X) Levels

NaSS	(%)	Copolymerizable Surfactant Concentration	0.5
WPC	(%)	Water Pre-charge	55.0
NaDOSS	(%)	Conventional Surfactant Concentration	4.0
NaCl	(%)	Neutral Salt Concentration	None
NaPS	(%)	Initiator Concentration	0.26
TDDM	(%)	Chain Transfer Agent Concentration	None
TMPTA	(%)	Polyfunctional Monomer Concentration	None
WBPT	(C)	Polymerization Water Bath Temperature	80
AGR	(rpm)	Agitation Rate	160
MAT	(hrs)	Continuous Monomer Addition Time	2.5
MPC	(%)	Monomer Pre-Charge for Seed Formation	10:0
MPCHT	(min)	Monomer Pre-Charge Initiation Hold Time	40

Where: (%) = percent by weight on total active monomer (i.e., ethyl acrylate + methacrylic acid + urethane functional macromonomer) except for MPC which is percent of total monomer mix.

Results and Discussion

The twelve independent variables, their levels, and the physical characteristics of the emulsions and neutralized aqueous solutions produced are compiled in Table III. The reproducibility of the physical properties measured for control HASE polymer (X) are summarized in Table II. It is reasonable to assume that the statistical data obtained for polymer (X) are fairly representative of the other formulations produced (e.g., for some of the more important properties: exotherm peak time +-5%, particle size +-10%, latex viscosity +-7%, and aqueous solution viscosity +-9%. The following is a discussion of some of the more significant findings for each of the independent variables examined.

Copolymerizable Surfactant. This series was conducted early in the experimentation to establish a control formulation with low grit and plating levels. Sodium styrene sulfonate (NaSS) has been used to help stabilize film forming emulsion polymers by augmenting the surface charge density (31). As disclosed in U.S. 4,801,671 (32) and confirmed here, use of an unsaturated aromatic sulfonic acid such as sodium styrene sulfonate is effective in

List II - Test Methods and Experimental Data Recorded

[a] Polymerization Start Temperature - the reactor temperature at time of NaPS pre-charge.

[b] Polymerization Exotherm (Exo) Temperature - the magnitude of exotherm for the monomer pre-charge seed formation after initiation (i.e., [b] = exotherm peak temperature - [a]).

[c] Polymerization Temperature Continuous (Cont) - the polymerization temperature during the midpoint of the continuous monomer addition.

[d] Monomer Pre-Charge Exotherm Peak Time (min) - the elapsed time from initiator pre-charge to the exotherm peak of the monomer pre-charge which is an indicator inhibition or retardation.

[e] Monomer Pre-Charge Exotherm Peak (%) - the exotherm peak time [d] divided by the monomer pre-charge hold time x 100%.

[f] Grit and Plating Level (%) - the total residue collected (plating plus filtered grit) as a percent of total monomer, after drying at 150 degrees C for 30 min in a convection oven.

[g] Latex Solids (%) - determined in triplicate on 0.5 g samples in an aluminum dish in a convection oven at 150 degrees C for 30 min.

[h] Latex Particle Size (nm) - by dynamic light scattering on a Coulter N-4 particle size analyzer (cumulants method).

[i] Latex Viscosity (mPa.s) - the Brookfield RVT viscosity obtained using an appropriate spindle at 20 rpm in a pint jar.

[j,k] 1% Aqueous Solution Viscosity (mPa.s) - the 10 and 100 rpm viscosities respectively for 1% total weight component solids in deionized water on a Brookfield RVT viscometer.

[l] Viscosity Index (VI) - the log of the 10 rpm Brookfield viscosity divided by the log of the 100 rpm Brookfield viscosity (i.e., log [j] / log [k]).

[m] Relative 1% Solution Clarity - rating scale is 0-5 with 0 being perfectly clear and 5 opaque. Solution clarity is important for some commercial thickener applications such as personal care products (e.g., shampoos) and oil field flooding operations.

[n] THF Solubility (%) - determined using 1 g dried latex solids in 100 ml THF, mixed 24 hrs at 25 degrees C, and centrifuged at 15000 rpm for 30 min. Solids were determined on an aliquot of supernatant by ASTM D2369-81 after filtering through a 0.45 micron filter.

reducing reactor grit and plating in HASE polymers. The data also show that the amount of NaSS must be optimized to minimize latex viscosity which may become excessive. Since NaSS is highly ionized, the apparent particle size increase with increasing stabilizer concentration may have been due to increased hydrophilicity resulting in increased particle swelling. Although a minimum was observed in solution viscosity at intermediate NaSS concentration, the viscosities observed may have been suppressed in the emulsions containing NaSS due

TABLE II - Processing and Physical Characteristics of Control HASE Polymer (X)

Control Polymer Code	Reactor: Polymerization Temp Start	Exo	Cont	MPC Exo Peak	MPC Exo Peak	Grit and Plating
(Units)	(C)	(C)	(C)	(min)	(%)	(%)
[Footnotes]	[a]	[b]	[c]	[d]	[e]	[f]
X-1	77.8	1.9	79.0	21	52.5	0.072
X-2	77.7	2.0	79.2	19	47.5	0.017
X-3	77.7	2.0	79.0	19	47.5	0.017
X-4	77.7	1.6	79.0	21	51.3	0.017
X-5	77.7	2.5	79.5	23	57.5	0.040
X AVG	77.7	2.0	79.1	20.5	51.3	0.033
X STD	0.0	0.3	0.2	1.7	4.1	0.024
X STD%	0.1	16.2	0.3	8.1	8.1	74.1

Control Polymer Code	Latex Emulsion Properties: Solids	pH	PS	Visc
(Units)			(nm)	(mPa.s)
[Footnotes]	[g]		[h]	[i]
X-1	29.67	3.08	326	164
X-2	29.48	3.10	382	-
X-3	29.60	3.12	325	182
X-4	29.63	3.24	306	192
X-5	29.27	3.05	-	177
X 1-5 AVG	29.53	3.12	335	179
X STD	0.16	0.07	33	12
X STD%	0.55	2.3	9.8	6.5

Control Polymer Code	1% Aqueous Solution Properties: Viscosity 10rpm	100rpm	VI	Clarity 0 Best	THF Sol
(Units)	(mPa.s)	(mPa.s)		(Rel)	(%)
[Footnotes]	[j]	[k]	[l]	[m]	[n]
X-1	3740	1136	1.169	3	-
X-2	3440	1120	1.160	2	-
X-3	3440	1320	1.133	3	-
X-4	3400	1084	1.164	3	-
X-5	3580	1154	1.161	4	41
X AVG	3520	1163	1.157	3.0	41
X STD	141	92	0.014	0.7	-
X STD%	4.0	7.9	1.2	23.6	-

Note: For explanation of [Footnotes] and (Units) see List II. For reaction conditions used in Control formulation (X) see Table I.

TABLE III - Processing and Physical Characteristics of HASE Polymer Variable Sets (Part A1)

Polymer Code	Independent Variable Type	Independent Variable Level	Reactor Polym Temp Start	Reactor Polym Temp Exo	Reactor Polym Temp Cont	Reactor MPC Exo Peak	Reactor MPC Exo Peak	Reactor Grit and Plat
(Units)			(C)	(C)	(C)	(min)	(%)	(%)
[Footnotes]			[a]	[b]	[c]	[d]	[e]	[f]
A-1	NaSS (%)	0.00	78.7	1.5	79.2	22	55	1.65
A-2	"	0.25	77.5	1.9	78.5	25	63	1.37
X-AVG	"	0.50	77.7	2.0	79.1	21	51	0.03
A-3	"	0.75	77.6	2.1	78.5	19	48	0.00
A-4	"	1.00	78.0	1.4	79.2	16	40	0.00
B-1	WPC (%)	55	77.9	1.4	77.2	16	40	100.00
B-2	"	108	76.7	2.3	77.0	18	45	0.00
B-3	"	162	77.5	2.0	78.3	21	53	0.00
X-AVG	"	216	77.7	2.0	79.1	21	51	0.03
C-1	NaDOSS (%)	2.0	77.6	1.4	78.5	20	50	0.00
X-AVG	"	4.0	77.7	2.0	79.1	21	51	0.03
C-2	"	6.0	78.4	1.9	79.5	21	53	0.00
C-3	"	8.0	77.6	1.7	78.5	24	60	0.00
C-4	"	10.0	77.5	2.1	79.5	20	50	0.00
X-AVG	NaCl (%)	0.0	77.7	2.0	79.1	21	51	0.03
D-1	"	0.1	77.2	2.1	79.0	27	68	0.05
D-2	"	0.5	77.8	1.3	78.9	33	83	2.05
D-3	"	1.0	77.5	1.6	79.0	40	100	100.00
E-1	NaPS (%)	0.07	77.6	1.2	78.6	46	115	0.00
E-2	"	0.13	77.2	1.0	78.3	36	90	0.00
X-AVG	"	0.26	77.7	2.0	79.1	21	51	0.03
E-3	"	0.39	76.9	2.0	78.6	17	43	0.00
E-4	"	0.52	77.5	2.2	78.7	16	40	0.00
E-5	"	0.78	77.1	2.6	79.2	18	45	1.01
X-AVG	TDDM (%)	0.0	77.7	2.0	79.1	21	51	0.03
F-1	"	0.1	77.6	1.7	78.8	25	63	0.00
F-2	"	0.2	77.7	1.3	79.0	22	55	0.00
F-3	"	0.4	77.4	2.2	79.5	25	63	0.03

Note: For explanation of Independent Variables see Table I.
For explanation of [Footnotes] and (Units) see List II

Continued on next page

TABLE III - Continued (Part A2)

Polymer Code	Independent Variable		Reactor					
			Polym Temp			MPC Exo Peak	MPC Exo Peak	Grit and Plat
	Type	Level	Start	Exo	Cont			
(Units)			(C)	(C)	(C)	(min)	(%)	(%)
[Footnotes]			[a]	[b]	[c]	[d]	[e]	[f]
X-AVG	TMPTDA (%)	0.00	77.7	2.0	79.1	21	51	0.03
G-1	"	0.05	77.3	2.0	78.5	18	45	0.00
G-2	"	0.20	77.8	1.8	78.7	20	50	0.00
G-3	"	0.40	77.4	2.1	78.9	21	53	0.25
G-4	"	0.60	77.9	2.5	79.6	22	55	0.16
H-1	WBPT (C)	70.0	67.4	7.0	69.8	127	318	1.12
H-2	"	75.0	72.7	2.0	74.0	38	95	0.00
X-AVG	"	80.0	77.7	2.0	79.1	21	51	0.03
H-3	"	85.0	83.0	1.6	84.2	13	33	0.00
H-4	"	90.0	85.9	2.0	88.0	12	30	0.10
I-1	AGR (rpm)	80	77.5	1.7	78.6	18	45	0.10
X-AVG	"	160	77.7	2.0	79.1	21	51	0.03
I-2	"	240	77.2	1.4	78.3	25	63	0.00
I-3		320	77.9	2.0	79.1	33	83	0.03
I-4	"	400	78.0	2.3	79.1	59	148	0.50
J-1	MAT (hrs)	1.5	77.6	1.9	79.2	23	58	0.68
J-2	"	2.0	77.9	2.1	79.2	19	48	0.00
X-AVG	"	2.5	77.7	2.0	79.1	21	51	0.03
J-3	"	3.0	77.5	1.6	78.3	21	53	0.00
J-4	"	4.0	77.5	1.8	78.3	23	58	0.00
K-1	MPC (%)	0.0	77.0	2.8	78.6	37	93	0.01
X-AVG	"	10.0	77.7	2.0	79.1	21	51	0.03
K-2	"	20.0	77.5	3.1	78.0	21	53	0.00
K-3	"	30.0	77.5	4.6	78.6	22	55	0.02
L-1	MPCHT (min)	0	77.8	4.0	79.2	28	-	0.03
L-2	"	10	78.0	2.9	78.8	26	260	0.03
L-3	"	20	78.0	2.0	79.0	22	110	0.00
L-4	"	30	77.2	2.0	78.1	23	77	0.00
X-AVG	"	40	77.7	2.0	79.1	21	53	0.03

Note: For explanation of Independent Variables see Table I.
For explanation of [Footnotes] and (Units) see List II

TABLE III - Continued (Part B1)

Polymer Code	Independent Variable Type	Level	Latex Emulsion Properties Solids	pH	PS	Visc
(Units)			(%)		(nm)	(mPa.s)
[Footnotes]			[g]		[h]	[i]
A-1	NaSS (%)	0.00	29.4	3.3	183	27
A-2	"	0.25	30.1	3.2	271	45
X-AVG	"	0.50	29.5	3.1	335	179
A-3	"	0.75	30.1	3.3	346	2075
A-4	"	1.00	29.7	3.2	460	7350
B-1	WPC (%)	55	-	-	-	-
B-2	"	108	29.8	3.3	428	1226
B-3	"	162	30.0	3.3	333	844
X-AVG	"	216	29.5	3.1	335	179
C-1	NaDOSS (%)	2.0	30.2	3.2	278	42
X-AVG	"	4.0	29.5	3.1	335	179
C-2	"	6.0	30.0	3.3	288	840
C-3	"	8.0	30.1	3.3	350	4610
C-4	"	10.0	29.9	3.1	-	14120
X-AVG	NaCl (%)	0.0	29.5	3.1	335	179
D-1	"	0.1	30.1	3.2	321	97
D-2	"	0.5	29.3	3.2	354	39
D-3	"	1.0	-	-	-	-
E-1	NaPS (%)	0.07	29.8	3.8	325	660
E-2	"	0.13	29.6	3.5	543	844
X-AVG	"	0.26	29.5	3.1	335	179
E-3	"	0.39	29.8	3.0	293	85
E-4	"	0.52	30.1	2.9	342	52
E-5	"	0.78	29.1	2.3	-	28
X-AVG	TDDM (%)	0.0	29.5	3.1	335	179
F-1	"	0.1	30.4	3.4	434	100
F-2	"	0.2	30.0	3.2	331	122
F-3	"	0.4	29.4	2.9	-	185

Note: For explanation of Independent Variables see Table I.
For explanation of [Footnotes] and (Units) see List II.

Continued on next page

TABLE III - Continued (Part B2)

Polymer	Independent Variable		Latex Emulsion Properties			
Code	Type	Level	Solids	pH	PS	Visc
(Units)			(%)		(nm)	(mPa.s)
[Footnotes]			[g]		[h]	[i]
X-AVG	TMPTDA (%)	0.00	29.5	3.1	335	179
G-1	"	0.05	29.8	3.3	318	206
G-2	"	0.20	28.8	3.3	342	69
G-3	"	0.40	30.2	3.2	287	77
G-4	"	0.60	29.8	3.0	-	65
H-1	WBPT (C)	70.0	29.6	3.2	220	31
H-2	"	75.0	30.0	3.3	398	1743
X-AVG	"	80.0	29.5	3.1	335	179
H-3	"	85.0	29.6	3.2	262	61
H-4	"	90.0	29.8	3.2	234	54
I-1	AGR (rpm)	80	30.0	3.4	297	260
X-AVG	"	160	29.5	3.1	335	179
I-2	"	240	30.0	3.3	332	144
I-3		320	29.7	3.1	482	233
I-4	"	400	30.1	3.2	160	151
J-1	MAT (hrs)	1.5	29.9	2.8	333	90
J-2	"	2.0	29.2	3.2	300	128
X-AVG	"	2.5	29.5	3.1	335	179
J-3	"	3.0	29.7	3.2	307	190
J-4	"	4.0	29.9	3.3	354	293
K-1	MPC (%)	0.0	29.7	3.2	153	264
X-AVG	"	10.0	29.5	3.1	335	179
K-2	"	20.0	29.6	3.3	268	130
K-3	"	30.0	30.0	3.3	243	120
L-1	MPCHT (min)	0	30.1	3.3	241	375
L-2	"	10	30.2	3.3	349	1188
L-3	"	20	30.1	3.3	373	1244
L-4	"	30	30.1	3.3	355	448
X-AVG	"	40	29.5	3.1	335	179

Note: For explanation of Independent Variables see Table I.
For explanation of [Footnotes] and (Units) see List II.

TABLE III - Continued (Part C1)

Polymer Code	Independent Variable Type	Independent Variable Level	1% Aqueous Solution Properties: Viscosity 10rpm	Viscosity 100rpm	Viscosity VI	Clarity 0 Best	THF Sol
(Units)			(mPa.s)	(mPa.s)		(rel)	(%)
[Footnotes]			[j]	[k]	[l]	[m]	[n]
A-1	NaSS (%)	0.00	5000	1338	1.18	2	-
A-2	"	0.25	4140	1250	1.17	2	-
X-AVG	"	0.50	3520	1162	1.16	3	41
A-3	"	0.75	4240	1182	1.18	3	-
A-4	"	1.00	4740	1250	1.19	2	-
B-1	WPC (%)	55	-	-	-	-	-
B-2	"	108	3740	1086	1.18	4	-
B-3	"	162	3760	1150	1.17	4	-
X-AVG	"	216	3520	1162	1.16	3	41
C-1	NaDOSS (%)	2.0	1660	530	1.18	4	-
X-AVG	"	4.0	3520	1162	1.16	3	41
C-2	"	6.0	4420	1360	1.16	2	-
C-3	"	8.0	6580	1786	1.17	2	-
C-4	"	10.0	8260	2220	1.17	3	46
X-AVG	NaCl (%)	0.0	3520	1162	1.16	3	41
D-1	"	0.1	3180	1018	1.16	3	-
D-2	"	0.5	2560	858	1.16	5	-
D-3	"	1.0	-	-	-	-	-
E-1	NaPS (%)	0.07	4340	1260	1.17	5	-
E-2	"	0.13	5240	1476	1.17	5	-
X-AVG	"	0.26	3520	1162	1.16	3	41
E-3	"	0.39	2360	786	1.16	2	-
E-4	"	0.52	2140	762	1.16	1	-
E-5	"	0.78	1392	556	1.15	1	51
X-AVG	TDDM (%)	0.0	3520	1162	1.16	3	41
F-1	"	0.1	1580	750	1.11	0	-
F-2	"	0.2	260	214	1.04	0	-
F-3	"	0.4	30	58	0.84	0	100

Note: For explanation of Independent Variables see Table I.
For explanation of [Footnotes] and (Units) see List II.

Continued on next page

TABLE III - Continued (Part C2)

Polymer Code	Independent Variable Type	Independent Variable Level	1% Aqueous Solution Properties: Viscosity 10rpm	Viscosity 100rpm	Viscosity VI	Clarity 0 Best	THF Sol
(Units)			(mPa.s)	(mPa.s)		(rel)	(%)
[Footnotes]			[j]	[k]	[l]	[m]	[n]
X-AVG	TMPTDA (%)	0.00	3520	1162	1.16	3	41
G-1	"	0.05	3680	1118	1.17	4	-
G-2	"	0.20	2380	770	1.17	5	-
G-3	"	0.40	1000	370	1.17	5	-
G-4	"	0.60	430	204	1.14	5	33
H-1	WBPT (C)	70.0	460	202	1.16	5	-
H-2	"	75.0	9420	2358	1.18	5	35
X-AVG	"	80.0	3520	1162	1.16	3	41
H-3	"	85.0	1840	672	1.15	4	-
H-4	"	90.0	1182	508	1.13	2.5	45
I-1	AGR (rpm)	80	3200	1000	1.17	3	-
X-AVG	"	160	3520	1162	1.16	3	41
I-2	"	240	4280	1272	1.17	2	-
I-3		320	7220	1917	1.17	3	42
I-4	"	400	5400	1516	1.17	2	-
J-1	MAT (hrs)	1.5	4100	1284	1.16	2	-
J-2	"	2.0	4110	1249	1.17	3	-
X-AVG	"	2.5	3520	1162	1.16	3	41
J-3	"	3.0	2860	932	1.16	4	-
J-4	"	4.0	2600	874	1.16	3	-
K-1	MPC (%)	0.0	5360	1470	1.18	2	-
X-AVG	"	10.0	3520	1162	1.16	3	41
K-2	"	20.0	2260	718	1.17	5	-
K-3	"	30.0	1520	512	1.17	5	-
L-1	MPCHT (min)	0	3580	1030	1.18	5	-
L-2	"	10	6680	1678	1.19	4	-
L-3	"	20	7520	1898	1.18	2	-
L-4	"	30	5300	1500	1.17	3	-
X-AVG	"	40	3520	1162	1.16	3	41

Note: For explanation of Independent Variables see Table I.
For explanation of [Footnotes] and (Units) see List II.

to a salt effect (Na+ ion and aqueous phase polymerization of NaSS forming polyanion). The potential for salt effect will be discussed further in the Conventional Surfactant and Neutral Salt set sections.

Water Pre-Charge. With the exception of the final latex viscosity which decreased significantly with increasing initial level of water added (final solids content was constant at 30% by weight), there was little effect of water pre-charge concentration on other physical properties. The lowest initial water amount used resulted in total coagulation late in the polymerization (at 90% of continuous monomer addition). This was probably not due to excessive solids concentration since the solids level at this point in the polymerization was similar to the other runs in the set.

Conventional Anionic Surfactant. The apparent effects of increasing surfactant (sodium dioctylsulfosuccinate) concentration during polymerization were an increase in emulsion viscosity and aqueous solution viscosity of the neutralized terpolymer. Although the surfactant employed is a highly ionized organic salt with appreciable water solubility, the observed effect was opposite of that found for increasing neutral salt (NaCl) concentrations. The importance of surfactant in the emulsion polymerization process is very well known, and its effect on some of the previously mentioned physical and chemical properties of the polymer produced could account for some of the results observed. However, since surfactants are associative species, their presence can either increase (co-thickening) or decrease viscosity of an associative polymer solution depending on concentration (25,33,34,35).

To determine if surfactant co-thickening was at least a partial explanation for the observed effect, sufficient NaDOSS was post-added to the aqueous solution of the control HASE thickener X-5 (4% NaDOSS on active monomer) to provide a total surfactant concentration equal to that of of the C-4 emulsion (10% NaDOSS). The associative effects of the surfactant appeared to account for about half of the observed viscosity increase (See Table IV).

Neutral Salt. The apparent effect of increasing salt (NaCl) concentration during the emulsion polymerization process was a decrease in the aqueous solution viscosity of the neutralized terpolymer. The result may have been due to one or more physical or chemical factors including changes in molecular weight, monomer sequencing, the amount aqueous phase polymerization, particle morphology, particle swellability or due to a simple electrostatic effect (ionic strength) reducing the hydrodynamic volume of the polyanion in solution. This latter phenomenon is the well known polyelectrolyte effect, where at low solution concentrations the polymer coil expands due to the electrostatic repulsion of the fixed charges on the poly-ion (36). The expansion or contraction of the coil is a function of the neutral salt concentration. To determine how much of the observed viscosity reduction was due to salt effect, a post-addition of NaCl was made to the aqueous solution of the X-5 control formulation (the formulation without salt) to give a total NaCl concentration equal to emulsion D-2 (0.5% NaCl on total monomer). The viscosity of the post-add salt solution was similar to that made with salt during polymerization suggesting a substantial salt effect (See Table IV).

Initiator. The general effect of increasing initiator (NaPS) concentration was a modest decrease in emulsion and aqueous solution viscosities. Solution clarity

TABLE IV - Evaluation of Salt Effect in HASE Polymers Containing Neutral Salt or Conventional Surfactant

Polymer Code	Salt	Salt Concentration Original (%)	Post-Add (%)	Total (%)	Solution Viscosity 10 rpm (mPa.s)	100 rpm (mPa.s)
X-5	NaCl	0	0	0	3580	1154
D-2	NaCl	0.5	0	0.5	3180	1018
X-5	NaCl	0	0.5	0.5	3204	1076
X-5	NaDOSS	4	0	4	3580	1154
C-4	NaDOSS	10	0	10	8260	2220
X-5	NaDOSS	4	6	10	6430	2136

Note: (%'s) are based on total active monomer.

also improved. Although these observations might suggest the anticipated reduction in molecular weight that is frequently observed with increasing initiator concentration, an evaluation for salt effect (free sodium and persulfate ions) was not conducted nor was the effect of polymer bound persulfate determined. Compared to the control formulation, the increased THF solubility for the sample containing the highest level of initiator is consistent with molecular weight reduction.

Chain Transfer Agent. Increased aqueous solubility with decreasing molecular weight has previously been observed for ASE polymers using optical density measurements (20,23). In our study, a sharp decrease in aqueous solution viscosity and a corresponding improvement in solution clarity were observed with increasing levels of TDDM chain transfer agent. The HASE polymer containing the highest level of TDDM (F-3) also had the highest THF solubility of all samples tested. These observations are consistent with an expected reduction in molecular weight. Since emulsion H-3 was completely soluble in THF, a molecular weight determination was conducted relative to polystyrene standards: Mn 17000, Mw 82000, Mz 160000. Because of low solubility and the questionable validity of the results that would be obtained, molecular weight determinations were not conducted on the other emulsion samples.

Polyfunctional Monomer. Low levels of polyethylenically unsaturated monomers are commonly included in ASE and HASE polymerization recipes to lightly crosslink the emulsion particles. The desired effect is conversion from a mostly alkali-soluble emulsion to a mostly alkali-swellable emulsion with a resulting increase in the apparent aqueous solution viscosity (1). When these emulsions are diluted and swollen with alkali, clear microgel solutions are formed whose particle swelling diameters may be hundreds of times those of the original particle dimensions. The thickening efficiencies of swellable ASE polymers (and presumably sellable HASE polymers as well) are very dependent on the level of polyfunctional monomer present (2,19). Too little polyfunctional monomer produces low crosslink density resulting in excessive solubility and insufficient swelling. Too much produces polymer that is too tightly crosslinked resulting in particles which exhibit little or no swelling. An optimum level of crosslinker, therefore, exists (usually less than about 1%) which will provide maximum solution viscosity.

The effect of all levels of TMPTA crosslinker in the present study was a reduction in aqueous solution viscosity. One possible explanation is that the optimum crosslink density had already been exceeded in the control emulsion containing no TMPTA. Two potential sources of polyfunctional monomer theoretically are generated during synthesis of the urethane-functional associative macromonomer which is a reaction product of a,a-dimethyl meta-isopropenyl benzyl isocyanate (American Cyanamide trade name m-TMI) and an alkylaryl ethoxylate (nonionic surfactant). m-TMI can react with trace moisture present in the nonionic surfactant during synthesis to form an amine (Reaction 1). The amine can subsequently react with more m-TMI to form a difunctional urea monomer (Reaction 2).

$$\underset{\text{(m-TMI)}}{H_2C{=}C(CH_3){-}R{-}N{=}C{=}O} + H{-}O{-}H \longrightarrow H_2C{=}C(CH_3){-}R{-}NH_2 + CO_2 \quad (1)$$

$$H_2C{=}C(CH_3){-}R{-}NH_2 + \underset{\text{(m-TMI)}}{O{=}C{=}N{-}R{-}C(CH_3){=}CH_2} \longrightarrow H_2C{=}C(CH_3){-}R{-}NH{-}CO{-}NH{-}R{-}C(CH_3){=}CH_2 \quad (2)$$

Nonionic surfactants also frequently contain some ethylene glycol and polyethylene glycol (PEG) which is formed due to the presence of trace moisture during ethoxylation (Reaction 3). The m-TMI can react with both ends of the ethylene glycol or PEG forming difunctional urethane monomer (Reaction 4).

$$n\ H_2C\overset{O}{—}CH_2 + H{-}O{-}H \longrightarrow HO{-}(CH_2{-}CH_2{-}O)_n{-}H \quad (3)$$

$$HO{-}(CH_2{-}CH_2{-}O)_n{-}H + 2\ \underset{\text{(m-TMI)}}{H_2C{=}C(CH_3){-}R{-}N{=}C{=}O} \longrightarrow$$

$$H_2C{=}C(CH_3){-}R{-}NHCOO{-}(CH_2{-}CH_2{-}O)_n{-}CONH{-}R{-}C(CH_3){=}CH_2 \quad (4)$$

The moisture content of the NP-50 surfactant used to prepare the associative macromonomer was 0.15% by Karl-Fisher titration. The PEG content was determined to be about 1% by an accepted extraction technique (37). Based on the levels of moisture and PEG in the surfactant, both types of difunctional monomer (Reactions 2 and 4) would be expected to be present in the macromonomer. Because the analysis is difficult and imprecise, no attempt was made to determine the amount of crosslinking monomer present; however, since the amount of macromonomer was same in each HASE formulation, the amount of difunctional monomer was also a constant.

Polymerization Temperature. With the exception of emulsion polymer H-1 which was made at 70 degrees C, increased polymerization temperature resulted in the expected reduction in aqueous solution viscosity with an overall improvement in solution clarity. This effect is similar to that found for increasing levels of chain transfer agent again suggesting molecular weight reduction. Polymer H-1 did not follow this trend presumably due to the low rate of free radical generation at 70 degrees C which resulted in a low rate of monomer conversion. Evidence for this was the absence of an exotherm peak during the seed formation stage.

Agitation Rate. An increase in agitation rate generally increased solution viscosity with a maximum observed at 320 rpm. One possible interpretation of this observation is that a more uniform polymer composition is produced due to improved monomer transfer from droplets. Because of its molecular size and low water solubility compared to MAA and EA, the macromonomer would be expected to be generally poor mobility. Increased rate of agitation could facilitate the process. However, another observation was a maximum in the exotherm peak time with increasing agitation rate. The maximum in solution viscosity may be related to the latter effect, and further support for this conclusion is provided later.

Monomer Addition Time. Of the polymerization variables Morgan and Jensen (8,9) studied, a reduced monomer addition rate was reported to be most important for improved copolymer homogeneity as characterized by solution clarity. In our study, an increase in monomer addition time had little effect on solution clarity or emulsion viscosity but did moderately decrease aqueous solution viscosity. The more important variable with respect to monomer addition was the amount of monomer in the pre-charge for seed formation.

Monomer Pre-Charge. Solution viscosity and clarity increased with decreasing monomer pre-charge up to and including polymer K-1 made without any monomer pre-charge (i.e., without a seeding technique). A free monomer build-up occurred in this sample, however, and an exotherm peak of about 3 degrees C was observed after 37 min continuous monomer addition. The resulting reactor temperature trace was, therefore, similar to many of the other polymerizations.

Monomer Pre-Charge Initiation Hold Time. On examination of the data for the first 11 independent variable sets, there appeared to be some relationship between exotherm peak time and the start of the continuous monomer addition. Maximums in solution viscosity were clearly obtained in several of these variable sets (Initiator Concentration, Polymerization Temperature, Monomer Pre-Charge, and Agitation Rate) when the exotherm peak was close to the start of continuous monomer addition. To investigate this apparent effect further, a final set of experiments was prepared varying the start of monomer addition from 0-40 min after monomer pre-charge initiation. Although the results may have been coincidental, the relationship was again observed. Maximum viscosity was obtained on emulsion L-3 where the polymerization peak was within two min of the start of monomer addition. Starting the monomer addition earlier or later reduced solution viscosity. Data extracted from Table III demonstrating this effect are compiled in Table V and Figure 1.

Conclusions

Several of the HASE independent processing variables examined showed strong effects on latex stability during processing, on the final latex emulsion properties, or on the aqueous solution properties upon neutralization. For example, strong inverse relationships were observed for reactor grit and plating as a function of the copolymerizable surfactant (NaSS) concentration, and for aqueous solution viscosity as a function of both the crosslinking monomer (TMPTA) and the chain transfer agent (TDDM) concentrations. These results were mostly predictable (i.e., NaSS increased stability of the particles by increasing surface

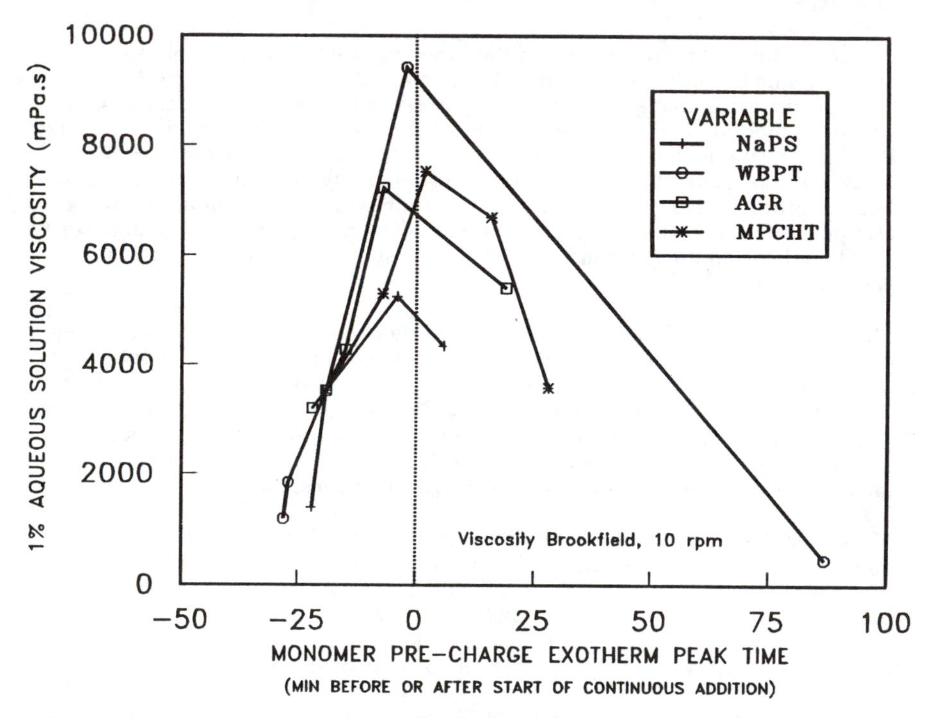

Figure 1. Relationship between the 1% alkaline aqueous solution viscosity (Brookfield, 10 rpm) of the final HASE latex and the monomer pre-charge exotherm peak time relative to the start of the continuous monomer addition. Data plotted are for the following independent variable sets: sodium persulfate initiator concentration (NaPS), water bath polymerization temperature (WBPT), agitation rate (AGR), and monomer pre-charge hold time (MPCHT). The monomer pre-charge exotherm peak time is a dependent variable.

TABLE V - Effect of Monomer Pre-Charge Initiation Hold Time / Exotherm Peak Time Ratio on Aqueous Solution Viscosity

Polymer Code	Independent Variables: Type	Amount	* MPCHT	Dependent Variable: MPC Exo Peak	[d] - MPCHT	[d]/ MPCHT * 100	1% Aqueous Solution Viscosity
(Units)			(min)	(min)	(min)	(%)	(mPa.s)
[Footnotes]				[d]		[e]	[i]
E-1	NaPS (%)	0.07	40	46	6	115	4340
E-2	"	0.13	40	36	-4	90	5240
X-AVG	"	0.26	40	21	-19	51	3520
E-3	"	0.39	40	17	-23	43	2360
E-4	"	0.52	40	16	-24	40	2140
E-5	"	0.78	40	18	-22	45	1392
H-1	WBPT (C)	70.0	40	127	87	318	460
H-2	"	75.0	40	38	-2	95	9420
X-AVG	"	80.0	40	21	-19	51	3520
H-3	"	85.0	40	13	-27	33	1840
H-4	"	90.0	40	12	-28	30	1182
I-1	AGR (rpm)	80	40	18	-22	45	3200
X-AVG	"	160	40	21	-19	51	3520
I-2	"	240	40	25	-15	63	4280
I-3	"	320	40	33	-7	83	7220
I-4	"	400	40	59	19	148	5400
K-1	MPC (%)	0.0	40	37	-3	93	5360
X-AVG	"	10.0	40	21	-19	51	3520
K-2	"	20.0	40	21	-19	53	2260
K-3	"	30.0	40	22	-18	55	1520
L-1	MPCHT (min)	0	0	28	28	-	3580
L-2	"	10	10	26	16	260	6680
L-3	"	20	20	22	2	110	7520
L-4	"	30	30	23	-7	77	5300
X-AVG	"	40	40	21	-19	53	3520

* The end of the initiation hold period for monomer pre-charge is also the starting point for continuous monomer addition.

Note: For explanation of Independent Variables see Table I.
For explanation of [Footnotes] and (Units) see List II.

charge density, TMPTA crosslinked the particles excessively reducing their ability to swell appreciably, and TDDM decreased molecular weight resulting in a reduction in hydrodynamic volume). Among the strong direct relationships observed were increased latex viscosity as a function of NaSS concentration, and increased latex and aqueous solution viscosity as a function of conventional surfactant (NaDOSS) concentration. These results appear to be related to surface charge density with respect to the acidic latex and to the well known surfactant co-thickening effect for the aqueous solution.

Perhaps the most interesting, yet unexpected result obtained which was observed in several variable sets, was a relationship between aqueous solution viscosity and a common dependent processing variable (start of continuous monomer addition relative to the timing of the observed exotherm peak during seed formation). Maximum solution viscosity was obtained when these two events coincided. At this point the polymerization process is assumed to be very active and substantially uninterrupted with the introduction of additional monomer. We believe that under these conditions a more homogeneous copolymer with respect to molecular weight, monomer sequencing or balance of soluble and swellable polymer is produced maximizing thickening efficiency.

Acknowledgment

The encouragement and financial support for these studies by the Polymer Development, Pioneering Research, and Thickener Research and Development Groups of DeSoto Inc. is gratefully acknowledged.

Literature Cited

1. Shay, G. D. In *Polymers in Aqueous Media: Performance Through Association*; Glass, J. E., Ed.; Advances in Chemistry Series No. 223; American Chemical Society: Washington, D.C., 1989; Chapter 25.
2. Fordyce, D. B.; Dupre', D.; Toy, W. *Offic. Dig.* 1959, *31*, 284-302.
3. Fordyce, D. B.; Dupre', D.; Toy, W. *Ind. Eng. Chem.* 1959, *51* , 115.
4. Guziak, I. F.; Maclay, W. N. *J. Appl. Polym. Sci.*, 1963, *7*, 2249-2253.
5. Matsumoto, T.; Shimada, M. *Kobunshi Kagaku* 1965, *22*, 172.
6. Muroi, S. *Kogyo Kagaku Zasshi* 1966, *69*, 1551.
7. Fordyce, R. G.; Ham, G. E. *J. Am. Chem. Soc.* 1947, *69*, 695-696.
8. Nishida, S. Ph.D. Thesis, Lehigh University, 1980.
9. Nishida, S.; El-Aasser, M. S.; Klein, A.; Vanderhoff, J. W. In *Emulsion Polymerization*; ACS Symposium Series No. 165; American Chemical Society: Washington, DC, 1981; 291-314.
10. Greene, B. W. *J. Colloid Int. Sci.* 1973, *43* (2), 449-461.
11. Hen, J. *J. Colloid Interface Sci.* 1974, *49*, 425.
12. Vijayendran, B. R. *J. Appl. Polym. Sci.* 1979, *23*, 893-901.
13. Bassett, D. R.; Hoy, K. L. In *Polymer Colloids*, Vol. II; Fitch, R. M., Ed.; Plenum: New York, 1980, 1-25.
14. Egusa, S.; Makuuchi, K. J. *J.Colloid Int. Sci. 1981, 79, 350-364.*
15. Rogers-Moses; P. J.; Schaller, E. J. *Amer. Paint Coat. J.* 1984, 54-58.
16. Shay, G. D.; Rich, A. F. *J. Coat. Tech.* 1986, *58* (732), 43-53.
17. Muroi, S. *J. Appl. Polym. Sci.* 1966, *10*, 713-729.

18. Loncar, F. V.; El-Aasser, M. S.; Vanderhoff, J. W. *Polym. Mater. Sci. Eng.* 1985, *52*, 299-303.
19. Verbrugge, C. J. *J. Appl. Polym. Sci.* 1970, *14*, 897-909.
20. Muroi, S. *Kogyo Kagaku Zasshi* 1966, *69*, 1551.
21. Evani, S.; Rose, G. D. *Proc. of the ACS Div. of Polymeric Mater.: Sci. and Eng. Mtg.* Fall 1987, *57*, 477-481.
22. Sonnabend, L. F. U.S. Patent 4,384,096, 1983.
23. Verbrugge, C. J. *J. Appl. Polym. Sci.* 1970, *14*, 911-928.
24. Sperry, P. R.; Thiebeault, J. C.; Kostansek, E. C. *Proc. 11th Int. Conf. Org. Coat. Sci. Technol.*, Advances in Org. Coat. Sci. and Tech. Series; Patsis, A. V., Ed., Vol. 9, 1985, 1-11.
25. Schaller, E. J. *Surface Coatings Australia* 1985, *22* (10), 6-13.
26. Karunasena, A. and Glass, J. E. *Proc. of the ACS Div. of Polymeric Mater.: Sci. and Eng. 1987, 56*, 624-628.
27. Karunasena, A. and Glass, J. E. *Proc. of the ACS, Div. of Polymeric Mater.: Sci. and Eng. 1987, 57*, 632-636.
28. Morgan, L. W.; Jensen, D. P. *Makromol. Chem.* 1985, Suppl. 10/11, 59-67.
29. Morgan, L. M.; Jensen, D. F.; Weiss, C. S. *Proc. of the ACS Div. of Polymeric Mater.: Sci. and Eng. Mtg.* Fall 1987, *57*, 689-693.
30. Shay, G. D.; Eldridge, E.; Kail, E. U.S. Patent 4,514,552, 1985.
31. Juang, M. S.; Krieger, I. M. *Polym. Prep., Div. Polym. Chem.*, American Chemical Society 1975, Vol. 16, *1*, 120-124.
32. Shay, G. D.; Kravitz, F. K.; Brizgys, P. V.; Kersten, M. A. U.S. Patent 4,801,671, 1989.
33. Thiebeault, J. C.; Sperry, P. R.; Schaller, E. J. In *Water Soluble Polymers*; Glass, J. E., Ed.; Advances in Chemistry Series No. 213; American Chemical Society: Washington, DC, 1986; 375-389.
34. Chang, C. J.; Stevens, T. E. European Patent 13836, 1980.
35. Witiak, D.; Dupre', J. U.S. Patent 4,529,773, 1985.
36. Oosawa, F. In *Polyelectrolytes*, Marcel Dekker Inc., 1971, Chapter 1.
37. Private communication with surfactant supplier, 1989.

RECEIVED October 10, 1990

ASSOCIATING POLYMERS

Chapter 8

Rheological Behavior of Liquid-Crystalline Polymer–Polymer Blends

D. Dutta and R. A. Weiss[1]

Polymer Science Program and Department of Chemical Engineering, University of Connecticut, Storrs, CT 06269–3136

This paper reviews the literature on the rheological behavior of liquid crystalline polymer/polymer blends. The advantages of blending a LCP with a thermoplastic polymer is that the LCP serves as a processing aid and can also form a reinforcing phase. Rheology plays an important role in determining the morphology and hence the mechanical properties of these blends. This review also discusses the slow behavior of neat liquid crystalline polymers.

In recent years, numerous studies of blends containing a liquid crystalline polymer (LCP) have been published. The main objective of the majority of these was to develop a fibrillar morphology of the dispersed LCP phase that could act as a reinforcement and improve the mechanical properties of the host polymer. For blends of thermotropic LCPs and thermoplastics, both phases are liquids at processing temperatures. As a consequence, the viscosity increases usually associated with filling thermoplastics with conventional reinforcing fibers such as glass or graphite do not occur. In fact, because of the anisotropic nature of an LCP melt, the melt viscosities of blends with flexible chain thermoplastics are often lower than that of the thermoplastic. Thus, blending LCPs into thermoplastic polymers can result in improved processability and reduced power consumption. A second consequence of blending LCPs into thermoplastic polymers is the development of a rigid, reinforcing LCP phase when the processed melt is cooled and solidified. In some cases, the orientation of the LCP phase generated during processing may be retained after solidification due to the long relaxation times of LCPs. In this way, one can preserve a microfibrillar morphology of a dispersed LCP phase in the fabricated blend, and the mechanical properties of such <u>self-reinforced blends</u> can approach those of fiberglass-reinforced plastics. The ability to develop these morphologies is influenced to a great extent by the rheology of the LCP and the rheology of the blend.

[1]Corresponding author

0097–6156/91/0462–0144$06.00/0

The purpose of this article is to review the published literature on the rheology of liquid crystalline polymer blends. A short section on the flow behavior of thermotropic LCPs is presented before dealing with the rheological and mechanical properties of LCP/polymer blends.

RHEOLOGY OF THERMOTROPIC LIQUID CRYSTALS

In thermotropic liquid crystals the mesophase transitions are brought about by changing temperature. Porter and Johnson (1,2) reviewed the rheology of liquid crystals (LC). The general viscosity behavior of a thermotropic LC as a function of temperature is represented schematically in Figure 1. A decrease in the viscosity, associated with the transformation from an isotropic melt to an anisotropic mesophase is usually observed when temperature is lowered. The viscosity again increases as the temperature of the nematic mesophase is decreased. In the mesophase the molecules orient along the flow direction and slide past one another, while in the isotropic melt, they can be highly entangled. The orientation of the molecules is responsible for the lower viscosity of the nematic mesophase relative to the isotropic melt.

Baird (3) and Wissbrun (4) reviewed the literature on LCP rheology. Onogi and Asada (5) proposed that the viscosity vs. shear rate behavior for LCPs can be represented by three distinct regions: 1. a shear thinning region at low shear rate 2. a Newtownian (plateau) region in an intermediate shear rate region and 3. a power-law shear thinning region at high shear rate (Figure 2).

Very few sets of data show all the three regions in a single polymer. However, in his review article, Wissbrun (4) was able to identify the three flow regions by analyzing the published data of a number of authors for both thermotropic and lyotropic LCPs.

An important characteristic of polymeric liquid crystals is that they have longer relaxation times compared with flexible coil polymers (4). For example, Suto et al. (6) studied the birefringence decay of hydroxypropyl cellulose (HPC) and ethyl cellulose (EC) after cessation of shear flow and found that whereas the birefringence decayed on the order of seconds for flexible coil polymer melts like polystyrene and polypropylene, the LCPs had relaxation times of the order of ten minutes or more. Similarly, Jackson (7) reported that polyethylene terephlhalate (PET)/parahydroxybenzoic acid (PHB) copolyesters have much longer terminal relaxation times than PET. In these polymers, the relaxation times increased as the PHB content in the copolyester increased due to decreased chain flexibility. Jerman and Baird (8) proposed that for LCP melts, two relaxation times were important: 1. for the stresses and 2. for the orientation. While these phenomena were related through the stress-optical law for conventional flexible coil polymers, they were independent for LCPs. The orientational relaxation time was longer than the stress relaxation time, and as a consequence, the orientation of thermotropic LCPs achieved during processing was retained in the solid state more easily than for flexible chain polymers.

The type of deformation (shear or extension) plays an important role in determining the orientation, texture and morphology of

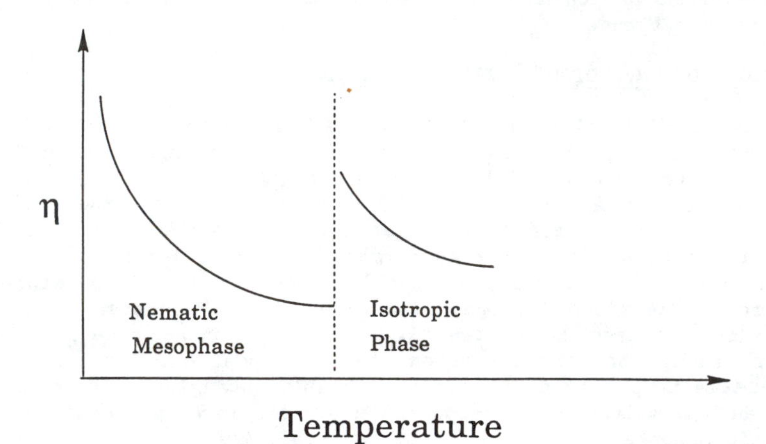

Figure 1. Schematic representation of the viscosity (η) versus temperature relationship for a nematic liquid crystal.

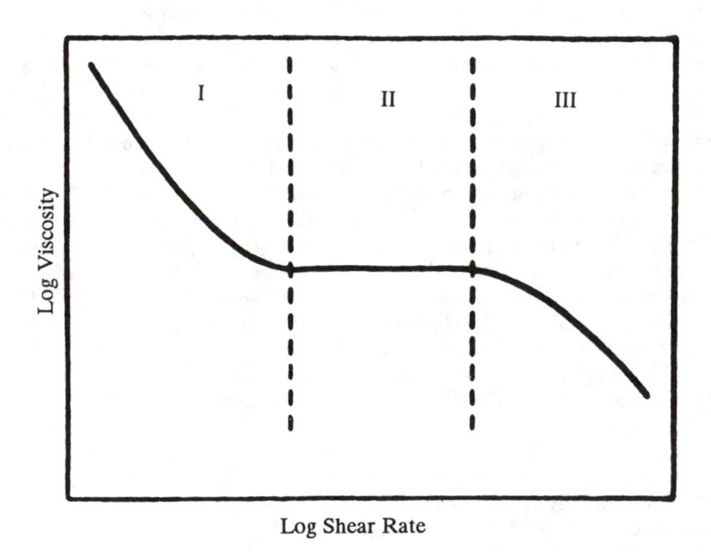

Figure 2. Schematic of the three distinct regions of flow behavior for thermotropic LCPs. (Reproduced with permission from reference 5. Copyright 1980 Plenum.)

an LCP (3). Ide and Ophir (9,10) showed that in extensional flow, the rod-like domains of an LCP were stretched and the molecules aligned along the flow direction. Shearing did not orient the molecules if the LCP domains were stable. For injection molded LCPs, the skin region was generally highly oriented because of extensional flow near the walls, but the core regions that experienced mainly shear flow were relatively unoriented. Baird et al. (11) reported similar results and obtained electron micrographs that showed that the highly oriented skin layer had a fibrous texture while the core layer that experienced shear flow did not.

Polymer melt extrudates generally tend to swell as a consequence of recoverable stress related to the deformation in the die entrance. In contrast, liquid crystalline PET/PHB copolymers exhibited negligible extrudate swell and in some cases, contraction of the extrudate was reported (8,12-13). Jerman and Baird (8) reported that extrudate swell increased with increasing temperature. The negligible extrudate swell was attributed to yielding of the LCP melt, similar to what is observed in fiber-filled isotropic melts. Another possible explanation involved the correspondence of the extruded swell with the negative first normal stress difference (N_1) that has also been reported for LCPs.

RHEOLOGY OF LCP/POLYMER BLENDS

LCPs can be used as processing aids by blending with thermoplastic polymers. The LCP phase is preferentially oriented in the direction of flow such that oriented LCP domains translate without entanglements. In this way, they can be viewed as a lubricant for the polymer melt, which reduces the effective viscosity of the blend. The lowering of viscosity not only reduces the energy consumption during processing but facilitates the filling of large and complex molds.

The temperature at which the blend is melt processed plays an important role in determining the effectiveness of the LCP as a processing aid. According to Cogswell et al. (14-16), the temperature range at which the thermoplastic polymer is melt processed must overlap the nematic temperature zone of the LCP.

Most studies of LCP/polymer blends did find a lowering of viscosity by the addition of an LCP. Other factors such as the shear rate, temperature and melt morphology also affected the viscosity of the blend. In the published studies of LCP/polymer blends, rheological characterization has relied mainly on capillary viscometry, though several researchers also used cone and plate and parallel plate rheometers to get data at lower shear rates. More important than the different shear rate regimes investigated, however, is the fact that in capillary viscometry the shear region in the capillary is preceeded by extensional flow in the entrance. This can perturb the blend melt morphology and yield significantly different viscosity results than one would achieve in shear alone.

One liquid crystalline copolyester that has been used extensively in blends is the copolyester of 6-hydroxy-2-naphthoic acid (HNA) and parahydroxybenzoic acid (PHB) (17-25). Several grades of this LCP have been commercialized by Hoechst-Celanese Corp. under the trade name Vectra and by Imperial Chemical Industries Ltd.

under the name Vitrex . Some of these grades also contain terephthalic acid (TA) and hydroquinone (HQ) comonomers. Nearly all researchers have observed that the addition of this LCP results in a decrease of viscosity of the blend.

Siegmann and coworkers (17) studied blends of HNA/PHB LCP with an amorphous polyamide and observed a large reduction in viscosity, as measured with a capillary viscometer, with the addition of as little as 5 wt. % LCP. The melts exhibited non-Newtonian behavior, which could be described by a power-law constitutive equation over a limited shear rate region. Swaminathan and Isayev (19) observed similar behavior for blends of this type of LCP and polyether sulfone (PES).

James et al. (24) and Froix et al. (25) also reported a reduction of viscosity of polyether sulfone (PES) due to addition of PHB/HNA LCP. The flow curves of blends containing up to 20 wt. % LCP resembled that of pure PES while the 50% LCP blends showed shear thinning behavior that was similar to that of the pure LCP. James and co-workers (24) also observed a four-fold decrease in viscosity with the addition of just 2% LCP.

Malik et al. (22) found that blends of HNA/PHB LCP with polycarbonate (PC) had lower viscosities and were more shear thinning than PC. Solid-state relaxation measurements indicated that the relaxation modulus also increased with the addition of LCP. Isayev and Modic (18) studied similar blend systems and observed a cross-over point in the flow curves of the LCP and PC, Figure 3. The cross-over can be taken as the point where the viscosity ratio of the neat components is approximately equal to unity. Maximum fibrillation of the LCP during flow occurred at this cross-over point.

Kohli et al. (23) investigated blends of PC and a lower melting point LCP based on HNA/PHB/TA/HQ. Blend compositions ranging from 5 wt % LCP to 80 wt % LCP were investigated and compared with the pure components. In contrast to the result of Isayev and Modic (18), they found that the viscosity decreased with increasing LCP content over the entire range of composition. (Figure 4). The shape of the flow curves were determined by the rheology of the polymer that made up the continuous phase. At low LCP content the PC constituted the continuous phase while a phase inversion occurred around 40-50 wt. % LCP content. Above this composition the flow curves were similar to that of the LCP. The blends were pseudoplastic and no Newtonian region was observed.

Blend studies with the copolyester of PET and PHB as the LCP component also showed a reduction in viscosity by the addition of LCP. Blizard and Baird (26) investigated blends of PET/PHB LCP with Nylon 66 and polycarbonate. Dynamic oscillatory and steady shear data showed a significant reduction of viscosity with the addition of the LCP, Figure 5. Similar results were also reported by Acierno et al. (27) for blends of PET/PHB and PC and by Zhuang et al. (28) for blends of PET/PHB with PC, PET or polystyrene (PS).

Several researchers have found that for certain blends the viscosity did not decrease monotonically with increasing LCP content (29-37). This was attributed to factors such as the melt morphology and temperature. Chung (29,30) studied blends of HNA/PHB LCP and Nylon 12 and found a minimum in the viscosity at 10 wt. %

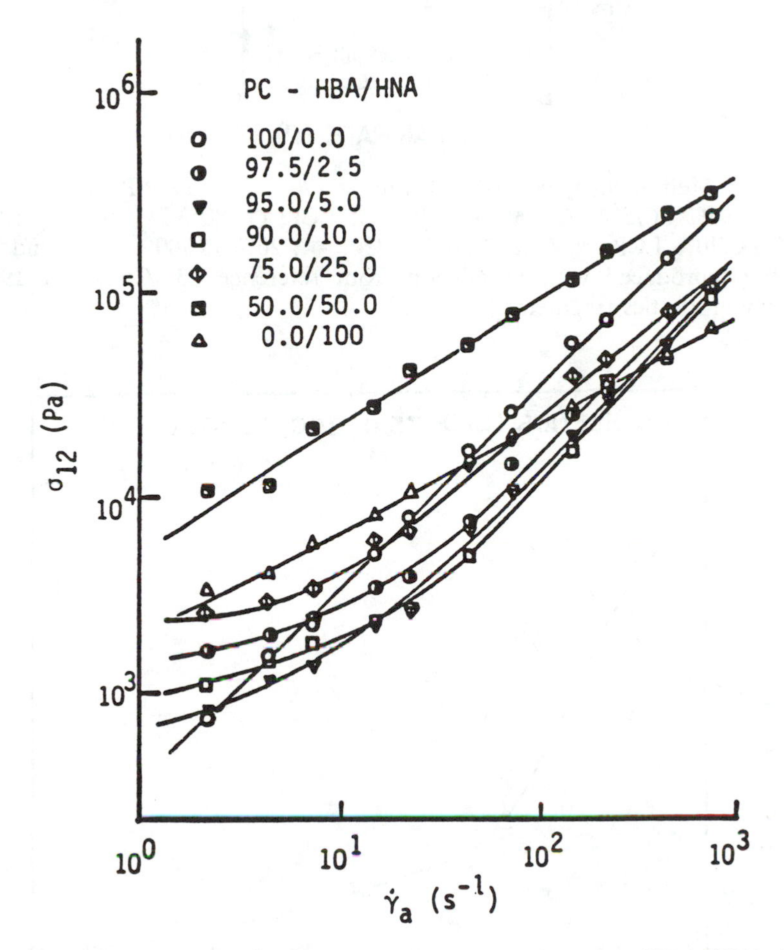

Figure 3. Shear stress (σ_{12}) versus shear rate data for HBA/HNA–PC blends showing the crossover point of the LCP and PC flow curves at γ = 44/s. (Reproduced with permission from reference 18. Copyright 1987 Society of Plastics Engineers.)

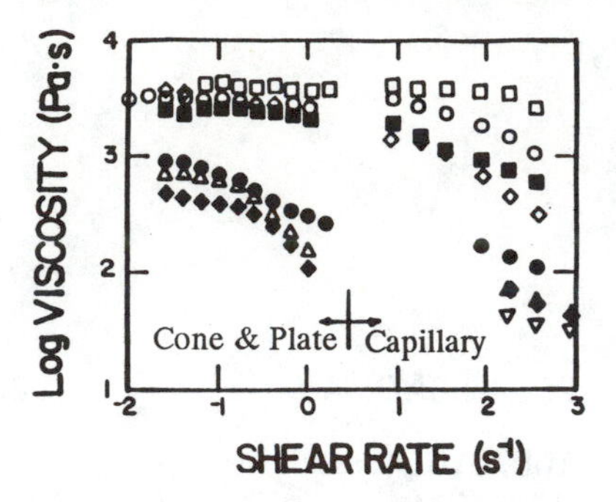

Figure 4. Melt viscosity versus shear rate at 270 °C for LCP/PC blends: □, PC; ○, 95% PC/5% LCP; ■, 90% PC/10% LCP; ◇, 80% PC/20% LCP; ●, 60% PC/40% LCP; Δ, 40% PC/60% LCP; ◆, 20% PC/80% LCP; and ∇, LCP. (Reproduced with permission from reference 23. Copyright 1989 Society of Plastics Engineers.)

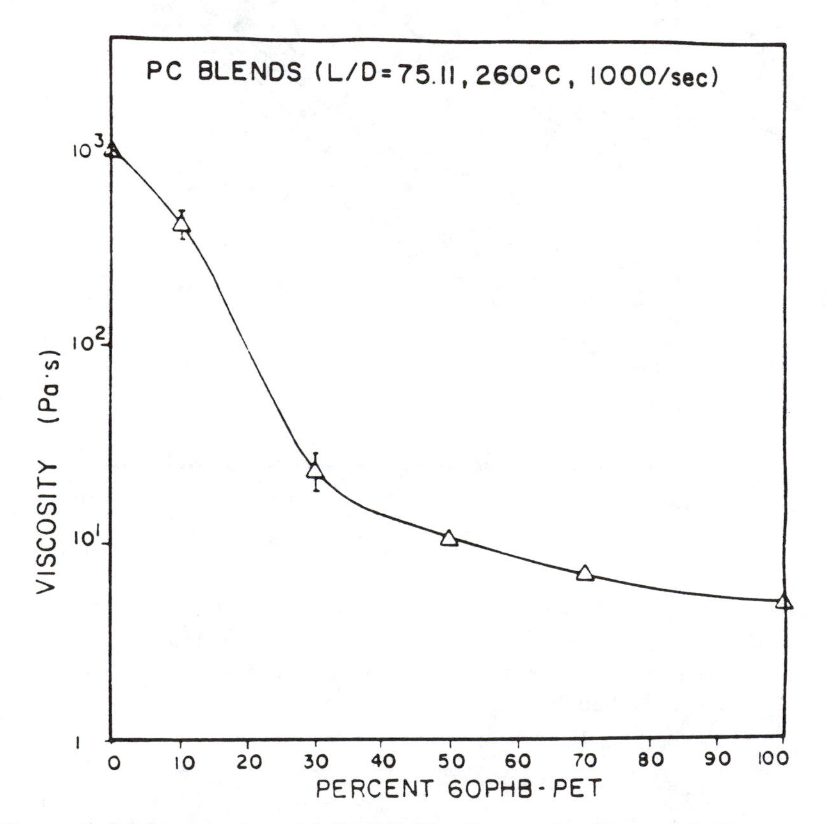

Figure 5. Melt viscosity of LCP/PC blends as a function of LCP content. (Reproduced with permission from reference 26. Copyright 1987 Society of Plastics Engineers.)

LCP concentration and a maximum at 20 wt. % LCP content. He proposed that below 10% LCP content, the LCP domains were well dispersed and in the Nylon 12 matrix and acted as a lubricant, but at 20 wt. % LCP an interconnected morphology formed, thereby increasing the viscosity of the melt.

Lorenzo et al. (31) found that temperature had a significant effect on the melt flow index (MFI) and complex viscosity η^*, of blends of PET/PHB and styrene-butadiene copolymer. When the blends were extruded below the melting point of the LCP, η^* was independent of the LCP content. But when the extrusion was done above the melting temperature of the LCP, MFI and η^* showed minima at 10 wt. % LCP concentration.

The viscoelastic behavior of blends of PET/PHB LCP and PC was investigated by Nobile et al. (32). At low shear frequency rates (below $0.3s^{-1}$) the complex viscosity η^* increased with increasing LCP content. At higher frequencies, however, η^* decreased with increasing LCP concentration.

Similarly, Weiss et al. (33-37) observed that the addition of a thermotropic LCP based on 4,4'-dihydroxy,α,α'-dimethylbenzalazine to PS raised the steady shear and dynamic viscosities at relatively low shear rates ($<1s^{-1}$), Figure 6. These data were obtained with a cone and plate rheometer in which the blends experienced only simple shear flow. They proposed that the tumbling and rotation of the phase-separated LCP domain resulted in the increase in viscosity. No significant deformation or extension of the LCP dispersed phase was observed, which is consistent with theoretical predictions for the simple shear flow of two-phase fluids.

At higher shear rates, Weiss et. al. (33-37) found that the viscosity decreased with increasing LCP content. In this case, the data were obtained with a capillary rheometer. The simple shear flow in the capillary was preceded by an extensional flow in the entrance region, and they concluded that the extensional flow deformed and aligned the LCP domains in the flow direction. This change in morphology during flow resulted in the decrease in viscosity at higher shear rates, as well as the development of the microfibriller LCP structure needed for a self-reinforcing blend.

Research on LCP/Polymer blends has dealt mainly with blends of thermotropic copolyesters and flexible thermoplastic polymer. DeMeuse and Jaffe (38), however, studied blends in which both the components were copolyesters based on PHB and HNA. The composition of the individual copolymers determined the rheological properties of the blends. When the composition of the copolymers were similar, the viscosity of the blend obeyed a simple rule of mixture relation,

$$\ln\eta = w_1 \ln\eta_1 + w_2 \ln\eta_2$$

where η is the viscosity of the blend, w_1 and w_2 are the wt. fractions of the two polymers and η_1 and η_2 are the viscosities of the individual for miscible blends. When the compositions of the copolymers were different, the viscosity of the blends followed an inverse rule of mixtures expression,

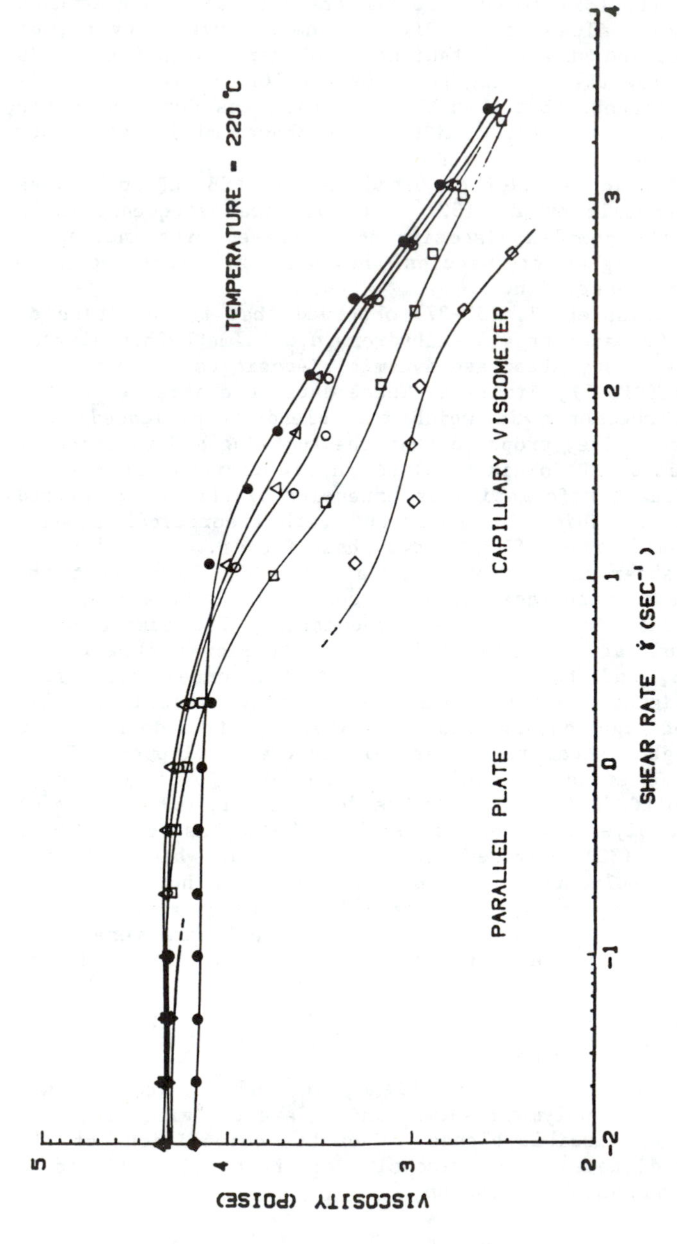

Figure 6. Viscosity versus shear rate for LCP/polystyrene blends: ●, polystyrene; Δ, 1.5% LCP; ○, 4.5% LCP; □ 10.0% LCP; and ◇, 100% LCP. Percentages denote weight percent LCP. (Reproduced with permission from reference 33. Copyright 1987 Society of Plastics Engineers.)

$$\frac{1}{\eta} = \frac{w_1}{\eta_1} + \frac{w_2}{\eta_2}$$

CONCLUSIONS

There has been considerable activity over the past 5 years in the research and development of liquid crystalline polymer/polymer blends. The two primary conclusions drawn from these studies were 1. that small to modest amounts of LCP added to a conventional thermoplastic polymer improve the melt processing of the thermoplastic and 2. when a suitable deformation history is employed, a microfibrillar morphology of the LCP-dispersed phase can result. Although it has been demonstrated that an extensional component of flow is required for developing the microfibrillar phase, very little fundamental understanding of the rheology of LCPs and their blends have resulted. Part of this is due to a lack of well characterized LCP samples for study and also to the strong influence of thermal history on the melt morpholgy and rheology of LCPs. Despite this shortcoming, self-reinforcing LCP/polymer blends are being tested in applications such as electronic substrates and injection molded parts. Many problems such as the anisotropy of molded and extruded parts and the adhesion between phases need to be overcome in order for these materials to be successful in these applications. In addition, the relationship between the blend rheology, processing conditions and final morphology and properties need to more thouroughly established.

REFERENCES

1. Porter, R. S.; Johnson, J. F. In Rheology; Academic Press: New York, 1969; Vol. 4.
2. Porter, R. S.; Johnson, J.F. J. Appl. Phys. 1963, 34, 51.
3. Baird, D. G. In Polymeric Liquid Crystals; Blumstein, A., Ed.; Plenum Press: New York, 1984.
4. Wissbrun, K.F. J. Rheol. 1981, 25, 619.
5. Onogi, S.; Asada, T. In Rheology:Astarita, G; Marrucci, G; Nicolais, L., Eds.; Plenum Press: New York, 1980.
6. Suto, S.; White, J. L.; Fellers, J.F. Rheol. Acta. 1982 21, 62.
7. Jackson, W. L. Macromolecules 1983 16, 1027.
8. Jerman, R. E.; Baird, D. G. J. Rheol. 1981 25, 272.
9. Ide, Y.; Ophir, Z. Polym. Eng. Sci. 1983 23, 792.
10. Ide, Y.; Ophir, Z. Proc. An. Tech. Conf. Soc. Plast. Eng. 1982 p 33.
11. Joseph, E. G.; Wilkes, G. L.; Baird, D. G. ACS Polym. Preprints 1984 25(2), 94.
12. Wissbrun, K. F. Polym. J. 1980 12, 163.
13. Sugiyama, H.; Lewis, D. L.; White, J. L. J. Appl. Polym. Sci. 1985 30, 2329.
14. Cogswell, F. N.; Griffin, B. P.; Rose, J. B. U.S. Patent, 4,386,174 1983.
15. Cogswell, F. N.; Griffin, B. P; Rose, J. B. U.S. Patent, 4,433,083 1984.

16. Cogswell, F. N.; Griffin, B. P.; Rose, J. B. U.S. Patent, 4,438,236 1984.
17. Siegmann, A.; Dagan, A.; Kenig, S. Polymer 1985 26, 1325.
18. Isayev, A. I.; Modic, M. J. Polym. Composites 1987 8, 158.
19. Swaminathan, S.; Isayev, A. I. Proc. ACS Div. of Polym. Mat. 1987 p 330.
20. Lee, B. Proc. An. Tech. Conf. Soc. Plast. Eng. 1988 p 1088.
21. Ko, C. U.; Wilkes, G. L. J. of Polym. Sci. 1989 37, 3063.
22. Malik, T. M.; Carreau, P. J.; Chapleau, N. Poly. Eng. Sci. 1989 29, 600.
23. Kohli, A.; Chung, N.; Weiss, R. A. Polym. Eng. Sci. 1989 29, 573.
24. James, S. G.; Donald, A. M.; Macdonald, W. A. Mol. Cryst. Liq. Cryst. 1987 153, 491.
25. Froix, M. F.; Park, M.; Trouw, N. U.S. Patent, 4,460,736, 1984.
26. Blizard, K. G.; Baird, D. G. Polym. Eng. Sci. 1987 27, 653.
27. Acierno, D.; Amendola, E.; Carfagna, C.; Nicolais, L.; Nobile, R. Mol. Cryst. Liq. Cryst. 1987 153, 533.
28. Zhuang, P.; Kyu, T.; White, J. L. Proc. Am. Tech. Conf. Soc. Plast. Eng. 1988 p 1237.
29. Chung, T. Plast. Eng. 1987 43, 39.
30. Chung, T. Proc. Am. Tech. Conf. Soc. Plast. Eng. 1987 p 1404.
31. Lorenzo, L.; Ahuja, S. K.; Chang, H. ACS Polym. Prepr. 1988 29(1), 488.
32. Nobile, M. R.; Amendola, E.; Nicolais, L.; Acierno, D.; Carfagna, C. Polym. Eng. Sci. 1989 29, 244.
33. Weiss, R. A.; Huh, W.; Nicolais, L. Polym. Eng. Sci. 1987 27, 684.
34. Weiss, R. A.; Huh, W.; Nicolais, L. Int. Conf. Liq. Cryst. Polym., 1987 Bordeaux, France, paper 8P8.
35. Weiss, R. A.; Huh, W.; Nicolais, L.; Kohli, A. Conference on Emerging Technologies in Materials, AIChE, August 1987 paper C01.50.
36. Weiss, R. A.; Huh, W.; Nicolais, L.; Yanisko, P. Proc. Reg. Tech. Conf. Soc. Plast. Eng. 1987, p 267.
37. Weiss, R. A.; Huh , W.; Nicolais, L. In High Modulus Polymers, Zachariades, A. E.; Porter, R. S., Eds., Marcel Dekker Inc.: New York, 1988 p 145.
38. DeMeuse, M. T.; Jaffe, M. Mol. Cryst. Liq. Cryst. 1988 157, 535.

RECEIVED October 23, 1990

Chapter 9

Shear-Thickening Behavior of Ionomers and Their Complexes

R. D. Lundberg and I. Duvdevani

Exxon Chemical Company, Linden, NJ 07036

Hydrocarbon solutions of Sulfonated Ethylene-Propylene-Diene Monomer Terpolymer (sulfo-EPDM) display dilatant (or shear thickening) rheological properties over a range of shear rates. The shear rate dependence is a function of molecular parameters (sulfonation level, counterion, molecular weight) and solvent polarity. These effects were investigated in xylene/alcohol solutions and it was found that measured dilatancy is restricted to a narrow band of shear stress beyond which the solutions behave as a shear-thinning fluid. The mechanism responsible for dilatancy is postulated to involve changes in the nature of associations under shear since solutions of non-associating polymers usually exhibit shear thinning behavior.

The influence of shear has also been explored with a new family of hydrocarbon soluble polymer complexes based on a combination of sulfonate ionomers and amine containing polymers. As compared with single ionomer solutions, such complexes offer much enhanced viscosities at low polymer concentrations, and visually dramatic dilatant solution behavior. Potential mechanisms for this shear thickening behavior are also discussed.

Most polymer solutions exhibit pseudo-plastic (or shear thinning) rheological behavior, while simpler, low molecular weight fluids, such as water or hydrocarbons, exhibit a Newtonian behavior. While Newtonian fluids are those whose viscosity does not depend on shear rate or on shearing time, shear thinning and shear thickening (dilatant) fluids have viscosities that can depend on shear rate or on time. Since the 1950's there have been a number of papers (1-4) which have described dilatant solution behavior of polymers primarily in aqueous solutions. One of the best known is that of

0097–6156/91/0462–0155$06.25/0

complexes based on polyvinyl alcohol and sodium borate in water as made by Burgoin (5) and Savins (6). They showed viscosity increases of 4 to 8 fold over a narrow shear rate range working with semi-dilute solutions. Similarly, shear thickening compositions of polyethylene oxide with either a petroleum sulfonate or an aldehyde resin were described by Ahearn (7) and Swenson (8) respectively. More recently, Peng and Landel (9) described such behavior with copolymers of methacrylic acid in hydrocarbon or water solutions. A recent paper (10) has provided a good general review on shear thickening behavior especially directed at suspensions of solid particles in Newtonian liquids. Although a substantial background on shear thickening systems exists, there have been very few papers on this subject directed at ionomers in non-aqueous solutions.

In recent years there has been an increased interest in ion-containing polymers due to their unusual bulk and solution properties. Some of the bulk work was described in the texts by Holliday (11) and by Eisenberg and King (12). A special symposium on ion-containing polymers was also held in 1978 and summarized in a monograph edited by A. Eisenberg (13).

Work on solution properties of ion-containing polymers has been published only recently. Much of this work was done in our laboratories by Lundberg et al (14-17) and in the University of Liege by Teyssie et al (18-20). The work by Lundberg et al involved salts of sulfonated polymers, mainly those based on polystyrene and ethylene-propylene-diene polymers (EPDM's) backbones. The work at Liege was mostly with neutralized carboxy terminated polymers (halato-telechelic) especially based on butadienes. In both cases the polymers in solution showed strong associations above the overlap concentration (C*) as evidenced by high viscosity and gelation. The associations could be balanced to exhibit unusual rheological effects. Unusual viscosity-temperature effects were observed when mixed solvents were used, whereby viscosity could increase with increasing temperature over a given temperature range (15-17, 21). Broze et al (20) of the Liege group reported dilatant behavior of a concentrated solution (10 g/dl) of a magnesium-dicarboxy-polybutadiene in decahydronaphthalene. Finally, a theoretical paper by Witten and Cohen (22) which were inspired by some of the observations described in this paper, discloses one mechanism for shear thickening behavior and accounts qualitatively for the observed dependence of viscosity on shear rate and on concentration.

This paper will summarize some of the viscosity/shear rate results we have obtained with solutions of metal neutralized sulfonates of EPDM (Sulfo-EPDM). In a second section, we will discuss similar behavior of polymer-polymer complexes of Sulfo-EPDM and vinyl pyridine copolymers. These systems constitute a new family of polymers with novel bulk and solution properties (23-24). Following this discussion we will provide a preliminary interpretation of these results.

Experimental

The zinc sulfonated ethylene propylene terpolymer (Zn Sulfo-EPDM) was prepared by techniques summarized in previous articles (25-26).

Typically, the Sulfo-EPDM samples employed in this paper contained 0.5 to 1.5 wt.% metal counterion and 0.3-0.9% sulfur for a sulfonate content of 10 to 30 milliequivalents per 100 gms. of polymer. Specific synthetic details as well as a discussion of neutralization chemistry for various metal salts of Sulfo EPDM have been previously described (25).

The styrene-4 vinylpyridine copolymers were synthesized via a free radical emulsion copolymerization process. A typical preparation route is as follows. Into a four neck flask equipped with an air driven stirrer and in an inert gas atmosphere, distilled water (120 mls), freshly distilled styrene (50 g), and 4-vinylpyridine (4.7 g), sodium lauryl sulfate (3.2 g), dodecylthiol (0.1 g), and potassium persulfate (0.2 g) were added and thoroughly mixed for two hours. The polymerization was conducted at 50°C for 5 hours. Subsequently, 3 mls of methanol containing 0.1% of hydroquinone was added and the polymer was precipitated in a large excess of acetone. The precipitate was filtered, suspended in methanol and vigorously agitated in a Waring blender to finely disperse the coagulated polymer. The polymer was again filtered and dried in a vacuum oven at 60°C for 24 hours. A 90% monomer conversion was typically achieved and the resulting polymer contained 1.07% nitrogen corresponding to 8 mole % 4-vinylpyridine incorporation. Depending on catalyst levels the viscosity average molecular weight ranged from 100,000 to a few million. The specific characteristics of these polymers have been described previously (23).

Solutions of Sulfo-EPDM's were prepared in xylene or xylene/alcohol at various concentrations using a magnetic stirrer. Polymer complex solutions were prepared by mixing Zn Sulfo-EPDM solutions with a solution of styrene-4 vinyl pyridine copolymer at various polymeric ratios. The specific polymers employed in preparation of the polymer complexes were Zn Sulfo-EPDM (0.59% Zinc, 0.31% Sulfur) and a styrene-4 vinyl pyridine copolymer (1.07% nitrogen) with $[\eta] = 2.5$, measured in xylene.

Viscosity vs. shear rate measurements were obtained with a Haake Viscometer (Rotovisco RV 100, CV 100 system) using a variety of sensors such as the ME-30 sensor system for the 0 to 300 sec^{-1} shear rate range and a ZB-30 sensor system for the 0-1000 sec^{-1} range. Measurement were usually done in the shear rate scanning mode. Some measurements were also done under a constant shear rate.

Results and Discussion

General Observations. Ionomer solutions can typically be more viscous than those prepared at the same polymer concentration for the non-ionic precursor. The addition of a polar cosolvent can markedly decrease this viscosity but as long as the ionomer concentration is above the overlap concentration of the precursor polymer (C>C*), there will normally be evidence of viscosity enhancement. Provided the base polymer molecular weight is sufficiently high, and the level of sulfonate functionality is on the order of 10 to 30 meq/100 grams of polymer, there is a distinct possibility of shear thickening behavior at concentrations of C>C*. While the quantification of this phenomenon requires sophisticated viscometers, the qualitative features are often quite apparent by very simple techniques. For example, rapid agitation of an ionomer solution - shaking a bottle - can result in a virtual gelation of the contents, which are then observed to relax and return to form a homogeneous liquid within a few seconds. Under extreme conditions, this tendency to form a temporary gel with agitation can be misinterpreted as a permanent phase separation, since a very slight agitation can create the appearance of gel. In this paper, we are only concerned with those shear effects which are temporary and reversible.

Sulfonate Ionomers in Low Polarity Diluents. The S-EPDM solutions employed in this study typically were based on xylene solvent and optionally contained low levels of alcohols (0.5 to 2 volume percent). The reduced viscosity-concentration profile of those solutions are illustrated in Figure 1 and compared with that for the starting EPDM backbone. Consistent with previously published data (16-17), it is evident that the Zn S-EPDM in xylene exhibits a higher reduced viscosity value at polymer concentrations >0.5 weight percent than the starting EPDM precursor. At lower concentrations, the reduced viscosity of the ionomer is less than that of the EPDM suggesting intramolecular association at low concentrations and intermolecular associations at $C>C^*$. The addition of 0.5 percent of C_{18} alcohol has a modest effect on viscosity at higher polymer concentrations while 2 percent of methanol virtually wipes out any evidence of intermolecular association. The reduced viscosity measured in 98/2 xylene/ methanol solutions is consistent with little or no ionic association, and it seems that this mixed solvent is a poorer one than xylene alone for EPDM (a somewhat lower $[\eta]$).

These data provide additional evidence to that already reported (16) that modest quantities of alcohols can reduce ionic association while larger quantities can virtually eliminate association. Thus, intermediate levels offer an approach to modulate association which is useful in our subsequent studies.

A typical shear stress/shear rate curve obtained with the Haake viscometer is shown in Figure 2. This Zn S-EPDM polymer is dissolved at 2 percent concentration in 99.5/0.5 xylene/hexanol cosolvent system. The shear rate range is from 0 to 30 sec^{-1} and the observed viscosity increases from 38 poise to 53 poise over this shear rate range.

An alternative cosolvent system is illustrated in Figure 3 and compared in viscosity-shear rate profiles to that of pure xylene, 2 percent EPDM in xylene, and 2 percent Zn-S- EPDM in xylene containing 0.5 percent stearyl alcohol. A shear rate range of 1 to ~3×10^4 sec^{-1} was studied. While the EPDM solution is modestly shear thinning at shear rates $>10^3$ sec^{-1} the ionomer behavior is quite different. As evidenced in Figure 1, the viscosity of S-EPDM at low shear rates is substantially higher than that of EPDM at the same concentration. At low shear rates the solution is virtually Newtonian; however, at shear rates of 100 sec^{-1} there is evidence of shear thickening, followed by a pronounced maximum in viscosity over the range of 10^2 to 10^3 sec^{-1}. Finally, above 10^3 sec^{-1} shear thinning behavior is observed. While

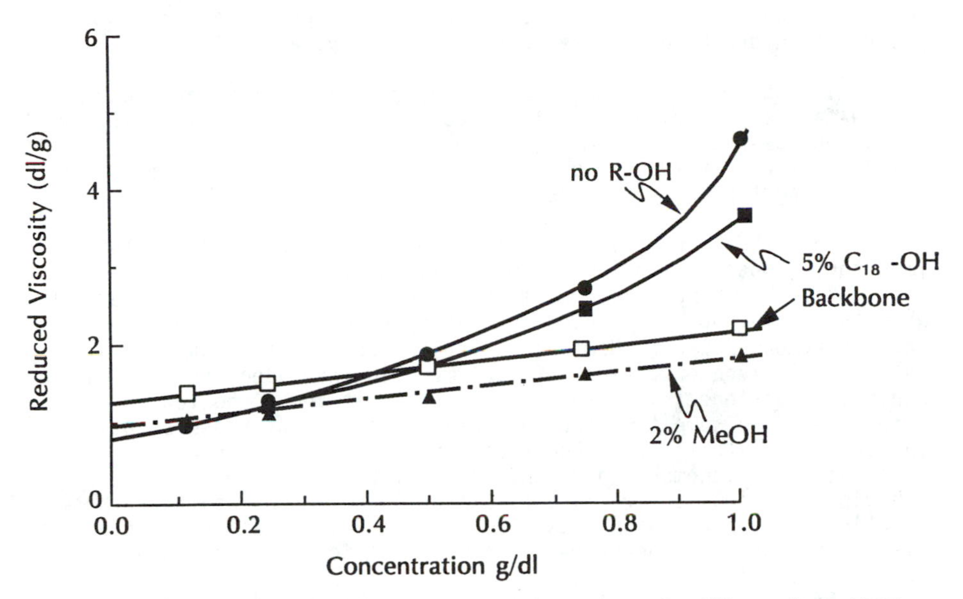

Figure 1. Reduced Viscosity-Concentration Profiles for EPDM Backbone and Zn-S-EPDM in Xylene and Xylene Plus Alcohol at 25°C.

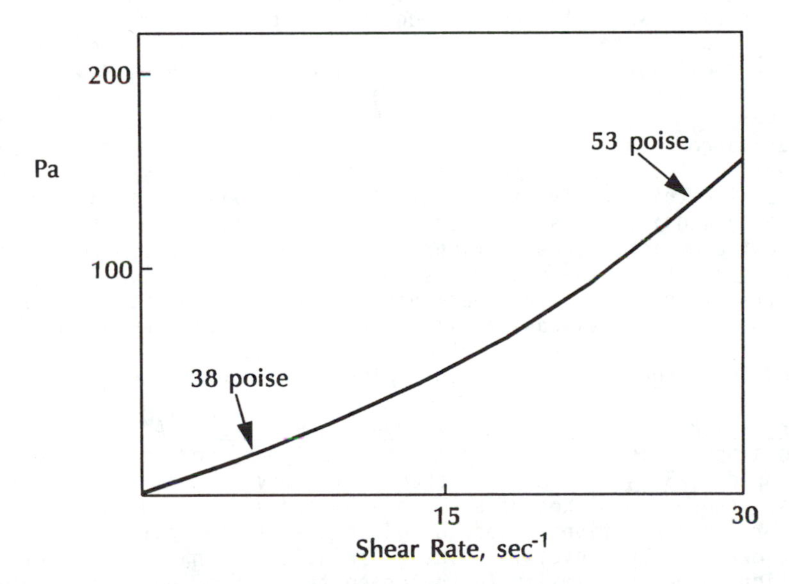

Figure 2. Shear Stress-Shear Rate Relationship for Zn-S-EPDM (2%) in Xylene Hexanol (99.5/0.5) at 25°C.

there is not a shear thickening profile that is "typical", Figure 3 does portray those general characteristics possessed by these materials in order to display dilatant behavior, as follows:

(1) At low shear rates, Newtonian behavior is generally observed.
(2) In a range of 50 to 10^3 sec^{-1} a maximum in viscosity is often observed.
(3) At shear rates beyond those at the viscosity maximum, shear thinning typically results.

While these observations are quite general, there is an important impact of polymer concentration, functional group level, and cation. Some evidence of this dependence is seen in Figure 4 when no cosolvent was used. At very low polymer concentrations, the solutions appear Newtonian over the shear rate range investigated. As concentration of polymer is increased, this polymer exhibits the onset of shear thickening at lower and lower shear rates. At the same time the low shear viscosity of these solutions increases substantially. Consequently, it may be inferred that the more viscous ionomer solutions are most prone to exhibit dilatant behavior at low shear rates. It should be emphasized that those solutions having viscosities >10,000 centipoise can approach gel-like behavior and therefore may not be sufficiently homogeneous to provide reliable viscosities.

Based on the data in Figure 4, it might be anticipated that a relationship between the maximum viscosity observed for a series of ionomer solutions and the shear rate at which the maximum is observed might exist. Some evidence in support of this is seen in Figure 5 where the peak viscosity is plotted as a function of the shear rate at which the peak viscosity is observed. While the data span a broad range of different S-EPDM samples, and solvent systems with and without cosolvents, it is apparent that there is a general trend to much lower shear rates as the peak viscosity increases. Moreover, the viscosity peak in this entire representation occurs at a narrow shear stress range of about one order of magnitude suggesting that shear thickening is destroyed at a critical shear stress for this family of polymers (the shear stress is the product of the viscosity and shear rate shown in Figure 5; the extreme left and extreme right lines in Figure 5 are constant shear stress lines which bracket all the observed shear stresses at peak). These data also reinforce the view that shear thickening behavior also exist for those lower viscosity solutions which appear Newtonian; however, the shear rate range at which this behavior is detectable is beyond the range easily accessible with our viscometers.

Shear Thickening of Polymer-Polymer Complexes. A new family of hydrocarbon soluble ionomer/aminated polymer complexes has been developed (23-24). As compared with conventional ionomer solutions, these complexes offer much enhanced viscosities at very low polymer concentrations, depending on polymer molecular weight and functional group levels. These solutions can display dilatant solution behavior similar to that seen for sulfonated ionomers, but the phenomenon is more dramatic and can occur with solutions of very low viscosity.

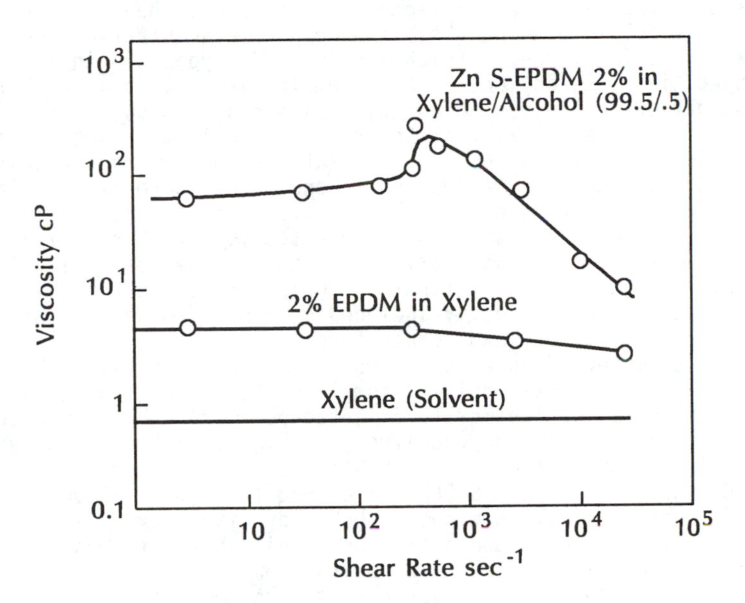

Figure 3. Viscosity-Shear Rate Relationships for EPDM and Zn-S-EPDM in Xylene or Xylene/Stearyl Alcohol at 25°C.

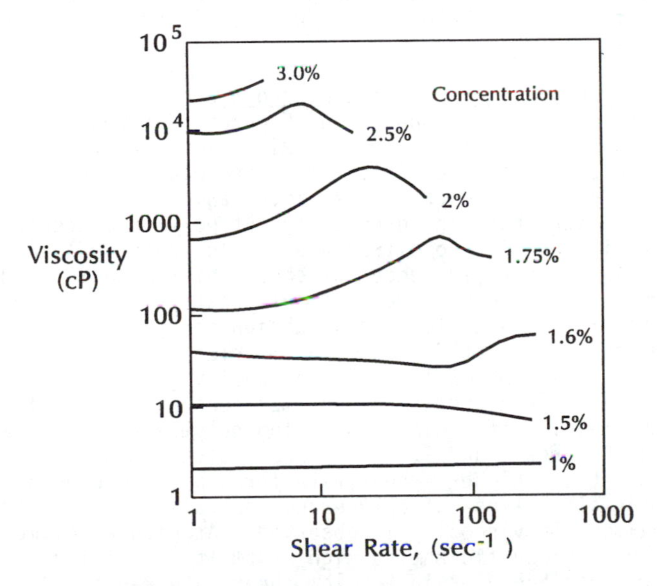

Figure 4. Viscosity-Shear Rate Relationships of Zn-S-EPDM in Xylene (no cosolvent) at Various Polymer Concentrations (Sulf. Content of 10 meq/100 gm).

There are a number of factors which affect shear thickening of polymer complexes. Among these are polymer concentration, sulfonate content, cation type, stoichiometry and type of vinyl pyridine employed in interaction. For example, 4-vinyl pyridine has been investigated in these studies; 2-vinyl pyridine shows little interaction. Similarly zinc as the sole cation is the subject of this study. Other cations (Cu, Ni, Co) interact much more strongly, and can lead to phase separation under conditions where zinc sulfonate-amine complexes remain dissolved. In order to investigate the effect of shear on the solution viscosity of these polymer complexes, shear rate was varied in the range of 0-1000 sec^{-1}. Parameters studied to determine their effect on the relation of shear rate with viscosity included sulfonate/amine stoichiometry, polymer concentration, and temperature.

For example, Figure 6 illustrates the viscosity-shear rate profiles for several polymer complexes in xylene having different stoichiometrics of ionomer to aminated polymer. The latter polymer (styrene-vinyl pyridine copolymer) exhibits Newtonian behavior at one-half weight percent concentration over the shear rate range up to 300 sec^{-1}. Zn S-EPDM (10 meq/100 gm. sulfonate level) exhibits essentially Newtonian behavior within experimental error at one-half percent concentration. With both of these component polymers the viscosities are quite low. However, upon blending these two components at various stoichiometric ratios of Zn/N, holding total polymer concentration constant, there is a marked dependence of viscosity on shear rate. Based on these amines it appears that an optimium stoichiometry exists for highly dilatant solutions, however, polymer concentration and temperature are also important factors. These factors will be discussed in the following sections.

Polymer Concentration and Zinc:Nitrogen Stoichiometry. Viscosity shear measurements are shown in Figures 7 through 10 for three different stoichiometries of the zinc ionomer to the aminated polymer wherein the viscosities of these solutions are examined as a function of shear over a shear rate range up to 300 sec^{-1}. The viscosity measurements in three different polymer concentrations at a high ratio of zinc to nitrogen are illustrated. It is apparent that no significant evidence of shear thickening is evident over the shear rate range examined. At alternate stoichiometries, such as Figures 8, 9 and 10, the situation is changed. For example, Figure 8 shows that at Zn:N stoichiometries close to 1 and at very low shear rates, below 15 sec^{-1}, viscosity goes through a maximum and then markedly decreases at polymer concentrations of 0.6-0.8%. At higher levels of amine containing polymer (Zn:N<1), as illustrated in Figures 8 and 9, it is apparent that shear thickening is manifested at a higher shear rate for a polymer concentration of 0.6%. In all cases of polymer concentrations of 0.5 to 0.8%, a clear maximum in viscosity is observed. At higher polymer levels, viscosities are sufficiently high that it is difficult to detect the shear rate behavior in the low shear rate range. As the level of aminated polymer is increased relative to ionomer (Zn:N<1) there is an increase in the shear rate range over which shear thickening

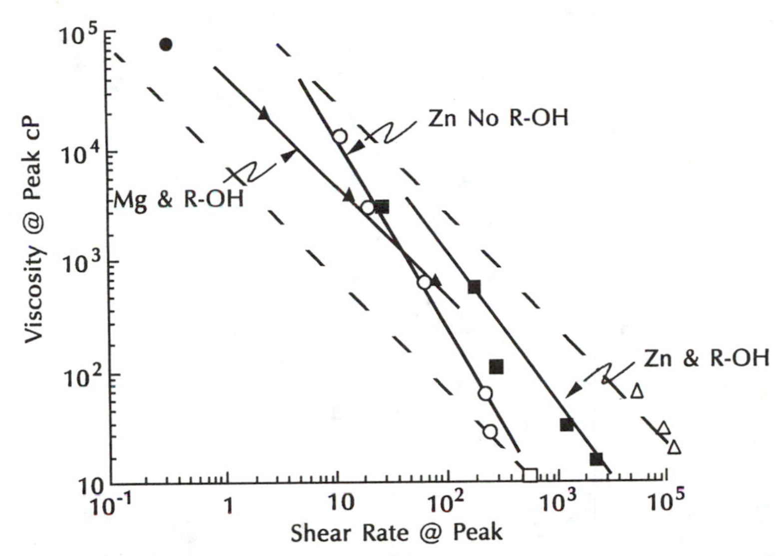

Figure 5. Critical Shear Thickening Limits for Zn-S-EPDM and Mg-S-EPDM in Xylene or Xylene/Alcohol at 25°C. The broken lines to the left and to the right of the data are lines of constant shear stress.

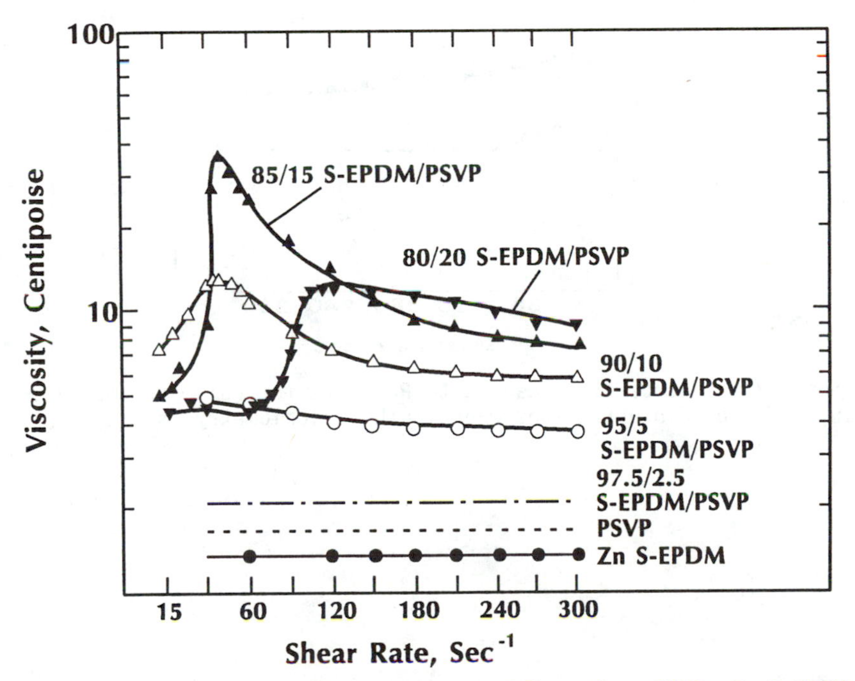

Figure 6. Viscosity-Shear Rate Profiles for PSVP, Zn-S-EPDM and Complexes in Xylene at 25°C.

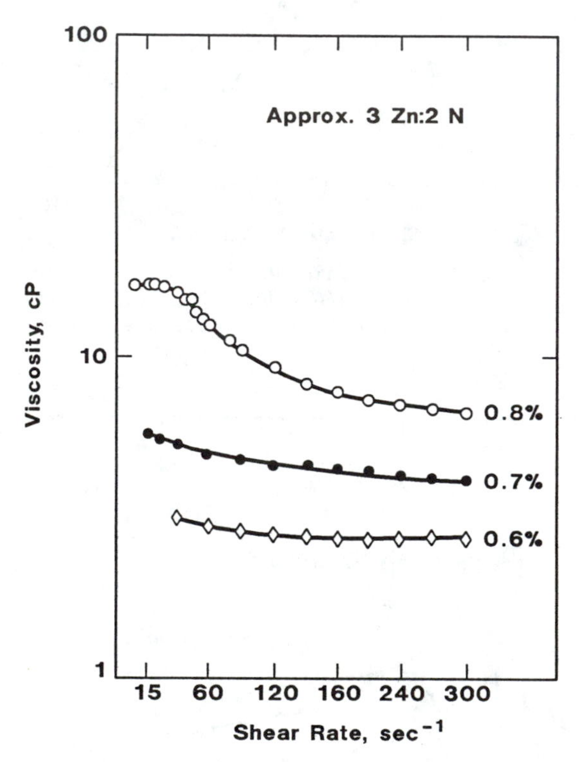

Figure 7. Viscosity Shear Rate Profiles for Polymer Complexes at Different Concentrations where Zn:N Stoichiometry is 3:2.

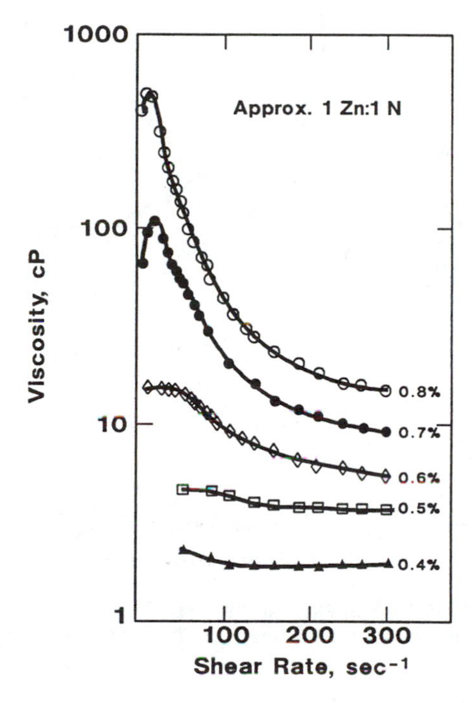

Figure 8. Viscosity Shear Rate Profiles for Polymer Complexes at Different Concentrations where Zn:N Stoichiometry is 1:1.

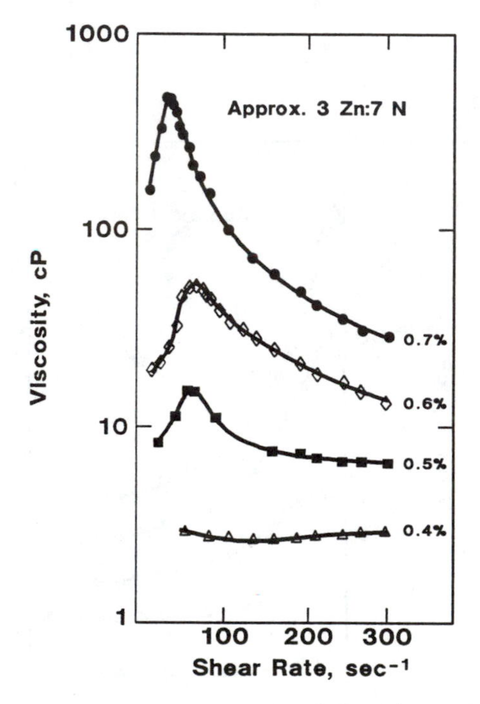

Figure 9. Viscosity Shear Rate Profiles for Polymer Complexes at Different Concentrations where Zn:N Stoichiometry is 3:7.

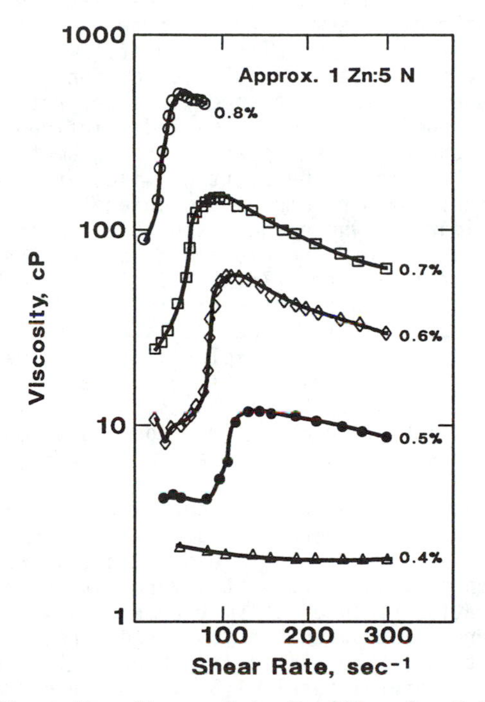

Figure 10. Viscosity Shear Rate Profiles for Polymer Complexes at Different Concentrations where Zn:N Stoichiometry is 1:5.

behavior is observed as shown in Figure 10 such that the phenomenon can now be observed in the shear rate as high as 100 sec^{-1} (at 0.5% polymer).

These studies clearly indicate that at very low polymer concentrations of (0.3-0.4%) there is little evidence of shear thickening behavior at the indicated shear rate range; however, at higher polymer concentrations when the stoichiometries are in a specific range for the two polymers making up the complex, shear thickening is observed. In many cases, apparent viscosities increase by nearly an order of magnitude, over a very narrow shear rate range.

It has been observed in a previous publication (23) that a maximum in the solution viscosity occurs near a stoichiometry of one zinc per nitrogen atom for these two interacting polymers. Those measurements were typically made at low shear rates on the order of 10 sec^{-1} or less. It is evident from the data in Figures 6 through 10 that the specific shear rate range at which viscosity measurements are made can markedly influence the location of the maximum suggesting an apparent stoichiometry. Another way to look at this behavior is by plotting viscosities at given shear rates for different polymer complex stoichiometries. Those results are shown in Figure 11 where the viscosities observed at a number of different shear rates are plotted as a function of ionomer to aminated polymer stoichiometry (Zn:N). It is apparent that a maximum is observed in the stoichiometry region of 40/60 Zn:N ratio for the two polymer systems at low shear rates. Depending on the shear rate over which this behavior is explored, the maxima in viscosity/stoichiometry can vary significantly with the shear rate at which the measurements are made. Consequently, any conclusions concerning zinc:amine stoichiometric interactions based on such data can be misleading unless the shear rates and polymer concentrations are clearly defined.

The data shown thus far were measured at shear rates from 0 to 300 sec^{-1}. Some studies were done over a wider shear rate range, up to 1000 sec^{-1}. Under these conditions, we were able to explore the influence of shear rate at several stoichiometries over a broader concentration range of polymer complexes and determine some trends which were not apparent at lower shear rates. For example, Figure 12 illustrates the viscosity-shear rate dependence for a polymer of Zn:N of approximately 1:5 over a concentration range of 0.4-0.7% weight percent. At low polymer concentrations, there is evidence of a small increase in viscosity as shear rate increases from 300 sec^{-1} to 600 sec^{-1}. At higher polymer concentrations, there is a marked difference in behavior with a sudden onset of shear thickening behavior near 100 sec^{-1}. These data would suggest that there is a change in the shear thickening mechanism that occurs at polymer concentrations in the range of 0.48 to 0.5% for this particular polymer complex. A similar result is observed at somewhat different stoichiometry (Zn:4N) and illustrated in Figure 13. Again, there is evidence in the relatively dilute concentration regime of 0.44% that shear thickening occurs at shear rates of about 300 sec^{-1}. As total polymer concentrations are increased, there is evidence of very sharp and sudden onset of shear thickening behavior which occurs at a much lower shear rate.

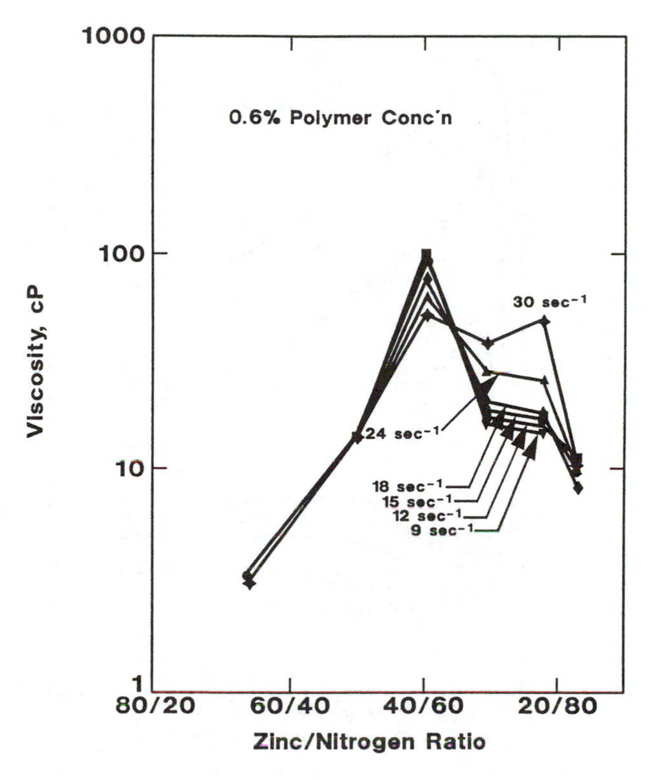

Figure 11. Solution Viscosity for Polymer Complexes at 0.6% Concentration as Function of Zn:N Stoichiometry Measured at Various Shear Rates at 25°C.

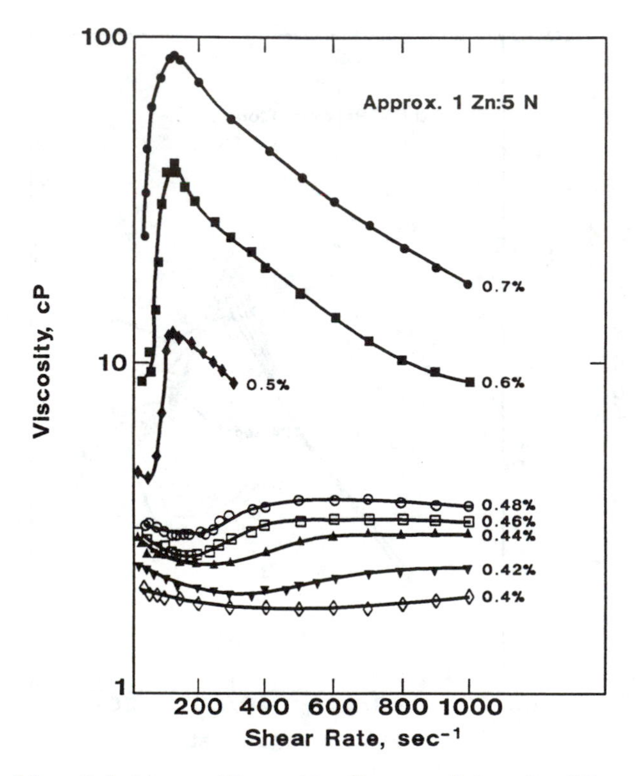

Figure 12. Solution Viscosity-Shear Rate Profiles at Zn:N Stoichiometry of 1:5 Measured at Various Polymer Concentrations at 25°C.

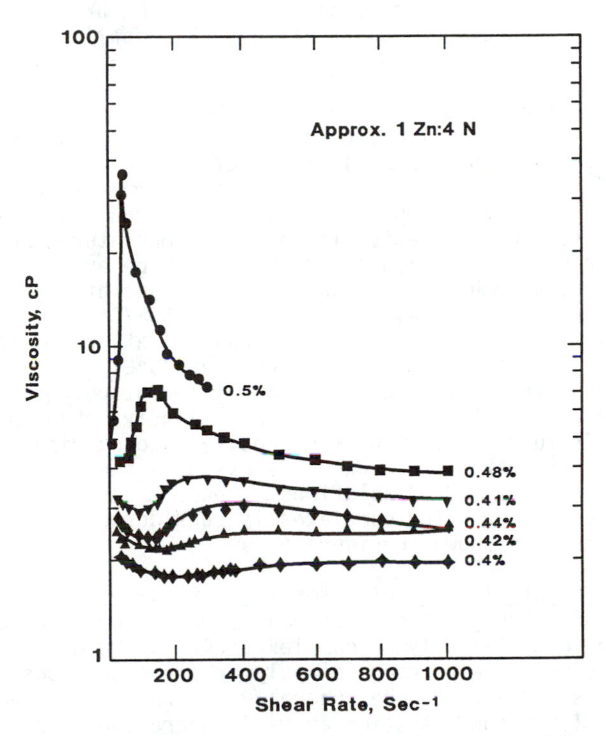

Figure 13. Solution Viscosity-Shear Rate Profiles at Zn:N Stoichiometry of 1:4 Measured at Various Polymer Concentrations at 25°C.

These data are consistent with the previous data in the onset of pronounced shear thickening taking place at approximately the same shear rate range observed in Figures 6 through 10.

We interpret the behavior in Figures 11 and 12 as being suggestive of a critical polymer concentration required for dilatant behavior which will vary for different stoichiometries of polymer complexes. When the polymer concentration for such systems becomes sufficiently high to permit substantial interpolymer interaction (corresponding to coil overlap or C*) very modest changes in shear rate can induce enhanced polymer associations. Under these conditions, slight shear rate changes can induce sudden increases in viscosity. This behavior is consistent to the proposed mechanism described in a previous publication (23) dealing with polymer complexes.

Effect of Temperature on Viscosity Shear Rate Relationships. Figure 14 illustrates viscosity-shear rate relationships for polymer complexes at Zn:N stoichiometry of 3Zn:7N as measured at 25, 50, and 75°C. The effect of increasing temperature is to lower viscosity across the complete shear rate range as might be expected. A second observation is that as temperature increases the shear rate range, over which shear thickening behavior is observed, is shifted to somewhat higher shear rate ranges at higher temperatures. Increasing the temperature from 25 to 75°C provides a peak viscosity which appears at about 40 sec^{-1} at the lower temperature and is shifted to about 150 sec^{-1} at 75°C. This behavior is generally consistent with other data at different polymer concentrations where shear thickening is sufficiently pronounced to be measured. It suggests that peak viscosity is occurring at a narrow range of shear stress as was discussed in the section on shear thickening in sulfo-EPDM solutions. However, it appears that the critical shear stress for complexes is considerably less than that for the ionomers as shown in Figure 5.

Interpretation of The Results for Polymer Complexes. Examination of Figures 6 through 14 suggest that shear thickening behavior is readily apparent for polymer complexes over a broad range in shear rates and over a wide range of polymer concentrations. However, these effects appear to be generally of two types: (a) one that occurs at polymer concentration above C* where the effects are very sudden in terms of the impact of shear rate on viscosity; and (b) a second effect which occurs generally in more dilute regimes where the effects are more subtle and the magnitude of the effects are much reduced - such as illustrated in Figure 12 at polymer concentrations <0.48%. We postulate that in order to have the dramatic changes in viscosity as a function of shear rate illustrated at higher polymer concentrations, it is a requirement that the total polymer concentration approach or exceed C*. Under these cases, small changes in shear rate can promote interpolymer complexation to a degree where these materials virtually gel. Under such conditions, agitating these fluids even to a slight degree, can create a gel-like appearance which takes a matter of seconds or minutes to relax and reform the homogeneous fluid, again. The effect at lower concentrations is such that much higher shear rate

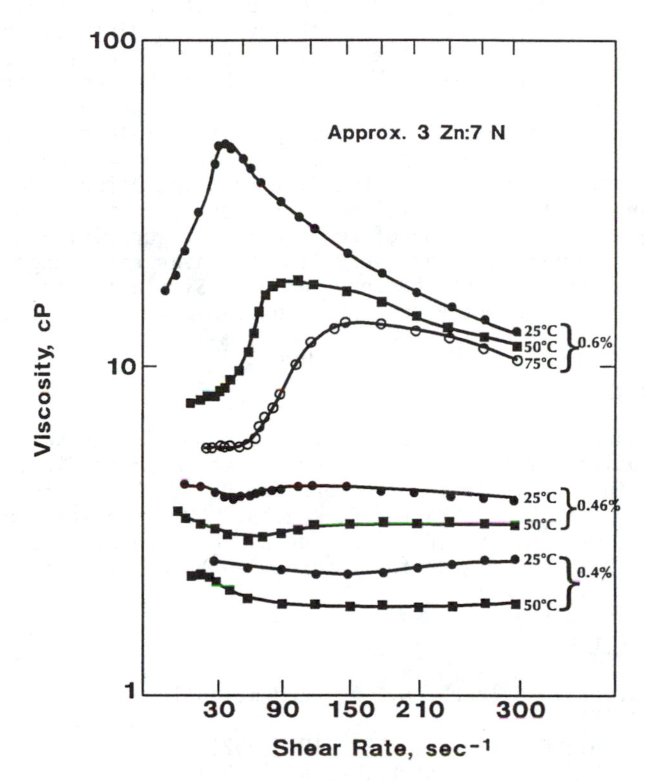

Figure 14. Viscosity-Shear Rate Behavior at Different Polymer Concentrations and Different Temperatures for Polymer Complexes where Zn:N is 3:7.

ranges are required to exhibit modest increases in viscosity. This latter behavior is one which occurs only when conditions promote maximum interpolymer association through orientation of polymer chains.

Other effects which moderate interactions between the zinc sulfonate and the amine groups, such as elevated temperature also defer the onset of shear thickening higher shear rates as is evident in Figure 14. Therefore, this behavior is illustrative of weaker interactions between the complexing groups and therefore any evidence of shear thickening behavior would be observed at higher shear rates. Similarly, the addition of a cosolvent, such as hexanol or other alcohols at modest levels can virtually eliminate any evidence of shear thickening. However, low levels of such alcohols tend to defer the onset of the shear thickening effects as well as reduce their magnitudes to a very marked degree.

In summary, solutions of the polymer complexes described in this paper can manifest pronounced shear thickening behavior. Those factors which reduce the interaction, between the polymer chains, or alters their lifetime sufficiently, can eliminate shear thickening behavior. Therefore, it is not surprising that these effects are observed over very limited shear rate ranges and the onset of the behavior can be very sudden. Similarly, these effects are expected to be very concentration-dependent as is clearly the case.

Acknowledgments

We have appreciated contributions from a number of workers who contributed polymer samples and/or specific measurements in the course of this study. Those contributors are Dr. D. G. Peiffer, Messrs. J. Wagensommer, M. Vieira and E. Kauchak. We especially value the efforts of Mr. R. R. Phillips for his many rheological measurements. Finally, we gratefully acknowledge the comments of Dr. G. Ver Strate.

Literature Cited

1. Eliassaf J., Silberberg, A. and Katchalsky, A., Nature, 176, 1119 (1955).
2a. Peterlin A., Kolloid Z., 182, 110 (1962).
 b. Ibid, J. Poly. Sci.-Letters, 2, 67 (1964).
 c. Burow, S. P., Peterlin, A. and Turner, D. T., Polymer, 6, 35 (1965).
 d. Bianchi, U. and Peterlin, A., Eur. Poly. J., 4, 515, (1968).
 e. Ibid, J. Poly. Sci.-A2, 6, 1011 (1968).
3. Raphalen, M. N. L. and Wolff, C., J. Non-Newtonian Fl. Mech., 1, 159 (1976).
4. Wolff, C., Adv. Coll. and Interface Sci., 17, 263 (1982).
5. Bourgoin, D. J., J. Chemie. Phys., 59, 923 (1963).
6. Savins, J. G., Rheol. Acta, 7, 87 (1968).
7. Ahearn, G. P., U.S. Patent 3,663,477 (1972).
8. Swenson, B. L., U.S. Patent 4,212,747 (1980).
9. Peng, S. T. J. and Landel, R. F., J. Appl. Phys., 52, 5988 (1981).

10. Barnes, H. A., Journal of Rheology, 33(2), 329 (1989)
11. Holliday, L., Ed., Ionic Polymers, Applied Science Publishers, London 1976.
12. Eisenberg, A. and King, M., Ion-Containing Polymers, Academic Press, New York 1977.
13. Eisenberg, A., Ed., Ions in Polymers, Advances in Chemistry Series No. 187, American Chemical Society, Washington DC 1980.
14. Lundberg, R. D. and Makowski, H. S., ibid, pp 21-36.
15. Lundberg, R. D. and Makowski, H. S., J. Poly. Sci.-Poly. Phys., 18, 1821 (1980).
16. Lundberg, R. D. and Phillips, R. R., J. Poly. Sci.-Poly. Phys., 20, 1143 (1982).
17. Lundberg, R. D., J. Appl. Poly. Sci., 27, 4623 (1982).
18. Broze, G., Jerome, R. and Teyssie, P., Macromol., 15, 920 (1982).
19. Ibid, 1300 (1982).
20. Broze, G., Jerome, R., Teyssie, P. and Marce, C., Macromol., 16, 996 (1983).
21. Tant, M. R., Wilkes, G. L., Storey, R. F. and Kennedy, J. P., Polymer Preprints, 25, 2 118, (1984).
22. Witten, T. A. and Cohen, M. H., Macromolecules, 18, 1915 (1985).
23. Lundberg, R. D., Phillips, R. R. and Peiffer, D. G., J. of Poly. Sci., Physics ED., 27, 245 (1989).
24. Peiffer, D. G., Duvdevani, I., Agarwal, P. K., and Lundberg, R. D., J. Polym. Sci., Polym. Letters ED., 24, 581 (1986).
25. Makowski, H. S., Lundberg, R. D., Westerman, L. and Bock, J., Ions in Polymers Ed by Eisenberg, A., ACS Series, Amer. Chem. Soc., Washington, D.C., 187, 3, 1980.
26. Agarwal, P. K., Makowski, H. S. and Lundberg, R. D., Macromolecules, 13, 1679 (1980).

RECEIVED September 10, 1990

Chapter 10

Hydrocarbon-Soluble Associating Polymers as Antimisting and Drag-Reducing Agents

Donald N. Schulz, K. Kitano, I. Duvdevani, R. M. Kowalik, and J. A. Eckert

Exxon Research and Engineering Company, Route 22 East, Annandale, NJ 08801

Polymers can modify not only a solution's (shear and elongational) viscosity, but its antimisting and drag reducing properties, as well. Associating polymers tend to be more potent solution rheology modifiers than conventional polymers. This paper describes the synthesis and solution properties of associating functional alpha olefin polymers. Specifically, carboxyl associating polymers are prepared by a special Ziegler Natta polymerization method which overcomes catalyst poisoning. The carboxyl polymers can self associate via H-bonding or they can be titrated with Lewis base containing polymers (e.g. styrene-vinyl pyridine) to form interpolymer complexes. The H-bonding polymers show enhanced shear and elongational viscosity compared with their nonfunctional counterparts. They also exhibit shear thickening rheology and antimisting activity. In turn, the interpolymer complexes exhibit enhanced viscosity, shear thickening rheology and shear-stable drag reduction.

Addition of polymers to a fluid can affect not only its shear and elongational viscosity, but its antimisting and drag reducing properties, as well. Antimisting (AM) is the ability of a fluid to resist break-up into minute droplets; drag reduction (DR) is the reduction of turbulence or the enhancement of flow of a fluid.

Conventional polymers build solution properties mainly by virtue of their high molecular weights and chain entanglements. For example, nonfunctional long chain poly (alpha olefins) are excellent pipeline drag reducing agents for crude oil because of their ultra high polymer molecular weight (10^6-10^7 daltons) and excellent hydrocarbon solubility(1). However, since shear stability also scales with polymer chain length, these materials are also shear sensitive.

0097–6156/91/0462–0176$06.00/0

On the other hand, associating polymers tend to be more potent solution rheology modifiers than conventional polymers. For example, sulfo EPDM ionomers and their interpolymer complexes build viscosity, antimisting, and/or drag reducing properties at lower polymer concentrations and molecular weights than conventional polymers. They also result in polymers which tend to be shear thickening and/or more shear stable(2-10). However, sulfo EPDM polymers are usually available only up to moderately high molecular weights (10^5 daltons) and can have hydrocarbon solubility and/or viscosity problems, depending upon the level and type of counterion.

Additionally, some associating carboxylate methacrylate functional hydrocarbon polymers (e.g. poly (t-butyl styrene-co-methyl acrylic acid) have been found to be mist suppression agents for jet fuel(11). Yet, these materials are generally only available with low to moderate molecular weights (10^4-10^5 daltons) because they are prepared by free radical polymerzation. Moreover, they require rather high levels of MAA to achieve mist reduction. Higher levels of MAA, in turn, reduce solubility in hydrocarbon and increase water sensitivity.

This paper combines the ultra high MW (10^6-10^7 daltons) capability and excellent hydrocarbons solubility of certain long chain poly (alpha olefins) with the associating capability of acid functionality. The preparation and the viscosity (shear and elongational), antimisting and drag reducing properties of certain acid functional polymers e.g. poly(1-octene-co-undecenoic acid) are described. These polymers self associate via H-bonding and form interpolymer complexes with basic polymers, e.g. poly(styrene-co-vinyl pyridine).

EXPERIMENTAL

Synthesis of Poly(1-Octene-co-methyl-10-undecenoate)

(a) Copolymerization of 1-Octene and methyl-10-undecenoate.

A 2-liter flask was charged with a mixture of n-heptane (480 ml), 1-octene (500ml), methyl-10-undecenoate (6.4 g), and diethyl aluminum chloride (72 mmole), and heated to 60°C.

The catalyst containing $TiCl_3$(described in U.S. Pat. No. 4,240,928) (2.0 g) was then added with n-heptane (20 ml). After stiring for 1 hour, the reaction was terminated with a small amount of isopropyl alcohol.

The polymer was precipitated and washed with isopropyl alcohol and vacuum dried at 60° to yield 87.9 g of colorless material. IR spectral analysis showed that the copolymer contained 0.8 mole percent of the methyl-10-undecenoate unit. Intrinsic viscosity of the material was found be about 4.3 dl/g in a decalin solution. The polymer Mn was $4.6x10^6$ by means of GPC.

(b) Acid hydrolysis of 1-octene-methyl-10 undecenoate copolymer (Polymer A1)

1-Octene-methyl-10-undecenoate copolymer similar to the one described in (a) above was converted to a respective copolymer having alkylenecarboxylic acid side chains in the general methods described below.

A solution of the ester copolymer (10 g) in xylene (500 g) was placed in a 2-liter flask and heated to 40°C. Concentrated sulfuric acid (20 ml) was then added. After stirring for one hour, the reaction mixture was cooled down and washed with a mixture of water and isopropyl alcohol three times. A white product was obtained by precipitationg from the solution with isopropyl alcohol. Further purifucation by reprecipitation and drying in a vacuum oven at 50°C gave a colorless rubbery polymer. Typcially, these polymers would have .02-.5 mole% acid functinality by IR and a Mn's of $2\text{-}5x10^6$ depending upon the starting ester and extent of hydrolysis.

(c) Base Hydrolysis

A flask was charged with a solution of poly(1-octene-co-methyl undecenoate) copolymer similar to the one described in (a) above (4.0 g) in 200 g THF and).82 g t-BuOK. The solution was heated to 50° -60°C. After one hour another 150 ml THF was added and 3.6 ml of 2N H_2SO_4 was added to neutralize the solution (Ph=5). After cooling, the polymer was precipitated in 600 ml of water/isopranol (1:1 vol./vol.). The polymer was filtered washed with water and isopranol, and dried to yield 4.0 g of product which has about 100% of the original ester groups hydrolyzed to carboxylate groups by IR. The viscosity of this polymer in Xylene (2%) was 19 cP at $30s^{-1}$.

(d) Acid Treatment (Polymer A2)

2.0 g of the polymer prepared according to Example 1(c) was dissolved in 100 g xylene. A 3 ml quantity of concentrated H_2SO_4 was added at room temperature. The batch was stirred for 1 hour at room temperature and subsequently precipitated in isopranol/water and dried under vacuum with heating. The polymer showed carbonyl and ester groups in the IR (75% COOH).

Systhesis of Styrene-Vinylpyridine Copolymer (Polymer B)

A representative example for the systhesis of styrene-4-vinylpyridine copolymer (SVP) is outlined below.

Into a 1-liter 4-neck flask the following ingredients were introduced:

100 g distilled styrene
6.4 g sodium lauryl sulfate
240 ml. distilled water
0.4 g potassium persulfate
9.4 g 4-vinylpyridine

The solution was purged with nitrogen gas for 10 minutes to remove dissolved oxygen. As the nitrogen gas pruge began, the solution was heated to 55°C. After 24 hours, the polymer was precipitated from solution with methanol. Subsequently, the resulting polymer was washed several times with a large excess of methanol and dried in a vacuum oven at 60°C. for 24 hours. Elemental analysis showed a nitrogen content of 1.13 wt.% percent which corresponds to 8.4 mole percent 4-vinyl pyridine.

Typical Preparation of Interpolymer Complexes (7,19)

Polymer A1 having acid functionalities and polymer B having base functionalities were separately dissolved in xylene at 1 weight percent concentration. Various ratios of these two solutions were then prepared. An example is given below:

Polymer A1, prepared by acid hydrolysis according to the procedure of Example 1(b), has a poly(1-octene) backbone with $-(CH_2)_8-COOH$ alkylenecarboxylic acid side groups randomly attached along the backbone. The carboxylic acid level is in the order of 0.02-0.5 mole percent. The average molecular weight is about 2 million based on an intrinsic viscosity in xylene of about 3.5. Polymer B, prepared according to the procedure of Example 2, is a copolymer of styrene and vinyl pyridine with a pyridine level of about 8 mole percent and viscosity average molecular weight of about 2 million. Mixtures of the xylene solution at 1% each were blended to form the interpolymer complexes and the resulting viscometrics measured.

Solution Viscometrics

Shear viscometrics were measured on a Haake viscometer. Elongational viscosities were measured by a tubeless siphon apparatus (14,15,20,21).

Drag Reduction Method

Drag reduction was evaluated by flowing polymer/xylene solutions through a 2.13 mm inside diameter stainless steel tube and measuring the resulting frictional pressure drops and flow rates. The flows were generated by loading a pair of stainless steel tanks (1 liter each) with a previously dissloved polymer/xylene solution, pressurizing the tanks with nitrogen gas (300kPa) and discharging the solution through the tube test section. Pressure drops were measured across a 50 cm straight segment of the tube with a pair of flush mounted tube wall pressure taps and a differential pressure transmitter. Flow rates were measured by weighing samples of the effluent liquid collected over measured time periods.

Flow rates in the drag reduction experiments ranged from about 12 to 25 g/s; these rates correspond to solvent Reynolds numbers from about 12,000 to 25,000 (solvent Reynolds number=mean flow velocity x tube diameter/solvent kinematic viscosity). Drag reduction was measured by comparing flow rates of the polymer/xylene solutions with the flow rates of the xylene solvent at equal pressure drops. Results were expressed as percent flow enhancement which is defined as

$$\text{Percent Flow Enhancement} = 100 \times \frac{\text{Flow Rate of solution} - \text{Flow Rate of solvent}}{\text{Flow Rate of Solvent}}$$

The sensitivity of the solutions to flow degradation was evaluated by recycling solutions through the system. Under these conditions flow enhancement values decrease on successive passes when flow degradation occurs.

RESULTS AND DISCUSSION

Synthesis of Associating Polymers

At the cornerstone of the associating polymers concept is the need for functional "hooks" on polymer chains that can associate or can be converted to associating groups. Most hydrocarbon associating polymers are synthesized via free radical(11) or post polymerization methods(21). Functional alpha olefins are difficult to prepare by direct polymerization because: (a) alpha olefin polymers are usually prepared by cordination (Zeigler-Natta) Catalysis and (b) functional groups (e.g. -OH,-NH2,-COOH) tend to compete with and poison the catalyst sites, as shown below.

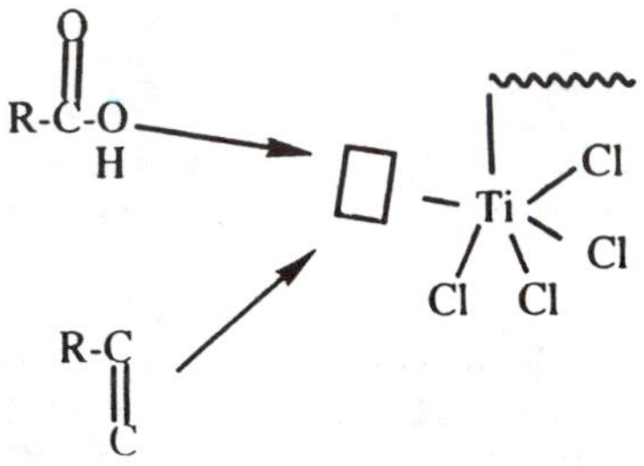

R-C(=O)-OH → □ — Ti(Cl)₄ ← R-C=C

We (12) and others(22,23) have found that functional hydrocarbon polymers can be prepared by Ziegler-Natta polymerization via the use of precomplexed ester monomers (Equations 1-3).

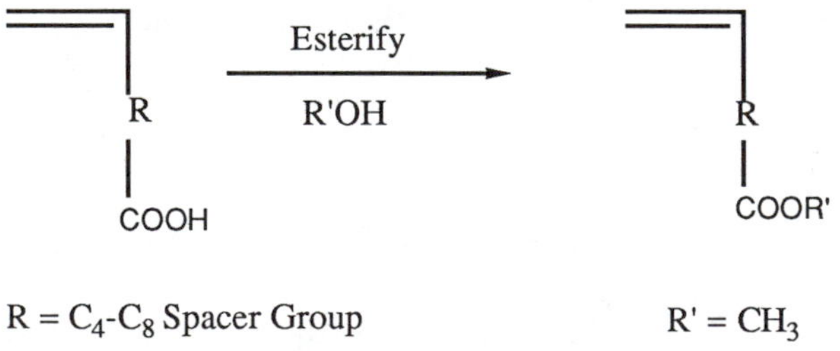

$$CH_2{=}CH{-}R{-}COOH \xrightarrow[R'OH]{\text{Esterify}} CH_2{=}CH{-}R{-}COOR' \quad (1$$

$R = C_4$-C_8 Spacer Group $\qquad R' = CH_3$

$$\text{CH}_2{=}\text{CH}{-}\text{R}{-}\text{COOR}' \xrightarrow{\text{AlX}_n\text{Y}_{n-3}} \text{CH}_2{=}\text{CH}{-}\text{R}{-}\text{C}(\text{OR}'){=}\text{O} \Rightarrow \text{AlX}_n\text{Y}_{n-3} \qquad (2$$

$$\text{CH}_2{=}\text{CH}{-}\text{R}{-}\text{C}{=}\text{O} \Rightarrow \text{AlX}_n\text{Y}_{n-3} \; + \; \text{CH}_2{=}\text{CH}{-}\text{R}'' \xrightarrow{\text{Work - up}} \sim\sim(\text{R}{-}\text{COOR}')\sim\sim(\text{R}'')\sim\sim \qquad (3$$

Specifically, carboxyl functional alpha olefins are reacted with alcohols to form the corresponding esters. The ester monomers have a more muted reactivity toward Ziegler-Natta catalysts than do carboxylic acid groups. The ester groups are further protected by precomplexation with aluminum compounds. Table I shows the stability of the ester-aluminum complexes as determined by IR. The order of stability is:

$$\text{TEAL} \ll \text{DEAL} < \text{EASC} < \text{EADC} < \text{AlCl}_3$$

Table I. Stability of Complexes

AlKYL Al (mol/mol Ester)		IR ABSORPTION (√C=0) Δ From Free Ester (cm^{-1}) at RT
TEAL	1.2	-85 (a)
TEAL	2.0	-85
DEAC	1.2	-100
EACS	1.2	-110
EADC	1.2	-130
$AlCl_3$(b)	1.0	-135

(a) @ 70°C (complex incomplete)
(b) Insoluble complex

The complex with TEAL is too weak to be useful; the complex with $AlCl_3$ is too strong and forms an insoluble complex. The aluminum complexes of intermediate reactivity are appropriate for masking. The resulting masked monomers are stable to Ziegler Natta catalysts and can be used as carriers of the carboxyl functionality. In our case, the ester monomer is compolymerized with a non functional alpha olefin; i.e. 1-octene.

The presence of the spacer group R is important for two reasons. First, it minimizes the amount of free radical side reactions. Second, the comonomer reactivity ratios are more favorable when R and R' are similar (Equation 3). Since R' is a C_6 group, R should preferably lie between 4-10 carbon atoms (23).

Once the ester copolymer is prepared it is easily converted to an acid copolymer by hydrolysis of a fraction or all of the ester groups to acid groups. The acid groups, in turn, can self associate via hydrogen bonding, can be neutralized to form ionomers or can be titrated with basic polymers (e.g. styrene-co-vinyl pyridine) to form interpolymer complexes. (Scheme A)

Solution Properties of Acid Functional Polymers

Above 1g/dl Concentration (~C*), the acid functional polymers show enhanced shear viscosity compared with the unhydrolyzed ester precursors presumably because of intermolecular self association via H-bonding. Below this concentration, the associating polymers actually show a lower vicosity then the nonassociating one because of intramolecular association (Figure 1). Such solution shear viscosity behavior is typical for associating polymers (5).

Furthermore, a $1.7x10^6$ dalton molecular weight acid functional polymer exhibits a measurable quiescent extensional viscosity in the tubeless siphon experiment(19,20) while a non associating polymer of similar molecular weight does not. This extensional viscosity can be denatured by the addition of H-bond breakers; e.g. stearic acid (Figure 2). These results are also prima facia evidence for association phenomena.

A tubeless siphon height at break has been shown to correlate with mist index and the higher the tubeless siphon height, the greater is the antimist effectiveness(26). Indeed, even an unsheared .3% w/w poly(1-octene-co-undecenoic acid) proves to be a

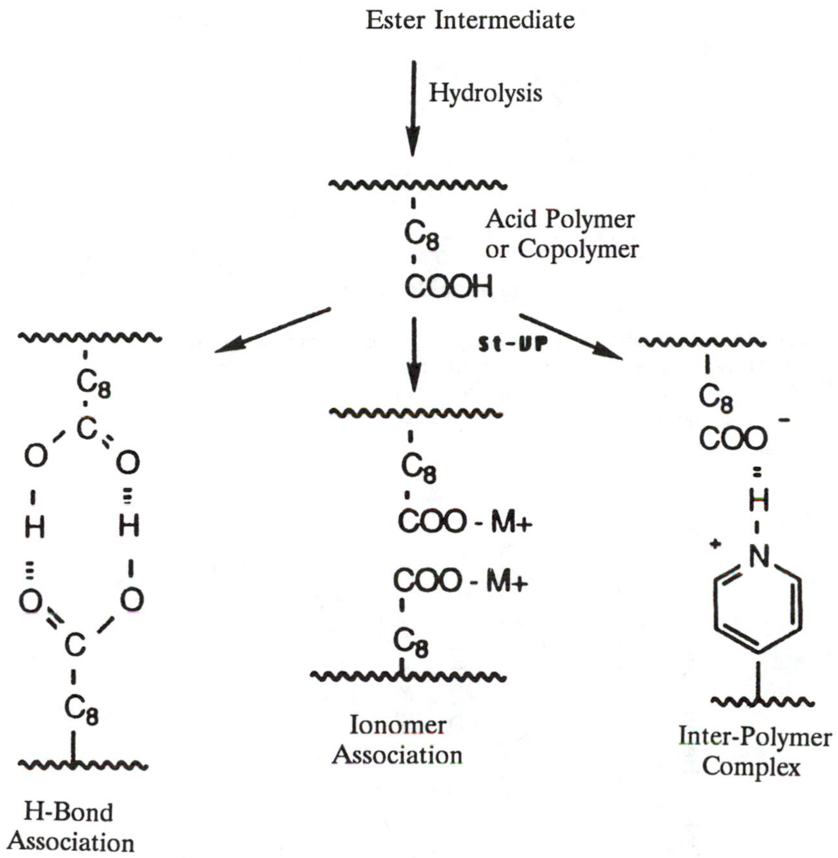

Ester Intermediate
Hydrolysis
C8
COOH
Acid Polymer
or Copolymer
St-VP
C8
C
O
O
H
H
O
O
C
C8
H-Bond
Association
C8
COO - M+
COO - M+
C8
Ionomer
Association
C8
COO -
H
+ N
Inter-Polymer
Complex

Scheme A

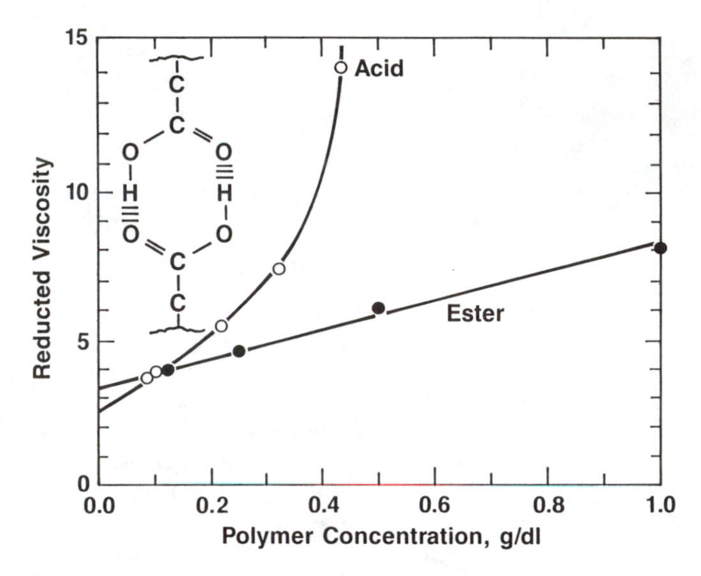

Figure 1. Reduced viscosity - concentration plots for an ester copolymer, i.e. poly(1-octene-co-methyl undecenoate), and the acid polymer formed by hydrolysis, i.e. poly(1-octene-co-unclecenoic acid).

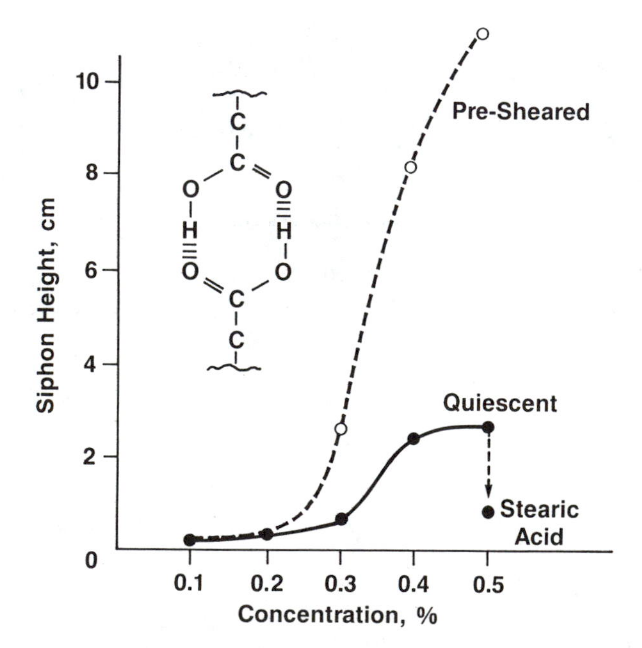

Figure 2. Plot of tubeless syphon height vs concentration for 1.7×10^{6} molecular weight poly(1-octene-co-undecenoic acid) in the quiescent state, as well as the effects of stearic acid denaturing and preshearing.

very effective mist suppression agent for jet fuel when atomized through a diesel injector pintle nozzle operating at about 10^4 KPa. An antimisting index(26) of 3 is achieved, which indicates the formation of droplets and threads rather than complete atomization to a fine mist. Further improvement in antimisitng activity can be achieved by adjusting polymer concentrations and levels of associating groups, as well as by preshearing. For example, preshearing the sample can increase the siphon height (and antimist activity) by 3-4x (Figure 2).

Solution Properties of Interpolymer Complexes

Besides self association, the acid functional polymers can be titrated with base containing polymers to form interpolymer complexes. For example, such complexes are formed by mixing poly(1-octene-co-undecenoic acid) or poly(1-octene-co-undecenoic acid/ester) with poly (styrene-co-vinyl pyridine). The viscosity reaches a maximum near the stoichometric equivalence point of the acidic and basic groups (In this case, a ratio of 90/10 acidic polymer/basic polymer)(Figure 3). Interpolymer complexes at or near the equivalence point are also shear thickening (Figure 4). Similar dilatant behavior has been observed for other interpolymer complexes and is consistent with a shift from intramolecular to intermolecular associations (9).

The interpolymer complexes also exhibit unusual drag reduction properties. Polymer drag reduction is the enhancement of flow or the reduction of drag for flowing fluids (e.g. flow in pipes, movement of ships, etc.). Polymers reduce the drag of flowing fluids presumably by limiting bursting and turbulence and extending the laminar flow regime. Drag reduction scales roughly with coil volume. Thus, it tends to increase with polymer chain length and the "goodness" of the solvent (27). Unfortunately, shear degradation also scales with polymer chain length (28).

We (7,19) have found that interpolymer complexes formed by reacting one polymer with acid groups and one polymer with basic groups can result in shear stable drag reducing agents. Figure 5 compares the drag reduction in xylene of 750 ppm of a 2 million MW interpolymer complex with 20 ppm of a 20 - 30 million MW nonfunctional polymer of the same backbone; i.e. poly(1-octene). Both have similar drag reducing ability at the beginning of the test. However, the convential nonassociating polymer shows a precipitous drop in flow enhancement after several passes in the pipe because of shear degradation, while the interpolymer complex is stable thoughout the duration of the test. Interpolymer complexes of sulfo - EPDM and amino polymers also exhibit similar phenomena (29), suggesting that the use of interpolymer complexes may be a general method for generating shear-stable drag reduction in fluids.

SUMMARY

This paper describes the synthesis and solution properties of associating functional alpha olefin polymers. Specifically, carboxyl polymers are prepared by masking the carboxyl functionality and using Zeigler-Natta polymerization. The carboxyl polymers can self associate via H-bonding or they can be titrated with basic polymers to form interpolymer complexes. The H-bonding

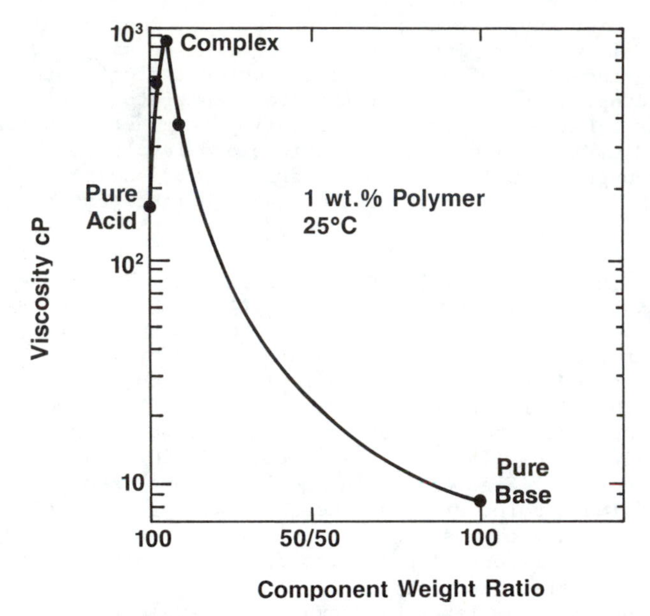

Figure 3. Plot of viscosity (in Cp) vs component ratio for interpolymer complexes of a 2.0×10^6 molecular weight acid functional polymer (acid content, 2.4%) and a 2×10^6 molecular weight basic functional polymer with 8.0% base content. The acid functional polymer is poly(1-octene-*co*-undecenoic acid). The basic functional polymer is poly(styrene-*co*-vinyl pyridine).

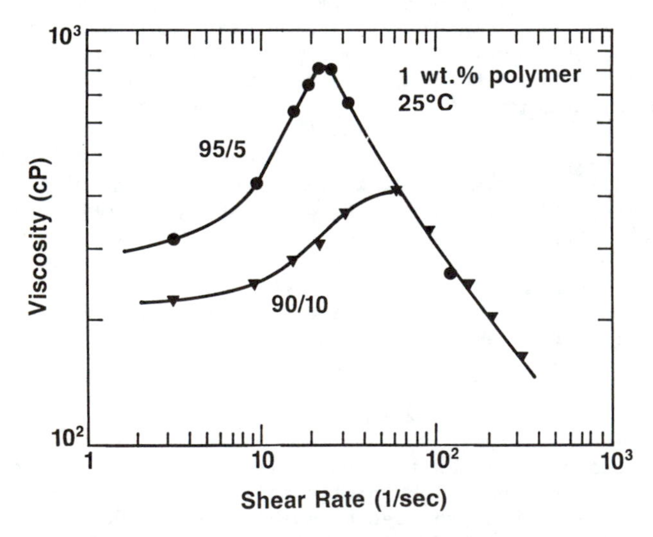

Figure 4. Plot of viscosity (in Cp) vs shear rate for the interpolymer complexes described in Figure 3.

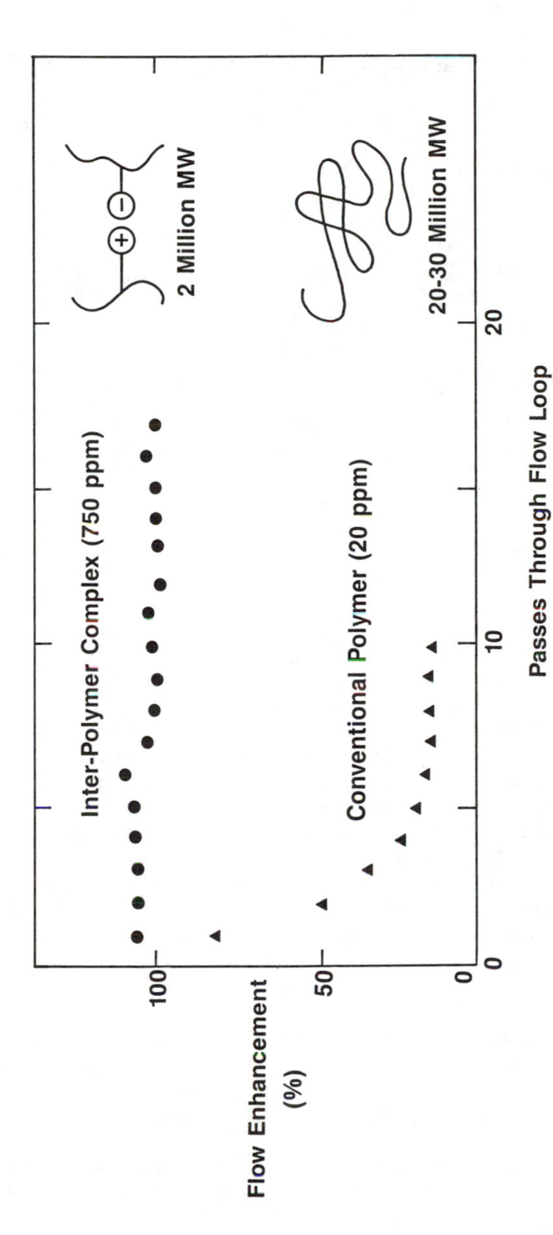

Figure 5. Plot of flow enhancement vs number of passes through a flow loop for 750 ppm of an interpolymer complex of poly(1-octene-*co*-undecenoic acid/ester) and poly(styrene-*co*-vinyl pyridene) (2 million MW) as compared with 20 ppm of a nonfunctional 20-30 million MW poly (1-octene).

polymers show enhanced shear and elongational viscosity compared with their nonfunctional counterparts. They also exhibit rheology and antimisting activity. In turn, the interpolymer complexes show enhanced viscosity, shear thickening rheology and shear-stable drag reduction.

LITERATURE CITED

1. Mack, M. P. , U.S. Patent 4, 433, 123 (Feb. 21, 1984)(Assigned to Conoco)
2. Lundberg, R. D.; Makowski, H. S., J. Polym. Sci.- Polym. Physics Ed. 18, 1821 (1980).
3. Lundberg, R. D.; Phillips, R. R., J. Polym. Sci. - Polym. Physic Ed. 20, 1143 (1982).
4. Lundberg, R. D., J. Appl. Polym. Sci., 27, 4623 (1982).
5. Lundberg, R. D.; Phillips, R. R.; Peiffer, D. G., J. Polym. Sci. - Polym. Physics Ed., 27, 245 (1989).
6. Peiffer, D. G.; Duvdevani, I.; Agarwal P. K.; Lundberg, R. D., J. Polym. Sci.- Polym. Lett. Ed., 24, 581 (1986).
7. Kowalik, R. M.; Duvdevani, I.; Peiffer, D. G.; Lundberg, R. D.; Kitano, K.; Schulz, D. N., J. of Non-Newtonian Fluid Mech., 24 (1987).
8. Lundberg, R. D.; Duvdevani, I.; ACS Div. of Polymeric Materials: Science and Engineering Prepr. 61, 259 (1989).
9. Lundberg, R. D.; Peiffer, D. G.; Duuvdevani, I.; U. S. Patent 4,536,539, (August 20, 1985). (Assigned to Exxon)
10. Duvdevani, I.; Peiffer, D. G.; Eckert J. A.; Lundberg R. D., U. S. Patent 4,516,982 (May 14, 1985) (assigned to Exxon).
11. Ashmond, J. U. S. Patent 4,002,436 (January 1, 1977) (Assigned to ICI).
12. Schulz, D. N.; Kitano, K.; Burkhardt, T. J.; Langer, A. W. U. S. Patent 4,518,747 (May 21, 1985) (assigned to Exxon)
13. Kitano, K.; Duvdevani, I.; Schulz, D. N.; U. S. Patent 4,742,135 (May 3, 1985) (assigned to Exxon).
14. Duvdevani, I.; Eckert, J. A.; Schulz, D. N.; Kitano, K.; U. S. Patent 4,523,929 (June 18, 1985) (assigned to Exxon).
15. Duvdevani, I.; Eckert, J. A.; Schulz, D. N.; Kitano, K.; U. S. Patent 4,586,937 (May 6, 1986) (assigned to Exxon).
16. Duvdevani, I.; Schulz, D. N.; Kitano, K.; Peiffer, D. G.; U. S. Patent 4,599,377 (July 8, 1986).
17. Kitano, K.; Duvdevani, I.; Schulz, D. N.; U. S. Patent 4,880,043 (November 14, 1989).
18. Duvdevani, I.; Schulz, D. N.; Kitano, K.; Peiffer, D. G.; U. S. Patent 4,621,111 (November 4, 1986) (assigned to Exxon).
19. Kowalik, R. M.; Duvdevani, Kitano, K.; Schulz, D. N., U. S. Patent 4,625,745 (December 6, 1986).
20. Peng, S. T. J.; Landel, R. F. in Rheology Volume 2: Fluids, Starita, G. A.; Marrucci, G.; Nicolais L, Eds. Plenum Prss, New York, 1980, p. 385.
21. Peng, S. T. J.; Landel, R. F.; J. Appl. Phys., 52, 5988 (1981).
22. Makowski, H. S.; Lundberg, R. D.; Westerman, L.; Bock, J, in "Ions in Polymers", Eisenberg, A. Ed., ACS Advances Series, Vol. 187, Am. Chem. Soc., Washington, D. C., 1980, Ch. 3.

23. Clark, K. J., U. S. Patent 3,492,277 (1970) assigned to ICI).
24. Purget, M. D., Ph.D. Thesis, University of Mass. (1984).
25. Purget, M. D.; MacKnight, W. J.; Vogel, O., Polym. Eng. Sci., 27, (19) 1461 (1987).
26. Chao, K. K.; Child, C. A.; Grens II, E. A.; Williams M. C., AICHE Journal 30 (1), 111 (1984).
27. Morgan, S. E.; McCormick, C. L. Prog. Polym. Sci. 15, 507-549 (1990).
28. Patterson, R. W.; Abernathy, R.H., J. Fluid Mech. 43, 689 (1970).
29. Kowalik, R. M.; Durderani, I.; Peiffer, D. G.; Lundberg, R. D., U.S. Patent 4, 516,128 (April 2, 1985).

RECEIVED November 9, 1990

Chapter 11

Association of Acrylamide–Dodecylmethacrylate Copolymers in Aqueous Solution

C. E. Flynn and J. W. Goodwin

Department of Physical Chemistry, University of Bristol, Cantock's Close, Bristol BS8 1TS, England

Association polymers are water soluble polymers which can associate intermolecularly in solution. A series of association polymers was made by copolymerizing acrylamide with dodecylmethacrylate using a surfactant as emulsifier. The solution properties were investigated by capillary viscometry, by using pyrene as a fluorescent probe, and by shear wave propagation. Experimental results indicated the presence of hydrophobic regions and association at all polymer concentrations. Above a critical hydrophobe concentration, a marked increase in the degree of association was apparent. Quantitative measurements of pyrene monomer and excimer fluorescence gave an average number of hydrophobes required to form a hydrophobic region as approximately 10. This number was the same as the same as that calculated from a network model by applying the theory of rubber elasticity to the modulus obtained from shear wave propagation.

Water soluble polymers that have been modified by the incorporation of hydrophobic groups have been investigated as associative thickeners in aqueous solutions and dispersions (1). These can be of widely varied molecular types, for example cellulosics with alkyl grafts (2), hydrophobically modified ethoxylated urethane (3), and more recently the copolymers of acrylamide with alkylacrylamides or alkylacrylates (4, 5). The desired thickening properties are achieved by the use of lower molecular weight polymers than would otherwise be possible by ability of the hydrophobic groups to associate intermolecularly, thereby creating a "cross linked" supermolecule or a network. This leads to other distinctive solution and rheological properties, for example "shear" degradation of the polymer solution can be reduced as the "cross links" are reversible and specific interactions with solutes such as surfactants can be introduced (6).

The total mole fraction of hydrophobic groups is generally extremely low, in order to maintain good water solubility. This can

0097–6156/91/0462–0190$06.00/0

make conventional chemical analysis of these association polymers difficult. With the copolymers of acrylamide with alkylacrylamides or alkylacrylates less than one mole percent of the alkyl monomer is needed to produce dramatic changes in the rheology of the solution. However in order to be able to establish relations between the measured solution properties and the molecular structure it is important to be able to determine the total amount of association occurring within such polymer solutions.

In this work two approaches have been taken in order to obtain an indication of the total amount of association with acrylamide dodecylmethacrylate copolymers. Addition of a fluorescent probe has been used to both identify the presence of hydrophobic regions and quantify the number of regions within solutions of low elasticity. For solutions of higher viscosity and elasticity that behave as polymer gels the network moduli have been used to quantify the degree of cross-linking and thus the amount of association.

Experimental

Polymer Synthesis. The polymers were made following the techniques described by Bock et al (5). Deoxygenated micellar solutions of dodecyl methacrylate solubilised by sodium dodecyl sulphate were prepared with acrylamide in the aqueous phase. Polymerisation was initiated by ammonium persulphate. To ensure removal of all traces of impurities and surfactant an extra step was introduced. Aqueous solutions were extensively dialysed against doubly distilled water. The resulting solutions were used as stock solutions for the experiments. Dilutions were prepared with aqueous sodium chloride and dilute aqueous sodium hydroxide to give a 0.1 M NaCl solution at pH 7. Salt solutions were used in order to minimise any polyelectrolyte effects that may be present due to hydrolysis of amide groups to carboxyl. All solutions were mixed by tumbling for at least 48 hrs to ensure equilibration and dosed with sodium azide (approximately 0.03 g/100 cm^3) to prevent biological degradation.

The polymers synthesised are given in Table I.

Table I. Polymer Synthesis

polymer code	Amounts added at Synthesis			
	acrylamide (mol/l)	dodecyl methacrylate ($x10^4$ mol/l)	SDS ($x10^2$ mol/l)	initiator ($x10^4$ mol/l)
1.1	0.362	38.3	9.76	2.10
0.7	0.354	23.7	3.35	2.48
0.4	0.326	11.6	3.17	2.34
0.2	0.325	7.2	3.16	2.40
0.0	0.354	0	3.46	2.42

Capillary Viscometry. The viscosities of dilute solutions were determined as a function of polymer concentration at 25° C by measuring the flow times through a Cannon-Fenske type capillary viscometer. As the shear rate varies both across the capillary and with total flow rate, the flow time can only be directly related to viscosity for Newtonian fluids. The dilute polymer solutions were found to behave as Newtonian fluids. Shear thinning was observed at high polymer concentrations in a rotational viscometer, but only at concentrations in excess of those used in the capillary viscometer. Intrinsic viscoity $[\eta]$ and Huggins coefficient k_H were determined by plotting the reduced viscosity versus polymer concentration.

$$\eta_{reduced} = \frac{\eta_{solution} - \eta_{solvent}}{\eta_{solvent} \times c} = [\eta] + [\eta]^2 k_H c$$

The Huggins coefficient was obtained from the initial linear slope in all cases.

Fluorescence Studies. Steady state fluorescence emission spectra were recorded on a Perkin Elmer model 3000 spectrofluorimeter. Measurements on each set of samples was taken on a single day to minimise machine variation. The excitation wavelength was 330 nm. Absorption spectra were recorded on a Perkin Elmer 552 spectrophotometer.

The fluorescent probe used was pyrene. It was introduced into each solution separately by drying a film of pyrene onto the walls of the sample vessel, adding the polymer solution and mixing by tumbling for 72 hrs. The polymer solutions were then transferred to a cuvette and deoxygenated by flushing with nitrogen for 30 min.

In one series of experiments the concentration of polymer was varied while the pyrene concentration was held constant and the fluorescence spectra recorded. In another series a stock solution of polymer was made up and the concentration of pyrene added was varied. The fluorescence and absorption spectra were measured quantitatively.

Shear Wave Propagation. The high frequency behaviour of the samples was evaluated using a Rank Shearometer.

The sample was constrained between two parallel plates. A shear pulse was applied and the time of propagation of the shear wave through the weak gel was measured as a function of the plate separation. The frequency of the shear wave was approximately 1200 rad s^{-1} and the strain was $< 10^{-4}$. Under these conditions the system behaved as a linear viscoelastic material. The shear wave rigidity modulus was calculated from the velocity of the shear wave,

$$\tilde{G} = \rho v^2$$

where ρ is the sample density and v is the measured wave velocity.

Results and Discussion

Fluorescence-Variation of Polymer Concentration. Pyrene is a molecule that is sparingly soluble in water and will partition strongly towards any available hydrophobic regions in solution. When it is excited at 330 nm it produces a characteristic fluorescence spectrum. Generally five peaks can be observed, as in Figure 1. If the pyrene is in a region where it experiences low polarity the intensity of the first peak in the spectrum is low. In a region of high polarity this peak is enhanced. The intensity ratio of the first to third peak I_1/I_3 can therefore be used as a sensitive probe of the dielectric properties or polarity of the region where the pyrene resides (7,8). These properties make it a useful probe for studying such phenomena as the micellisation of surfactants (9, 10) and conformational changes in polymers (11).

The change in the ratio I_1/I_3 as the polymer concentration is varied from 0.05 to 0.65 g/100 cm^3 is given in Figure 2. For the hydrophobe free homopolymer, polymer 0.0, the ratio was high and constant for all concentrations. This shows that there are no available hydrophobic sites to bind pyrene on the acrylamide backbone and that there is little specific interaction with pyrene.

With the two polymers 0.2, 0.4 there was an initial steady decrease in the intensity ratio I_1/I_3 as the concentration of polymer was increased. Above a threshold concentration I_1/I_3 remained constant at 1.08 a value which is close to that observed in surfactant micelles (7). This value is indicative of the presence of substantial hydrophobic regions in the solution above the threshold concentration, and in this respect is similar to the behaviour observed with surfactant systems above the critical micelle concentration (9). However the association process is different to micellisation. With micellisation I_1/I_3 maintains its aqueous value until the critical micelle concentration, when a step change to a lower value occurs. With polymers 0.2 and 0.4 a steady decrease from the aqueous value of 1.8 to 1.08 was observed. One explanation is that low concentrations of the hydrophobic regions exist below the threshold concentration and I_1/I_3 is a reflection of the relative amounts of hydrophobic and aqueous pyrene. Alternatively there could be a change in the polarity of the hydrophobic regions sampled by the pyrene as the polymer concentration increases.

These trends are similar to those observed by Siano et al (12) with a similar acrylamide-dodecylacrylamide copolymer using sodium 8-anilino-1-naphthalenesulphonate as a probe molecule. They are also consistent with the preliminary pyrene lifetime experiments in acrylamide-decylacrylamide copolymer solutions reported by McCormick and Johnson (13). This suggests that such behaviour is a general feature of this type of copolymer molecule.

Comparison with Capillary Viscometry. Plots of reduced viscosity versus polymer concentration for polymers 0.0, 0.2, 0.4 and 0.7 are given in Figure 3. Table II gives the values of the Huggins coefficient and intrinsic viscosity obtained.

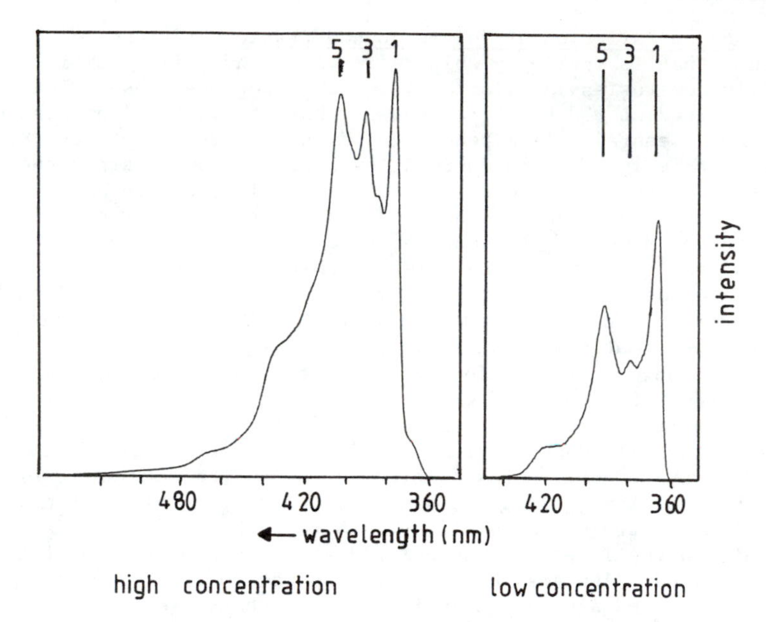

Figure 1. Example of pyrene fluorescence spectra at high and low polymer concentrations.

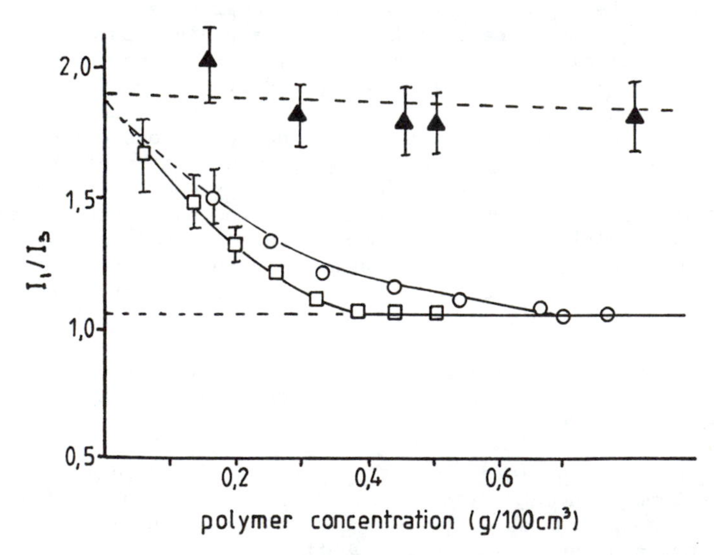

Figure 2. Change in I_1/I_3 ratio with concentration of polymers 0.4 (□), 0.2 (○), and 0.0 (▲)

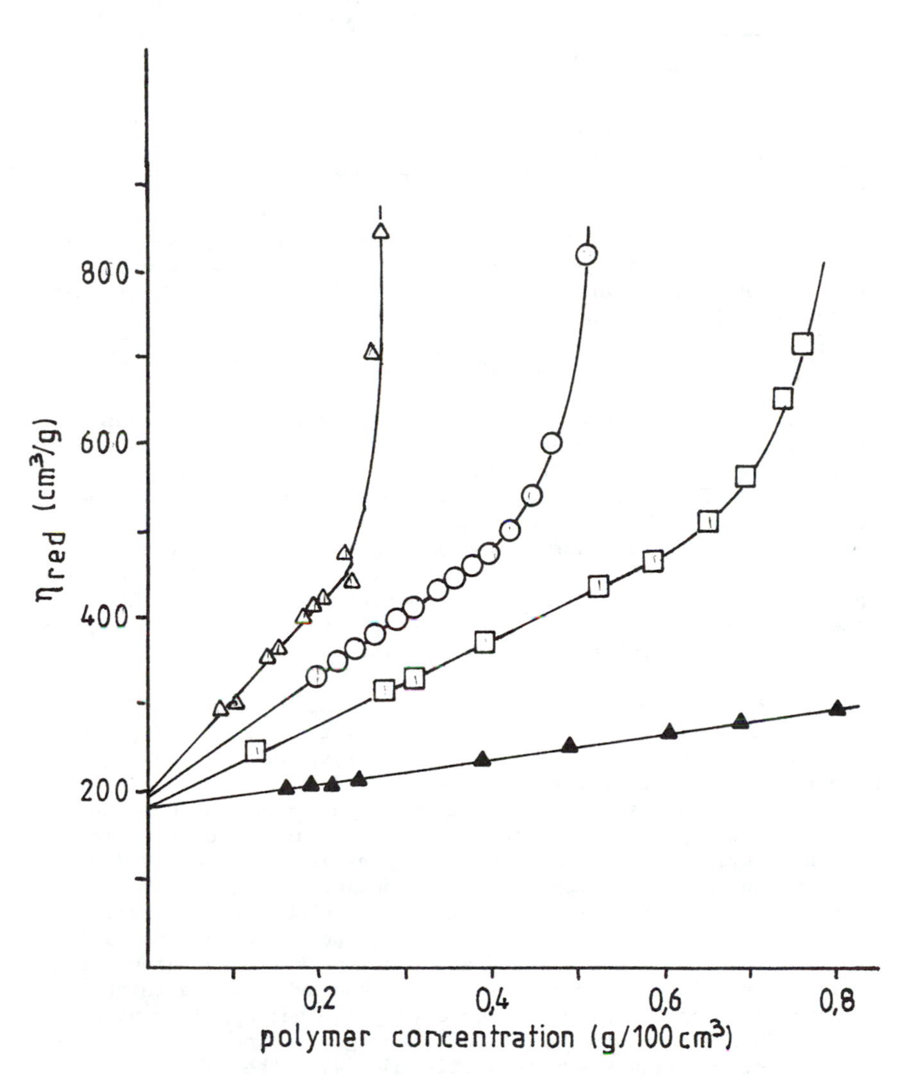

Figure 3. Reduced viscosity as determined by capillary viscometer, polymers 0.7 (△), 0.4 (○), 0.2 (□), and 0.0 (▲).

Table II. Dependence of Capillary Viscosity Constants on Hydrophobe Content

Polymer	Mole Fraction hydrophobe	Intrinsic viscosity (cm^3/g)	Huggins coeff.
1.1	0.011	NA	NA
0.7	0.0067	185	4
0.4	0.0036	195	2.2
0.2	0.0022	195	1.4
0.0	0.0	178	0.48

The four polymers have similar intrinsic viscosities, indicating a similar molecular size at high dilutions. A molecular weight can be calculated assuming that the hydrophobe polymers behave according to the Mark-Houwink equation with the coefficient and exponent that is used for polyacrylamide homopolymers (14), i.e.

$$[\eta] = 9.33 \times 10^{-3} \times M^{0.75}$$

However this can only be regarded as an estimation of the value due to the possibility of intra-molecular interactions causing a more compact polymer coil at low concentrations with such polymers.

The curve for each polymer has an initial linear slope. The Huggins coefficient of 0.48 calculated from this slope for the homopolymer of acrylamide, polymer 0.0, is similar to that expected with random coil polymers. For the polymers 0.2, 0.4 and 0.7 it increases with the mole fraction of the hydrophobic component used in the synthesis. This is consistent with the suggestion that high Huggins coefficient is indicative of intermolecular associations building up in solutions (15,16).

Above a "critical" concentraction, c_c, these polymers showed an upward curvature in the reduced viscosity. It is unlikely that the upturn is due to the onset of molecular overlap as the overlap concentration would be related to the molecular size, i.e. intrinsic viscosity, which was similar for all three polymers. However the value of c_c was approximately inversely proportional to the mole fraction of hydrophobe from synthesis; that is the total hydrophobe concentration at c_c was constant. If complete incorporation of hydrophobe at synthesis is assumed this hydrophobe concentration is approximately 0.2 mol/l. Thus a change in the nature of association and/or a more extensive degree of association is brought about at a "critical" hydrophobe concentration for these associative thickeners.

Comparison of the fluorescence data with the capillary viscometry data is made for polymers 0.2 and 0.4 in Figure 4 where the data are superimposed. c_c from viscometry coincides with the I_1/I_3 ratio becoming constant. This fits with the suggestion of a "critical" phenomenon in the nature of the hydrophobic association.

Fluorescence - Variation of Pyrene Concentration at Constant Polymer Concentration. In dilute solution pyrene exhibits the characteristic five peak monomer fluorescence spectrum. However when the concentration of pyrene is greatly increased an excimer is former, which has a fluorescent spectrum that is red shifted from that of the

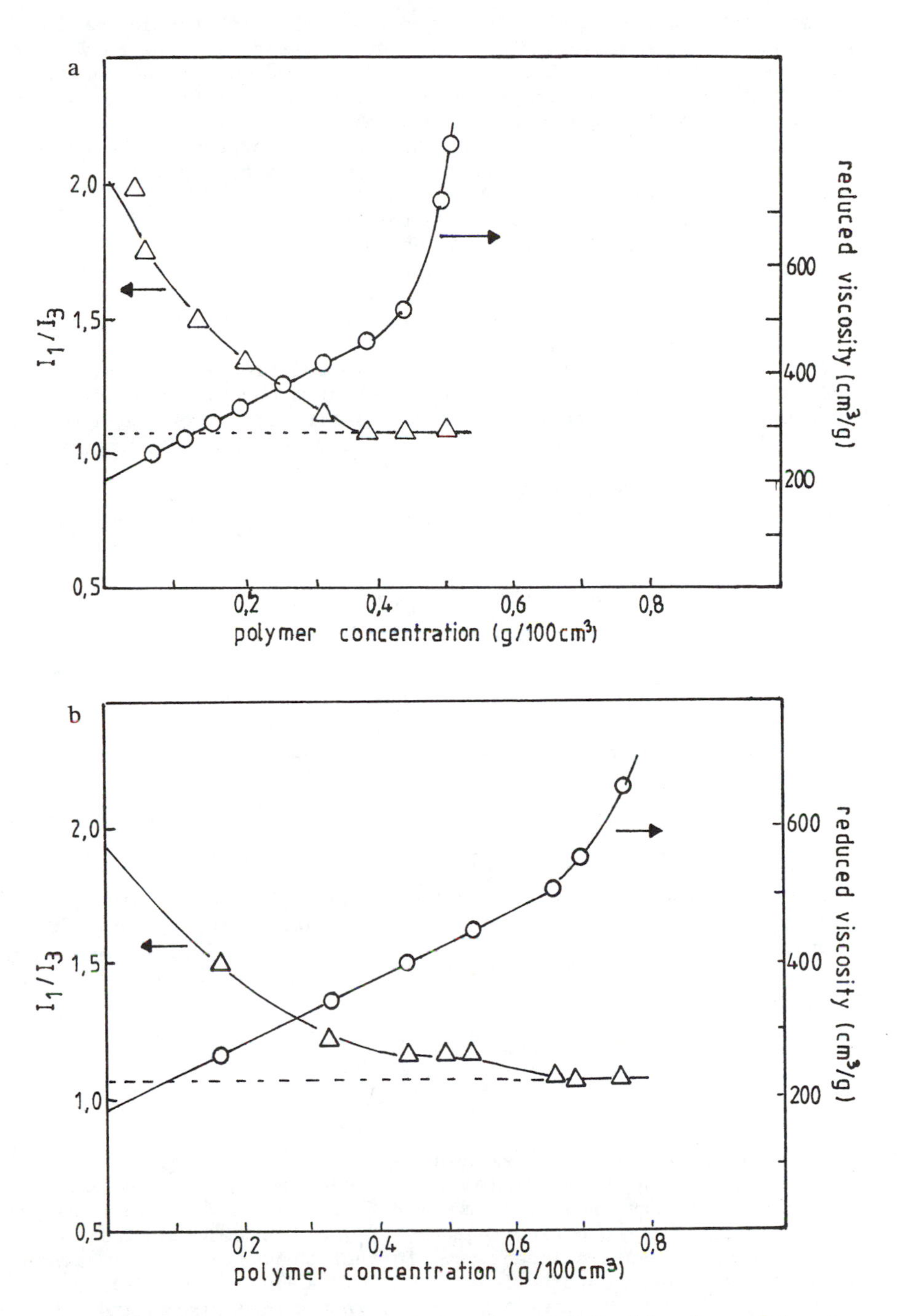

Figure 4. Comparison of reduced viscosity and I_1/I_3: (a) polymer 0.4 and (b) polymer 0.2.

monomer fluorescence (7,8). It was observed that substantial excimer formation was present when pyrene was equilibrated with the polymer solutions above c_c even though the overall pyrene concentration was low. This indicated a high effective pyrene concentration i.e. only a relatively small proportion of the total solution volume was available to the pyrene molecule. This would be expected from pyrene selectively partitioning into the small volume of hydrophobic regions produced by association.

Experiments were carried out where the concentration of polymer was held constant and the amount of pyrene was varied. As the concentration of pyrene was increased, the relative amount of excimer to monomer fluorescence increased. In effect a quenching experiment was performed where the monomer fluorescence was quenched by excess pyrene producing excimer.

As the solubility of pyrene in aqueous solution is extremely low it is almost exclusively solubilised at the hydrophobic regions. Therefore by changing the overall concentration of pyrene added to the solution, the expected number of pyrene molecules occupying any one region was altered. If it is assumed that the distribution of pyrene amongst hydrophobic regions is random and a Poisson distribution holds (17,18), then the probability that a region will be occupied by n pyrene molecules is,

$$\rho_n = \left(\frac{[P_{tot}]}{[H]} \right)^n \frac{\exp(-[P_{tot}]/[H]}{n!} \tag{1}$$

where $[P_{tot}]$ is the total pyrene concentration and [H] is the concentration of hydrophobic regions. Also ρ_n is given by the fraction of regions containing n pyrene molecules,

$$\rho_n = \frac{[H_n]}{[H]} = \frac{[P_n]}{n[H]} \tag{2}$$

where $[P_n]$ is the concentration of pyrene in regions containing n molecules and $[H_n]$ is the concentration of hydrophobic regions containing n pyrene.

Equating Equations 1 and 2 for the case of single occupancy gives

$$\ln \left(\frac{[P_1]}{[P_{tot}]} \right) = \frac{-[P_{tot}]}{[H]} \tag{3}$$

The formation of pyrene excimer is very efficient (8). If it is assumed that regions containing more than one pyrene will produce excimer fluorescence and pyrene monomer fluorescence will only be produced from singly occupied regions, $[P_1]$ can be determined from the monomer fluorescence intensity. In fact the monomer fluorescence should contain some contributions from regions that were multiply occupied and $[P_1]$ would therefore be somewhat overestimated and become a function of $[P_{tot}]$. In effect this would lead to [H] being underestimated from Equation 3.

The total pyrene concentration and the monomer pyrene concentrations in the polymer solutions were determined by comparison with calibration curves which were constructed using known concentrations of pyrene in propan-1-ol. Propan-1-ol was chosen as the I_1/I_3 ratio was similar to that in the polymer solutions above c_c, indicating a similar polarity of the pyrene environment. The total pyrene concentration was determined from the absorption spectra and monomer pyrene from the fluorescence spectra. Both the peak absorption at 335 nm (I_A) and the intensity of the third fluorescence peak (I_3) were found to be linear with pyrene concentration over the experimental range. The relevant peak heights can therefore be substituted directly into Equation 3.

Figure 5 shows the linear plots of $\ln (I_3/I_A)$ versus I_A. The gradients and concentrations of hydrophobic regions calculated using Equation 3 are given in Table III.

Table III. Concentration of Hydrophobic Regions Calculated from Fluorescence

Polymer	Conc. of solution (g/100 cm^3)	Conc. of hydrophobes (mol/dm^3)	Conc. of regions (mol/dm^3)	Ratio of hydrophobes to regions
0.4	0.32	1.64×10^{-4}	1.44×10^{-5}	11
	0.5	2.56×10^{-4}	2.18×10^{-5}	12
0.2	0.5	1.56×10^{-4}	1.73×10^{-5}	9

The concentration of hydrophobes calculated from synthesis assuming complete incorporation are also given in Table III. The concentration of regions increases proportionately with the total amount of hdyrophobe. Dividing the concentration of hydrophobes by the concentration of regions leads to a value of the average number of hydrophobes used in synthesis required to obtain one hydrophobic region. For both these polymers an average value of approximately 10 was obtained.

It is interesting to compare the results with fluorescence studies that were attempted with a different type of association polymer. An aqueous solution of a hydrophobically modified hydroxyethylcellulose (19) gave little or no excimer fluorescence. This suggests that the hydrophobic regions present cannot solubilise more than one pyrene. The regions may be small due to the low number of hydrophobic groups per polymer (3 or 4 per chain (19)) and the relatively high stiffness of the cellulosic polymer backbone resulting in low hydrophobe aggregation numbers. For the acrylamide copolymers a larger hydrophobic region is suggested as the data could be fitted assuming multiple occupancy of pyrene. A polymer structure which is consistent with this observation and the ratio of hydrophobes to hydrophobic regions is a flexible polymer chain containing short blocks of dodecyl methacrylate. Intermolecular association could then occur through interactions between these blocks.

Shear Wave Propagation. For the samples with highest mole percentage of hydrophobe, 0.7 and 1.1, fluorescence studies could not be performed due to the dramatic increases observed in viscosity. The solutions would not flow sufficiently to allow effective mixing and solubilisation of the pyrene. However these samples and also a more concentrated solution of polymer 0.4 have significant elastic moduli as determined by oscillatory measurements. An example of the variation of storage and loss modulus with frequency as measured using a Bohlin VOR Rheometer is given in Figure 6. The phase angle is below 10 throughout the frequency range and the solution is behaving as a weak viscoelastic gel. We can therefore model the solution as a swollen network using the theory of swollen rubber elasticity (20).

Provided there is no relaxation of stress within the network of the gel structure during the experimental timescale and that the deformation is small, the modulus of the network is directly related to the concentration of effective network linkages. In practice this requires measurement of the storage modulus at high frequency and low applied strains. The shearometer measures the wave rigidity modulus at 1200 rad s^{-1}. When there is little attenuation of the shear wave (which was observed with all the samples measured) the storage modulus is much greater than the loss modulus and the experimental wave rigidity modulus ($\tilde{G}$) can be taken as a good approximation to the high frequency limit of the storage modulus i.e. the network modulus (G_N).

$$\tilde{G} = G_N$$

From the theory of swollen rubber elasticity (20) the modulus of the network is given by

$$G_N = 2An_ERT \tag{4}$$

where R is the gas constant, T is the absolute temperature, n_E is the molar concentration of elastically effective network links and A is the front factor which has been given a value of 1-(2/link functionality) by Graessley (21). Assuming that one site on a polymer associates with only one other site, i.e. tetrafunctionality the "effective links" can be calculated using Equation 4. The results at a concentration fo 1.8 g/100 cm^3 for polymers 0.4, 0.7 and 1.1 are given in Table IV.

Table IV. Shear Wave Rigidity Modulus and Calculated Concentration of Elastically Effective Links for Polymers at 1.8 g/100 cm^3

Polymer	G_N(Pa)	n_E(mol/dm^3)
1.1	250	1.03×10^{-4}
0.7	180	7.4×10^{-5}
0.4	68	2.8×10^{-5}

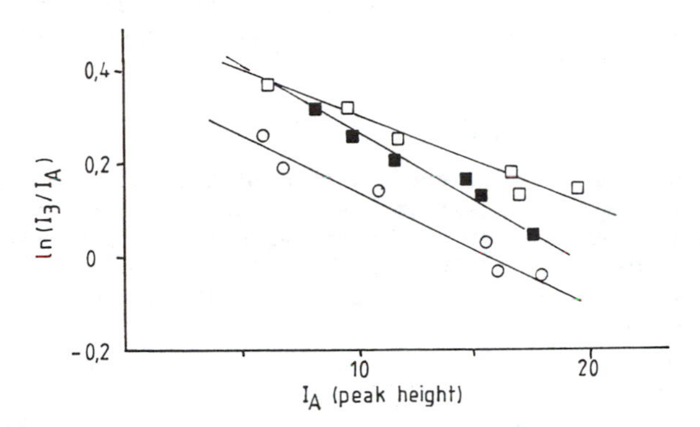

Figure 5. Pyrene fluorescence data.

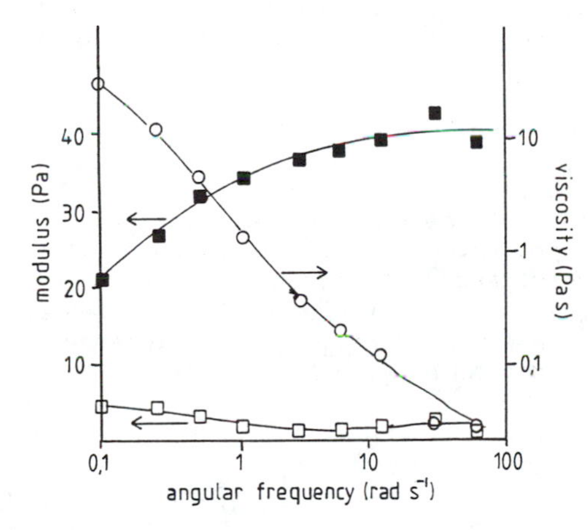

Figure 6. Storage (■) and loss (□) moduli and viscosity (○) for polymer 0.4 at 2.2 g/100 cm^3.

The number of "effective links", however, is not simply the number of associations present in the system. To convert the concentration of "effective links" to the concentration of actual links and thus to the concentration of hydrophobic associations requires assumptions about the form and completeness of the network.

The value of the wave rigidity modulus for polymers 0.4 and 0.7 in the concentration range from 0.5 to 3 g/100 cm^3 is given in Figure 7. There was a concentration below which the modulus was too small to be measured. Above this concentration there was a linear increase of modulus with concentration. The data can be fitted with a simple expression,

$$G_N = k(c-c') \tag{5}$$

where k and c' are the gradient and the intercept from Figure 7. The number of hydrophobes present in the solution is directly proportional to polymer concentration. This suggests that the number of effective links is directly proportional to the concentration of hydrophobes once above a critical concentration of polymer. Combing Equations 4 and 5

$$2An_E RT = k(c-c')$$

To convert the concentration of network links that are elastically effective n_E into the concentration of actual links or associations present in the polymer solution n_A, some corrections must be made. Firstly the number of associations required to form an infinite chain needs to be deducted i.e. one association per polymer chain. Only associations in excess of this number will lead to effective networking.

$$2An_E RT = k(c-c') = 2A(n_A-n_P)RT \tag{6}$$

where n_P is the polymer concentration.

In order for this simple linear relationship in concentration to hold we have to propose that associations can only build up above some critical concentration of polymer. If we suggest that there is a critical aggregation concentration of hydrophobes n_H' and that above this concentration extra hydrophobes will form associations containing on average N hydrophobes, Equation 6 gives

$$n_A-n_P = \frac{(n_H-n_H')}{N} - n_P = \frac{k(c-c')}{2ART} \tag{7}$$

where n_H is the total concentration of hydrophobes. The concentration of polymer molecules can be derived using the molecular weight determined from the intrinsic viscosity. The concentration of hydrophobes can be derived from the polymer concentration and the mole fraction of the hydrophobic monomer used in the synthesis. The calculated values of N and n_H' based on Equation 7 are given in Table V.

The network link concentration requires further corrections to account for the decrease of modulus by the formation of elastically ineffective loops from adjacent linking sites on the same polymer

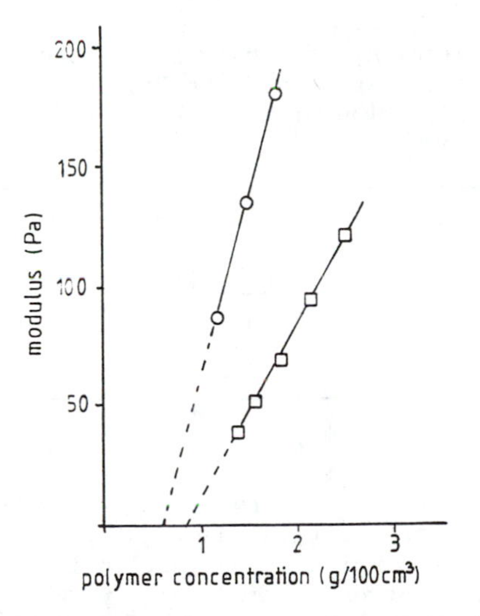

Figure 7. Wave rigidity modulus as a function of concentration for polymers 0.7 (○) and 0.4 (□).

chain and the increase of modulus due to physical entanglements of polymer chains. A model described in Reference (22) gives the network modulus as

$$G_N = 2A\left(\left[\frac{(n_H - n_H')}{N}(1-f_L) - n_P\right] + \frac{Ec^2}{T}\right)RT \qquad (8)$$

As polymer 0.0 had no measurable network modulus the contribution due to entanglements (Ec^2) can be neglected. The correction factor for the formation of elastically ineffective or closed loops $(1-f_L)$ (23) was calculated from data for the characteristic ratio of 11 for homopolyacrylamide chains (24). G_N was obtained as a function of polymer concentration using Equation 8 with adjustable values of the average number of hydrophobes per association, N, and the critical aggregation concentration of hydrophobes, n_H'. The values of N and n_H' that most closely fitted the data in Figure 7 are also given in Table V

Table V. Calculations of the Average Number of Hydrophobes per Association and the Critical Aggregation Concentration of Hydrophobes using data from Figure 6

Polymer	Equation 7		Equation 8	
	N	n_H' (mol/1)	N	n_H' (mol/1)
0.7	11	0.31	10	0.26
0.4	11	0.27	8	0.17

Both approximations require a finite concentration of unassociating hydrophobes before effective association can occur in order to fit the observed dependency of network modulus with concentration. The absolute value of the critical aggregation concentration of hydrophobes depends upon which approximation is considered, also the value is highly dependent on the molecular weight assumed for the polymers. However, in both cases it is of a similar order of magnitude to the "critical" hydrophobe concentration postulated from capillary viscometry of 0.2 mol/1.

The average number of hydrophobes per association was relatively independent of which approximation was used as well as the molecular weight and was 10 for both polymers. This value is in good agreement with that obtained with the more fluid samples using the fluorescence technique.

Conclusions

The solution properties of a series of acrylamide dodecylmethacrylate copolymers of similar intrinsic viscosity vary dramatically according to the level of hydrophobic monomer used in the synthesis. This can be attributed to the presence of hydrophobic associations between polymer molecules. Capillary viscometry studies show an increase in Huggins coefficient with levels of hydrophobe added and suggest the presence of a "critical" concentration above which the reduced viscosity rises rapidly due to associations. Fluorescent probe studies confirm the presence of hydrophobic regions and also show a "critical" concentration above which the average hydrophobicity experienced by the probe remains constant. Both observations are consistent with a change in either the nature or extent of association of hydrophobes in solution once a critical level has been attained.

By using the self-quenching effect of excimer formation by pyrene an estimate of the number of hydrophobic regions present in solution was obtained. This gave an average value of approximately 10 hydrophobes per association. For the solutions with highest mole percentage of hydrophobe, fluorescence studies could not be performed as the samples behaved as weak viscoelastic gels. Shear wave propagation experiments were made and the results fitted using the theory of swollen rubber elasticity. A "critical" concentration for association was required in order to give a good fit to the data. A resulting average aggregation number of approximately 10 hydrophobes per association was calculated, in very good agreement with that evaluated for the more fluid samples by fluorescence. A model for association that fits these data is the intermolecular association of two short blocks of hydrophobes to give an average aggregation number of 10 above a critical concentration of hydrophones in solution.

Acknowledgments

The support for this work by the Science and Engineering Research Council and B.P. International plc is gratefully acknowledged.

Literature Cited

1. Glass, J. E. In Water soluble polymers: Beauty with performance; Glass, J. E., Ed.; Advances in Chemistry 213; American Chemical Society: Washington, DC, 1986.
2. Landoll, L. M. J. Polym. Sci. Polym. Chem. Ed. 1982, 20, 443.
3. Karunasena, A.; Brown, R. G.; Glass, J. E. In Polymers in aqueous media; Glass, J. E., Ed.; Advances in Chemistry, 223; American Chemical Society: Washington, DC, 1989; pp 495-525.
4. Evani, S. U.S. Patent 4 105 649, 1984.
5. Turner, S. R.; Siano, D. B.; Bock, J. U.S. Patent 4 528 348.
6. Steiner, C. A. Polym. Prepr. 1985, 26(1), 224.
7. Thomas, J. K. The chemistry of excitation at interfaces; ACS Monograph No. 181; American Chemical Society: Washington, DC, 1984.
8. Birks, J. B. Photophysics of aromatic molecules; Wiley, NY, 1969.
9. Turro, N. J.; Yekta, A. J. Am. Chem. Soc. 1978, 100, 5951.
10. Aikawa, K.; Turro, N. J.; Yekta, A. Chem. Phys. Lett. 1979, 63, 543.
11. Deh-Ying, C.; Thomas, J. K. In Polymers in aqueous media; Glass, J. E., Ed.; Advances in Chemistry 223; American Chemical Society: Washington, DC, 1989; pp 325-343.
12. Siano, D. B.; Bock, J.; Myer, P.; Valint, P. L. ibid pp 425-436.
13. McCormick, C. L.; Brent Johnson, C. ibid pp 437-456.
14. Francois, J.; Sarazin, D. Polymer 1979, 20, 969.

15. Bock, J.; Siano, D. B.; Valint, P. L.; Pace, S. J. In Polymers in aqueous media; Glass, J. E., Ed.; Advances in Chemistry 223; American Chemical Society: Washington, DC, 1989; pp 411-424.
16. Gelman, R. A.; Barth, H. G. In Water soluble polymers; Glass, J. E; Advances in Chemistry 213; American Chemical Society: Washington, DC, 1986; pp 101-110.
17. Aikawa, K.; Turro, N. J.; Yekta, A. Chem. Phys. Lett. 1979, 63, 543.
18. Almgren, M.; Grieser, F.; Thomas, J. K. J. Am. Chem. Soc. 1979 101, 279.
19. Goodwin, J. W.; Hughes, R. W.; Lam, C. K.; Miles, J. A.; Warren, B. C. H. In Polymers in aqueous media; Glass, J. E., Ed.; Advances in Chemistry 223; American Chemical Society: Washington, DC, 1989; pp 365-380.
20. Flory, P. J. In Principles of polymer chemistry; Cornell University; Ithica 1953.
21. Graessley, W. W., Macromolecules 1975, 8, 186.
22. Goodwin, J. W.; Khidher, A. M. In Colloid and interface science, hydrosols and rheology; Kerker, M., Ed.; Academic Press, New York; 1976.
23. Walsh, D. J.; Allen, G.; Ballard, G. Polymer 1974, 15, 366.
24. Kulicke, W. M.; Kniewske, R.; Klein, J. Prog. Poly. Sci. 1982. 8, 373.

RECEIVED October 22, 1990

Chapter 12

Systems Approach to Rheology Control

Paul R. Howard, Edward L. Leasure, Stephen T. Rosier[1], and Edward J. Schaller

Rohm and Haas Company, 727 Norristown Road, Spring House, PA 19477

Nonionic hydrophobe-modified poly(oxyethylene) urethane (HEUR) associative thickeners have been designed possessing widely different but complementary viscosity-shear rate responses. Use of combinations of such thickeners provides excellent control of paint rheology without the need for additional cosolvents or surfactants in the formulation to obtain the desired high vs. low shear viscosity balance. Fundamental rheological characteristics of two such thickeners are presented, along with practical guidelines for their use in formulating latex paints.

Associative thickeners are water-soluble polymers containing hydrophobic groups which are capable of non-specific hydrophobic association, similar to surfactants. The multiple hydrophobic groups of the thickener molecules are thought to associate to form a network, although this may be an aggregation governed by mass action, rather than true micellar behavior.

Hydrophobe-modified poly(oxyethylene) urethane (HEUR) nonionic associative thickeners are typically built up from water-soluble poly(oxyethylene) segments joined by urethane groups (1-4). The diisocyanates used to link the water-soluble segments serve as internal hydrophobic groups, if the diisocyanate molecule is large enough and hydrophobic enough, or internal hydrophobes may be efficiently built up by reacting the diisocyanates with hydrophobic diols or diamines. Hydrophobic end groups may be incorporated by reacting hydrophobic alcohols, amines or acids with the diisocyanate groups, the resulting hydrophobic group effectively including the hydrophobic residue of the diisocyanate. Alternatively, hydrophobic monoisocyanates may be reacted with terminal poly(oxyethylene) chains. Excess diisocyanate may also be reacted with water to build up hydrophobic blocks; the overall hydrophobicity of the non-poly(oxyethylene) blocks is the operative factor. Although polyfunctional isocyanates or polyols may be used to construct branched thickener molecules, all of the HEUR thickeners commercially available today appear to have a linear structure. At this stage of the development of associative thickener chemistry, this type of structure, especially where end hydrophobic groups are much larger than

---

[1]Current address: Spectra Laboratories, Harrisburg, PA 17105

0097–6156/91/0462–0207$06.00/0

internal hydrophobes, also lends itself most readily to well controlled scientific investigations.

Effect of Hydrophobic End Groups

To determine the effects of the end hydrophobic blocks, poly(oxyethylene) urethane polymers were prepared by reacting four moles of 8,000 MW poly(oxyethylene) (Carbowax 8000, Union Carbide Corp.) with five moles of toluene diisocyanate. Intrinsic viscosity of the resulting polymer indicated an M_w of approximately 40,000. Since poly(oxyethylene) is not commercially available in this molecular weight range, it was not possible to compare the effects of the internal hydrophobic groups. A portion of the block copolymer was then reacted with an excess of dodecanol to provide hydrophobic end groups, and another portion with ethylene glycol methyl ether to provide (relatively) hydrophilic end groups. Figure 1 compares the aqueous solution viscosities of these two polymers. In spite of their similar molecular weights, the polymer with hydrophobic end groups provides viscosity far greater than that of the polymer with hydrophilic end groups. Since there is no dispersed phase for the associative thickener to adsorb onto, this enhanced viscosity must come from self-association.

These model poly(oxyethylene) urethane block polymers were used to thicken a 300 nanometer acrylic latex. A mixture of latex and thickener was prepared, adjusted to pH 9.0 with ammonium hydroxide and allowed to equilibrate for at least 60 hours at 25°C. The latex was separated by centrifuging, and the supernatant liquid was analyzed for free thickener. Figure 2 shows the experimental results, which clearly demonstrate adsorption of the hydrophobe-modified polymer on the latex particles. This thickener adsorbs rapidly at low concentrations and apparently saturates the latex surface at about 1% thickener in the continuous phase (5). In contrast, the non-hydrophobically modified poly(oxyethylene) block copolymer shows little adsorption. The hydrophobe-modified model thickener is similar in performance to the commercial thickener sold as Thickener QR-708 and Acrysol RM-825 (thickener A used in the experiments described below).

In further experiments, 40,000 M_w poly(oxyethylene) urethane polymers similar to those described above were capped with a series of straight chain alcohols ranging from C_6 to C_{14}. These were used to thicken the same 300 nm acrylic latex. As shown in Figures 3 and 4, increasing the carbon chain length of the hydrophobic end group markedly increased the latex adsorption and thickening characteristics. In this series, the low shear viscosity was most markedly affected by the changes in hydrophobe modification, while the high shear viscosity was less affected.

Formulating with Associative Thickeners

Since the hydrophobic association and adsorption exhibited by these thickeners is non-specific, it is greatly influenced by the presence of surfactants and water miscible organic solvents (5,6). Surfactants may compete for adsorption sites on particle surfaces, and can hinder or enhance associations between thickener hydrophobes, depending on the surfactant HLB. An appreciation of these interactions of the associative thickeners with other paint components forms the basis for the current approaches to formulating with them (7-11).

The commercial nonionic HEUR associative thickeners rely heavily on association to build low shear viscosity, and derive little from their relatively low M_w flexible backbones. They are therefore very responsive to formulation modifications that have the effect of decreasing association between thickener molecules or with dispersed particles. In addition to the choice of the thickener itself, the choice of latex can have a major effect

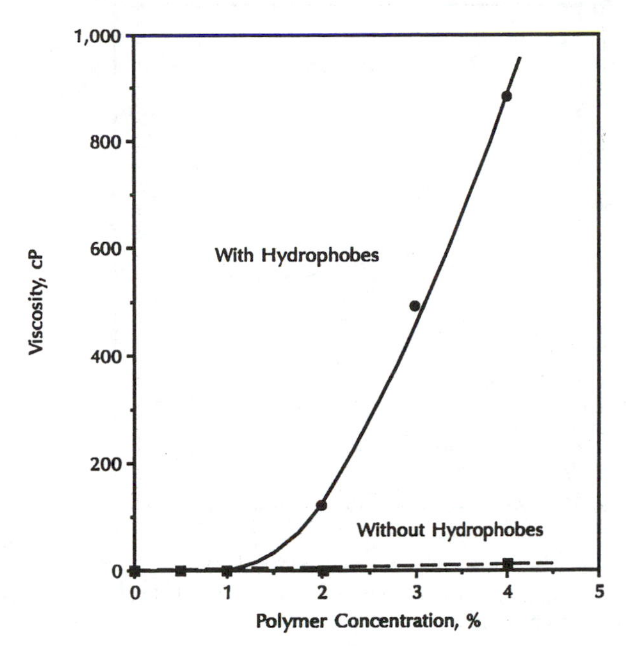

Figure 1. Effect of hydrophobic modification on thickening efficiency of poly(oxyethylene) urethane block copolymer in water.

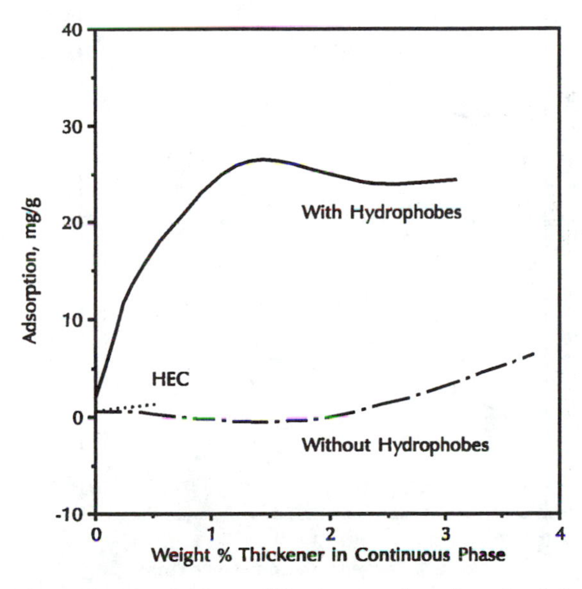

Figure 2. Effect of hydrophobic modification on adsorption of poly(oxyethylene) block copolymer on 300 nm acrylic latex.

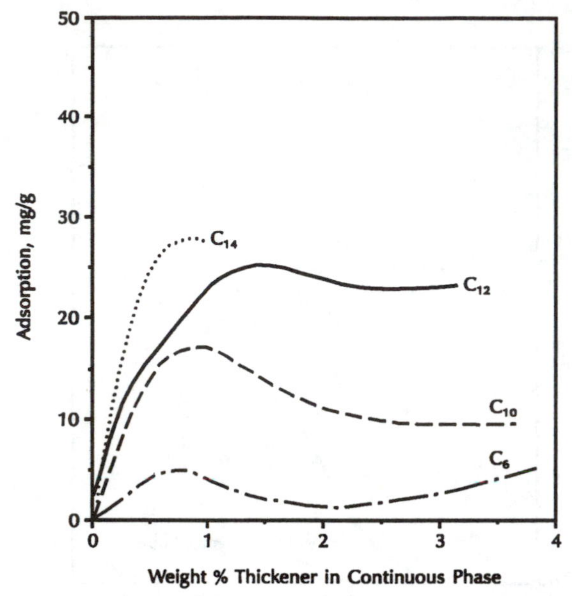

Figure 3. Effect of hydrophobic end group size in poly(oxyethylene) urethane block copolymers on adsorption on 300 nm acrylic latex.

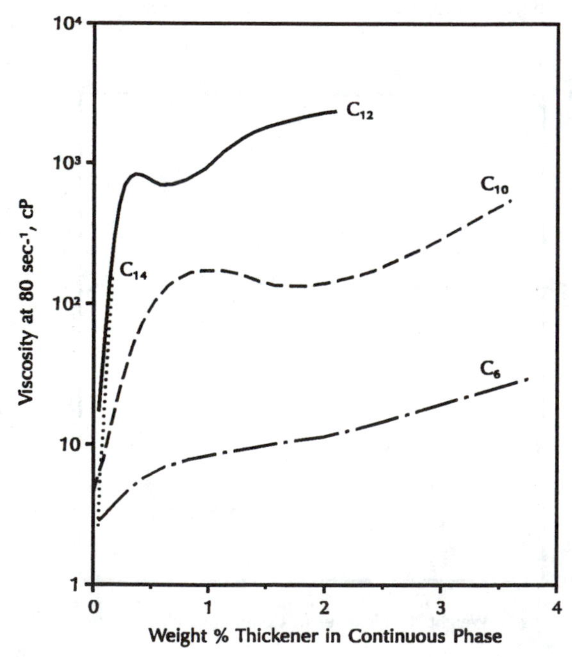

Figure 4. Effect of hydrophobic end group size in poly(oxyethylene) urethane block copolymers on viscosity (80 sec^{-1}) of 25% volume solids 250 nm acrylic latex.

on interactions with the associative thickener. Commercial latices vary widely in particle size and particle size distribution, and also in surface hydrophobicity, due to their composition and the surfactants and stabilizers used in their preparation. Type and particle size of pigments and extenders are not usually as important, since these tend to be more hydrophilic, and therefore do not form strong hydrophobic associations with the thickeners. However, the choice of dispersing agent and the level used often have important effects. Polymeric dispersants containing hydrophobic groups appear to enhance association with pigments, and may interact with the thickener hydrophobes in aqueous phase aggregates, while polyacid dispersants containing no significant hydrophobic groups often tend to interfere with realization of the usual associative thickener benefits. Water miscible solvents can modify the associations by changing the solvency of the aqueous continuous phase relative to the dispersed phase of the paint. Water insoluble coalescing agents have little effect, however.

There are often disadvantages to the use of surfactants or cosolvents to manipulate the rheology. Anionic surfactants are generally most efficient, but they must be used at low levels, since they may cause water sensitivity and foaming problems; nonionic surfactants will usually have less adverse effect on a weight-for-weight basis on foaming and water sensitivity, but because of their smaller effect on low shear viscosity, must be used at higher levels to achieve the desired viscosity suppression. The most popular choice with coatings formulators has usually been to use a water miscible cosolvent such as diethylene glycol monobutyl ether (DGBE); here, however, necessary solvent levels may exceed statutory limits, or may pose problems of excessive softening early in the life of the paint film. Propylene glycol or diethylene glycol methyl ether, often used as wet edge agents, can also be used to suppress low shear viscosity to some extent, but are less efficient than DGBE. All these approaches add to the cost of the paint formulation.

New HEUR Associative Thickeners

Since the cost of solvents and surfactants used to suppress low shear viscosity are especially high and their adverse effects on dry film properties are especially troublesome in low pigment volume concentration, high volume solids paints such as latex enamels and latex maintenance coatings, a new HEUR thickener was designed to cope with these problems. The thickener was balanced to provide high shear viscosity similar to that of the original products, but the hydrophobic modification was adjusted to give much lower association, thereby reducing the low shear viscosity to provide more nearly Newtonian rheology. This thickener, sold as Experimental Thickener QR-1001, has been successfully used for several years in European latex enamels and in latex maintenance coatings worldwide. By simultaneously reducing its cost and improving its thickening efficiency, recent process improvements have made this type of thickener more suited for general use in latex paints in combination with the more strongly associating type, making it possible to eliminate the need for cosolvents and surfactants as low shear viscosity suppressants. This new thickener, Acrysol RM-1020, is designated as Thickener B in the figures and tables that follow.

In order to determine the rheological differences between the strongly associating and more weakly associating HEUR thickeners, and to assess the performance of combinations of associative thickeners with different rheology profiles, a series of experiments were conducted in which each was used to thicken water, latex, pigment, and a simplified paint containing all three.

Aqueous Phase Thickening. The thickeners were mixed with water and allowed to equilibrate for three days. The rheology profiles of the samples were then measured with a Carri-Med controlled stress rheometer. The results for thickeners A and B are shown in Figures 5 and 6, respectively. Thickener A is Newtonian in water up to a shear rate of about 300 sec^{-1}. The viscosity increases monotonically as the concentration of thickener A is increased. Above 300 sec^{-1} shear rate, however, the viscosity of the thickener A aqueous solutions drops off rather abruptly. Considering that viscosity increases in water solutions are solely due to networks of intermolecular self-association, it appears that in thickener A solutions the network is partially broken down at shear rates above 300 sec^{-1}.

Thickener B (Figure 6) is much less efficient than thickener A in thickening water over the shear rate range accessible with our viscometer, but is still about three times more efficient than a poly(oxyethylene) urethane block copolymer without significant hydrophobic groups (Figure 1). Thickener B is also Newtonian throughout the measurable shear rate range, and no breakdown in viscosity is observed as in the case of thickener A. Probably thickener B, with its lower efficiency at low to intermediate shear rates and lack of shear induced viscosity breakdown, does not generate the same high level of network structure that is observed and subsequently broken down in thickener A.

Latex Thickening. Thickeners A and B have also been used to thicken a 250 nm latex of the type used in the experiments of Thibeault et al (5,6). In this case, the latex-thickener mixtures were made at 25 percent volume solids of latex, and sodium dodecyl benzene sulfonate surfactant was added at 0.4 weight percent on latex solids to prevent bridging flocculation due to excessive thickener adsorption on the latex (6). The rheology curves for the thickener-latex blends can be seen in Figures 7 and 8.

Thickener A (Figure 7) in the latex system is about 25 times more efficient than thickener B at low shear rates. At aqueous thickener concentrations up to 4%, the thickener A-latex system is essentially Newtonian at low shear rates, and shear thins at higher shear rates. However, in the presence of the latex, the onset of shear thinning occurs at a decade lower shear rate and the decrease in viscosity is more gradual. The major difference in the composition of the system is that it now contains latex particles which act as additional sites on which the thickener molecule hydrophobes can adsorb. Therefore, the viscosity is now a function of the thickener-thickener aggregate network junctions, thickener-latex network forming junctions, and volume solids thickening from the latex. The pure volume solids contribution to thickening (in the absence of latex-thickener interactions) is very slight, since measurements of the 25 percent volume solids latex in water showed a relative viscosity of only 2.5. In the five percent thickener solution, since the latex has the effect of increasing the viscosity 10 times, there is clearly an interaction. Also, in some way the presence of the latex or the latex-thickener junctions has both reduced the shear rate at which viscosity begins to decline, and moderated the sharpness of that decline compared to the behavior of thickener A in water.

At 4 to 5% thickener A in the aqueous phase of the thickener-latex system the viscosity appears to increase at the lowest measurable shear rates. This type of behavior is generally attributed to flocculation in classical colloid rheology, but the associative thickeners may act as flocculants or dispersants, depending on the thickener, surfactant and cosolvent concentrations present in the latex system. Our experiments were designed to avoid the volume exclusion flocculation regions mapped out by Sperry et al (6), but this merits further investigation.

Thickener B (Figure 8) shows essentially Newtonian behavior across the measurable shear rate range in the thickener-latex system, as it does in the thickener-water system. The viscosity of the latex system is 15 to 20 times higher than in the thickener-water system. In contrast to the 2.5 times viscosity increase due to the latex solids in water, this increase

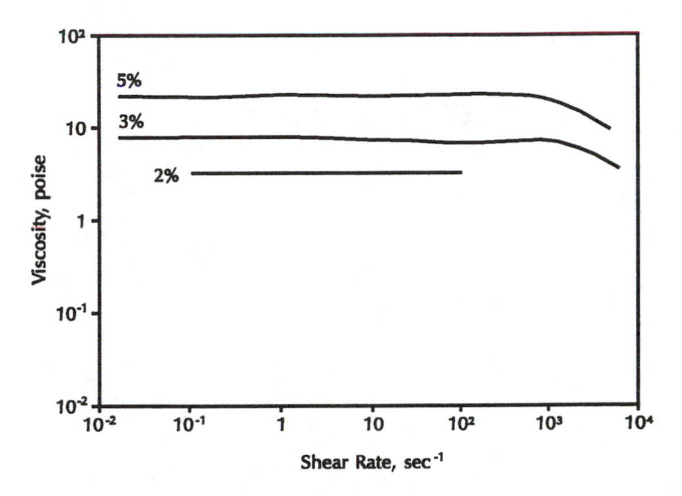

Figure 5. Effect of Thickener A concentration in water.

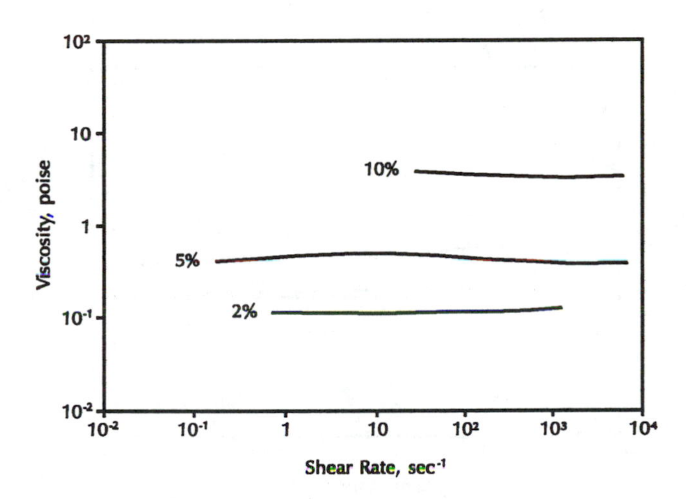

Figure 6. Effect of Thickener B concentration in water.

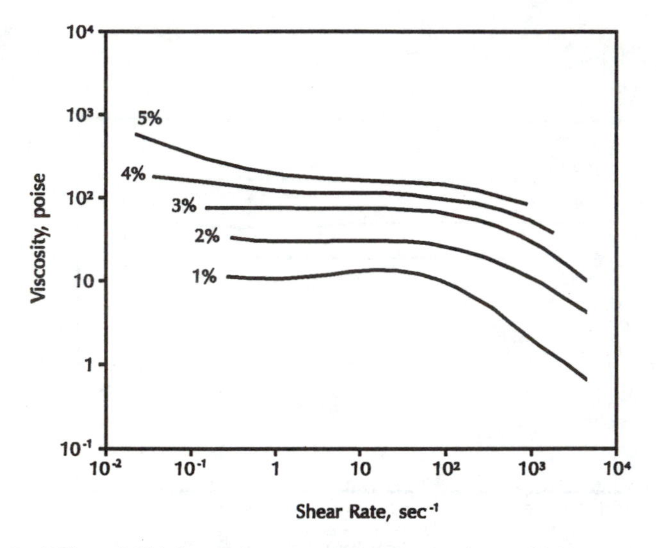

Figure 7. Effect of Thickener A concentration on rheology of 250 nm acrylic latex. (Thickener concentration expressed as weight of solid thickener on weight of aqueous phase.)

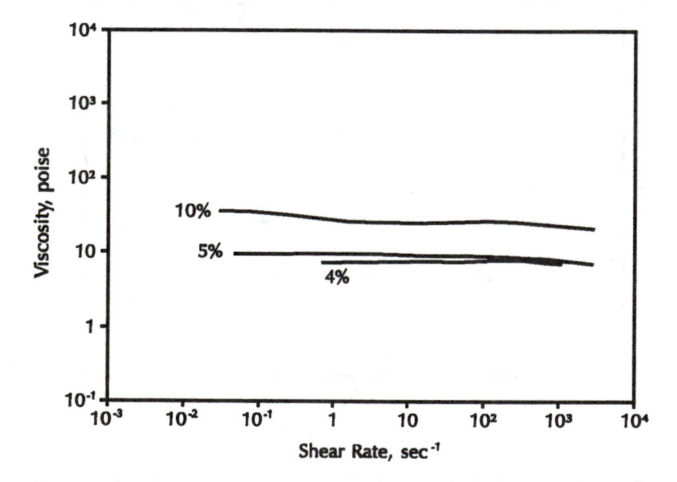

Figure 8. Effect of Thickener B concentration on rheology of 250 nm acrylic latex. (Thickener concentration expressed as weight of solid thickener on weight of aqueous phase.)

must also be due to thickener-latex interaction. However, this interaction has not altered the basic Newtonian behavior of thickener B.

Pigment Slurry Thickening. The two thickeners were also added to a pigment dispersion containing TiO_2 at 25 percent volume solids and a diisobutylene-maleic anhydride dispersant (Tamol 731) added at a level of one weight percent on pigment. The rheology curves for several concentrations of the thickeners in this system are shown in Figures 9 and 10. This system should be similar to the thickener-latex system in that the TiO_2 has a particle size of 200 to 250 nm. The viscosity of the unthickened system was 0.085 poise.

Each of the two thickeners exhibits very similar behavior in pigment and latex systems. However, practical paint experience has shown that the selection of dispersant can play a major role in the rheology of pigmented systems containing associative thickeners, and the dispersant used here is one that appears to enhance the pigment-thickener interaction. Thus, although the latex-thickener and pigment-thickener systems in this work show very similar results, we do not expect that this will always be the case as the pigment or dispersant is changed.

As noted earlier, an advantageous way to control the rheology of water borne paints is to use combinations of thickeners having different rheology profiles. The results presented above have detailed the rheology profiles of two HEUR associative thickeners in water, with latex, and with pigment. The latex and pigment dispersion were combined to create a model semi-gloss paint with 25 percent volume solids and 23 percent pigment volume concentration. The model paint consists of the same TiO_2, dispersant, latex, and surfactant used in the previous experiments. Portions of this model paint were thickened to a concentration of five weight percent thickener on continuous phase with thickener A and with thickener B. The thickened paints were then mixed so that the overall thickener level in the paint was maintained, but now with a 50/50 mixture of the two thickeners. The rheology curves for these three paints are shown in Figure 11. The two thickeners cause the paints to have rheology profiles that are predictable from the results in thickening the individual paint components. The model paint thickened with the blend of the two thickeners, as expected, has a rheology profile that is intermediate between those of the paints containing the individual thickeners. Note that in the latex, pigment dispersion and model paint systems, the viscosity of the systems containing thickener A show shear-thinning behavior above a shear rate of about 100 sec^{-1}, close to the shear rate range in which the commonly used Krebs-Stormer viscometer operates. Although we were unable to obtain good measurements for these model systems at 10,000 sec^{-1} with the Carri-Med viscometer, measurements on the ICI cone-plate viscometer showed the three model paints to be very similar in viscosity at this shear rate, generally considered to be representative of brushing shear rates.

Combinations of HEUR Thickeners

Since the new HEUR thickener B provides nearly Newtonian rheology in the shear rate range that is important to latex paint performance, its impact is seen primarily in the high shear viscosity. Although HEUR thickener A shows Newtonian behavior at low shear rates, it is decidedly shear-thinning in the shear rate range in which the common paint industry viscometers operate, i.e. from about 10 to 10,000 sec^{-1}. It is thus possible to adjust ICI cone-plate viscosity primarily with thickener B, and Krebs-Stormer viscosity primarily with thickener A (although thickener A will have some effect on ICI viscosity). This approach eliminates the need for cosolvent or surfactant to adjust the low shear viscosity, as would be the case if thickener A were used as the sole thickener.

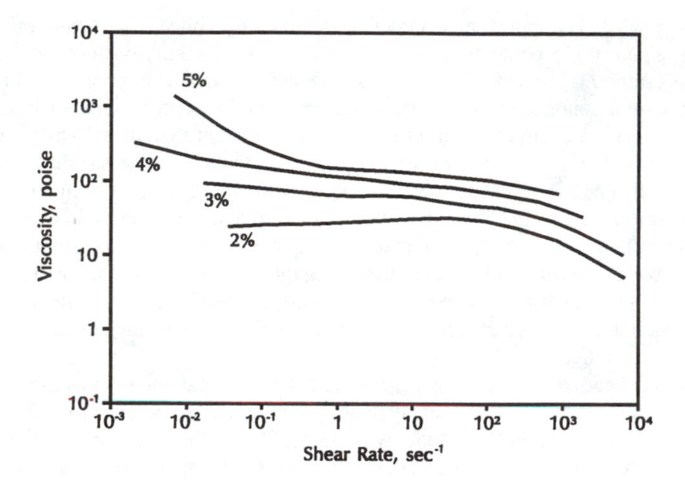

Figure 9. Effect of Thickener A concentration on rheology of 25% volume solids dispersion of enamel grade TiO_2, with 1% w/w diisobutylene/maleic anhydride dispersant solids on pigment solids. (Thickener concentration expressed as weight of solid thickener on weight of aqueous phase.)

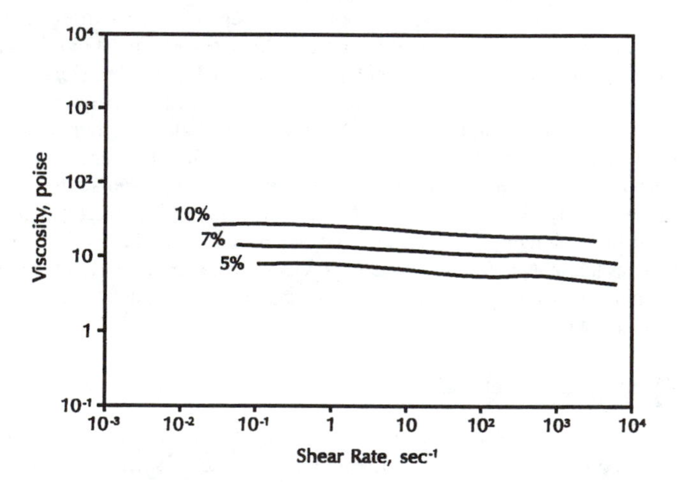

Figure 10. Effect of Thickener B concentration on rheology of 25% volume solids dispersion of enamel grade TiO_2, with 1% w/w diisobutylene/maleic anhydride dispersant solids on pigment solids. (Thickener concentration expressed as weight of solid thickener on weight of aqueous phase.)

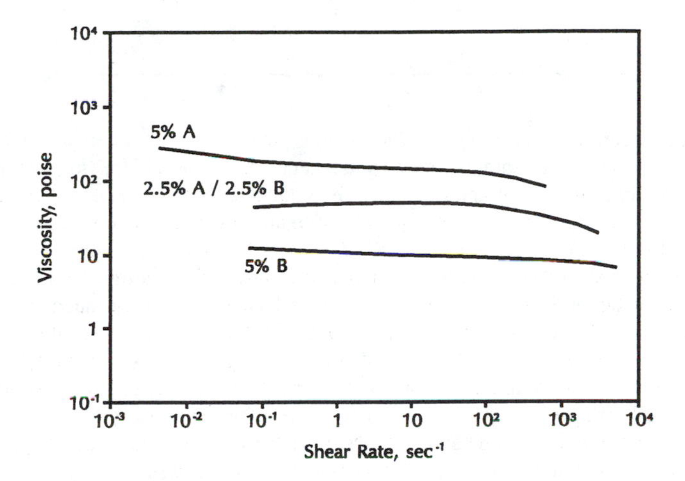

Figure 11. Comparison of model paints (25% volume solids, 23% pigment volume concentration) thickened with urethane thickeners A, B and a 50/50 combination of A and B. (Thickener concentration expressed as weight of solid thickener on weight of aqueous phase.)

Table I. Single HEUR thickener vs. HEUR thickener combination in an acrylic sheen paint

Composition and Properties	Paint 1	Paint 2
Thickener A, dry lb/100 gal	6.0	1.4
Thickener B, dry lb/100 gal	-	4.2
HEC, dry lb/100 gal	0.003	-
DGBE, lb/100 gal	31	-
Propylene glycol, lb/100 gal	-	33
Stormer viscosity, KU	90	89
ICI viscosity, Poise	1.5	1.5
Leveling	8+	8+
Sag	7	8+
Adhesion	9	9

Table I compares two acrylic sheen paints which differ only in the thickening strategy used. Paint 1 is formulated with thickener A, using DGBE to suppress the association so that enough of thickener A can be added to achieve an ICI cone-plate viscosity of 1.5 Poise; a very small amount of medium-high molecular weight HEC is added to control sag. Note that 31 lb. of DGBE per 100 gal. of paint is required to reduce the Krebs-Stormer viscosity to 90 KU. Paint 2 was formulated using a combination of thickener A and thickener B, and no DGBE (33 lb. propylene glycol was added per 100 gal. of paint to maintain freeze-thaw stability, but this has minimal effect on low shear viscosity). Note that the total of thickeners A and B used in paint 2 was slightly less than the amount of thickener A required in paint 1, and that better sag control was achieved without the need for HEC. This latter tendency has been noted in many formulations, and is likely due to the removal of the DGBE. Omission of the DGBE, plus the slightly lower total thickener level, resulted in some cost savings for paint 2, with even greater savings possible had the propylene glycol not been added.

Since the balance of low shear vs. high shear viscosity is affected by other formulation parameters than simply the choice of thickeners, the balance of thickener A vs. thickener B required to achieve any desired KU/ICI balance will depend on these considerations. Table II demonstrates the variation in total thickener level (A plus B) and percent thickener B required to hit a series of target viscosities ranging from 80 to 100 KU Krebs-Stormer viscosity and from 1.0 to 1.6 Poise ICI viscosity as the volume solids of the acrylic sheen paint is changed. Recall that thickener A is the more highly associative of the pair, and has its greatest effect at low shear rates. The lowest volume solids paints, with least surface area of dispersed solids available for association, require the greatest amount of thickener A. The higher the target Krebs-Stormer viscosity, the more thickener A required, all other things being equal. As the target ICI cone-plate viscosity is increased at a constant target Krebs-Stormer viscosity, more total thickener is required, and the percentage of thickener B required increases. Over the extremes of the ranges covered, the 30.4% volume solids paint formulated to 100 KU and 1.0 Poise ICI requires thickener A only, at 4.3 lb. per 100 gallons (2.4% on latex solids), while the 40.4% volume solids paint formulated to 80 KU and 1.6 Poise ICI requires only 3.7 lb. total thickener per 100 gallons (1.6% on latex solids), but 96% of that total is thickener B.

Table II. Thickener levels and ratios to achieve target rheologies in acrylic sheen paints

	1.0 Poise ICI		1.3 Poise ICI		1.6 Poise ICI	
	Total	% B	Total	% B	Total	% B
30.4% volume solids						
80 KU	5.4	52	7.1	68	9.1	80
90 KU	4.7	18	6.2	42	8.4	64
100 KU	4.3	0	5.5	16	7.7	47
35.4% volume solids						
80 KU	4.6	75	5.8	83	6.8	90
90 KU	4.3	56	5.4	67	6.3	76
100 KU	4.0	36	5.0	50	5.8	60
40.4% volume solids						
80 KU	2.8	88	3.2	99	3.7	96
90 KU	2.7	75	3.1	88	3.6	88
100 KU	2.6	59	3.0	80	3.6	80

Although cosolvents or surfactants are not required in order to adjust KU/ICI viscosity ratio when these two HEUR associative thickeners are used, system viscosity still is sensitive to such additives, and to the choice of latex. Four acrylic exterior house paint latices were screened in the sheen paint formulation at 37% volume solids in order to determine how much the total thickener level and ratio of thickener A to thickener B were affected. Target viscosities were 85 KU/1.3 ICI and 85 KU/1.6 ICI. The results presented in Table III show that the total thickener level varies over a threefold range, while the percentage of thickener B required varies from 63 to 92%, depending on the choice of latex and the desired high shear viscosity. The differences in the particle size and surfactant content of the latices are probably the reason for this effect, which is generally consistent with the practical experience gained with associative thickeners since their introduction.

Although use of the HEUR thickener pair provides the best application characteristics, thickener B can be used to adjust high shear brushing viscosity in paints where a conventional thickener is the primary contributor to low shear viscosity. Table IV

Table III. Effect of latex selection on thickener level and ratio in acrylic sheen paint

	1.3 Poise ICI		1.6 Poise ICI	
Latex	Total	% B	Total	% B
Chalk adhesion grade, 60.5%	4.4	63	5.1	65
High build chalk adhesion grade, 55%	2.7	73	3.3	86
High build grade, 50%	6.3	89	8.3	92
Small particle size grade, 46%	4.6	83	6.2	89

Table IV. Thickener B as a high shear viscosity improver with HEC in an exterior flat paint based on lobed latex

	HEC only	HEC + B	A + B
HEC, dry lb/100 gal	4.7	2.4	-
Thickener B, dry lb/100 gal	-	3.0	3.9
Thickener A, dry lb/100 gal	-	-	3.0
Stormer viscosity, KU	88	91	89
ICI viscosity, Poise		1.1	1.6
Film build	control	+18%	+22%
Leveling	4	6	8
Contrast ratio			
3 mil drawdown	0.97	0.97	0.97
brushout	0.92	0.96	0.96
Wet adhesion			
gloss alkyd	equal	equal	equal
chalky oil	equal	equal	equal

illustrates the use of thickener B to further increase the brushing viscosity of an HEC-thickened acrylic exterior flat house paint based on a novel lobed latex (12). The combination of HEC and thickener B can provide ICI cone-plate viscosity equal to that obtained with the combination of thickeners A and B, but not the leveling of the latter system. Nonetheless, the leveling is good enough to noticeably improve the hiding of the brushed coating, while adhesion is not adversely affected. This approach offers an attractive balance of cost and performance, especially in exterior flat paints, where leveling is not as critical as in interior paints.

In interior paints, thickener B can be used in conjunction with the lower cost hydrophobically-modified alkali soluble (HASE) or hydrophobically-modified hydroxyethyl cellulose (HMHEC) associative thickeners to obtain an attractive balance of film build, leveling and spatter resistance at costs competitive with conventional thickeners.

Summary

HEUR associative thickeners are strongly dependent on their specific hydrophobic adsorption characteristics in order to achieve efficient thickening and desirable rheology. Manipulation of these characteristics has allowed the design of HEUR thickeners which complement each other by providing varying degrees of shear thinning vs. Newtonian rheologies in the shear rate ranges of interest to the paint formulator. Combinations of these thickeners allow the formulator to achieve target high and low shear viscosities without the need for cosolvents or surfactants to alter the association characteristics of the thickeners. This has the advantages of simpler formulations, and often lower cost than the formulating techniques formerly employed.

Acknowledgments

We gratefully acknowledge the cooperation and assistance of Donald B. Larson and Jack C. Thibeault, who carried out much of the early work on applications of HEUR thickeners at the Rohm and Haas Company, in allowing us to quote from their unpublished work.

Literature Cited

1. Emmons, W. D.; Stevens, T. E. U.S. Patent 4 079 028, 1978.
2. Schimmel, K. F.; Christiansen, R.M.; Dowbenko, R. U.S. Patent 4 327 008, 1982.
3. Hoy, K. L.; Hoy, R. C. U.S. Patent 4 426 485, 1984.
4. Tetenbaum, M. T.; Crowley, B. C. U.S. Patent 4 499 233, 1985.
5. Thibeault, J.C., Sperry, P.R., and Schaller, E.J. In Water-Soluble Polymers; Glass, J. E., Ed.; Advances in Chemistry Series No. 213; American Chemical Society: Washington, DC, 1986; p 375.
6. Sperry, P.R., Thibeault, J.C., and Kostansek, E.C. In 11th Int. Conf. Org. Coatings Sci. and Tech., Proceedings; Patsis, A.V., Ed.; Adv. in Org. Coatings Sci. and Tech. Series No. 9; Technomic: Lancaster, PA, 1987; p 1.
7. Schaller, E. J. Surface Coatings Australia 1985, 22 (10), 6.
8. Glass, J. E.; Fernando, R. H.; England-Jongewaard, S. K.; Brown, R. G. J. Oil Colour Chemists Assoc. 1984, 67, 256.
9. Fernando, R. H.; Glass, J. E. J. Oil Colour Chemists Assoc. 1984, 67, 279.
10. Fernando, R. H.; McDonald, W. F.; Glass, J. E. J. Oil Colour Chemists Assoc. 1986, 69, 263.
11. Murakami, T.; Fernando, R. H.; Glass, J. E. J. Oil Colour Chemists Assoc. 1988, 71, 315.
12. Chou, C.S., Kowalski, A., Rokowski, J.M., and Schaller, E.J. J. Coatings Tech. 1987, 59, 93

RECEIVED August 22, 1990

Chapter 13

Steady-Shear and Linear-Viscoelastic Material Properties of Model Associative Polymer Solutions

R. D. Jenkins, C. A. Silebi, and M. S. El-Aasser

Department of Chemical Engineering and Emulsion Polymers Institute, Lehigh University, Bethlehem, PA 18015

Model associative polymers, whose structure consists of linear water–soluble poly(oxyethylene) backbones of number average molecular weight 16600–100400 with hydroxyl, dodecyl, or hexadecyl endgroups, are used to investigate the relationships between associative polymer structure and solution rheology. The steady shear viscosity profiles of solutions of model associative polymers with hexadecyl hydrophobic endgroups exhibit shear–thickening and shear-thinning regions that result from the extension and break-up of a dynamic association network under shear. These solutions are linearly viscoelastic with a well defined characteristic relaxation time constant, and exhibit an entanglement plateau in the storage modulus, which is interpreted as the pseudo–equilibrium modulus of the solution. The rheological properties of these solutions scale with a quantity we define as the molar density of association, which is calculated from the pseudo–equilibrium modulus of the solution. Physical interpretation of the rheological data yields a qualitative picture of the association mechanism.

Until the early 1980's, the predominant commercial rheology modifiers for latex paints were high molecular weight cellulose derivatives. However, undesirable application characteristics, such as roller spatter and the inability of cellulose derivatives to simultaneously produce the proper high shear and low shear viscosities in latex paints, prompted the development of associative polymers. Associative polymers improve the application characteristics of paints over those achieved through the use of cellulose derivatives by providing: better flow and leveling; and superior film build in pad, brush, or roller application; improved spray pattern and bubble release; reduced roller spatter; and the ability to simultaneously achieve these properties through the judicious use of surfactants and cosolvents. The literature contains a compilation of other advantages and comparisons between the performance of associative polymers and cellulose derivatives in paint formulations. (1-2)

0097–6156/91/0462–0222$06.00/0

As a first approach to understanding the association mechanism, researchers examined latex systems containing commercially available associative polymers. Unfortunately, the use of commercial associative polymers obscured the fundamental relationship between polymer structure and rheological properties because the polymers have a distribution of molecular weights and structures and are shipped dispersed in water/cosolvent mixtures. For example, Glass et al. (3-6) studied commercial polymers made by several different manufacturers to determine the effect of the following parameters on the rheological properties of coatings: polymer molecular weight and structure, latex type and size distribution, pigment volume concentration, and surfactant concentration. Their work focused on fully formulated coatings to highlight the numerous interactions possible among formulation components and polymers. Although they presented several complete viscosity profiles, they relied mainly upon changes in the low and high shear viscosities due to changes in dispersion formulation to monitor changes in the association network. In addition, the complexity of the interactions among the various formulation components and polymers hindered quantitative description of the relationship between polymer structure and coatings rheology.

Such studies on commercial associative polymers demonstrate the complexity of the technology, and motivate studies on model associative polymers to produce a fundamental understanding of the association phenomenon. In one of the first studies to use model polymers, Thibeault et al. (7) measured the adsorption isotherms and viscosity profiles in monodisperse latex dispersions of the following model polymers: a low molecular weight ethylene oxide based copolymer model associative polymer with large hydrophobe content; an ammonium salt of an anionic acrylic copolymer of high molecular weight with low hydrophobe content; and a (hydroxyethyl)cellulose non–associative polymer. They studied the effect of latex particle size, surfactant concentration, and cosolvent concentration, and observed that the desorption of the model polymer from the latex particle through the addition of surfactants and cosolvents coincided with a decrease in dispersion viscosity. They concluded that the associative polymer interacted with latex particles to physically cross–link the dispersion, and the shear–thinning viscosity profiles of the dispersion resulted from the disruption of the physical cross–links by shear forces. However, since they did not completely disclose the structure of the model polymer, no quantitative structure/property relationships could be drawn from the published data.

Realizing the need for studies on model polymers, several researchers used model associative polymers to examine the influence of polymer structure on solution properties. Bock et al. (8) synthesized copolymers of alkylacrylamide and acrylamide with weight average molecular weights of 3×10^6, and probed the effect of hydrophobe structure and electrolyte concentration on intrinsic viscosity and steady shear viscosity. Maerker and Sinton (9) used poly(vinyl alcohol) and polysaccharide with sodium borate at low pH as a model associating polymer system. They measured steady shear viscosity profiles and dynamic mechanical properties to investigate the flow structures induced by shear, and used nuclear magnetic resonance to determine number of association complexes in their solutions. Lundberg et al. (10) studied the viscoelastic properties of solutions that contained one of the following four hydrophobically modified urethane-ethoxylate model associative polymers: a linear polymer of number average molecular weight between 8700 to 24000 with either octyl, dodecyl, or nonylphenol hydrophobic endgroups, or a trimer with a number average molecular weight of 10500 and with nonylphenol hydrophobic endgroups. They concluded that linear polymers with large hydrophobes form an elastic network in solution. Some features seen in the rheological data presented by these studies, such as viscoelasticity and a maximum in the viscosity profile followed by shear thinning and elasticity, are qualitatively similar to features in our rheological data, which we discuss more fully later in this paper.

A review of the cited literature shows that a thorough investigation of the solution properties of model associative polymers of known structure, which vary systematically in hydrophobe length and molecular weight, should give considerable insight into the association mechanism. Our work studies the self–association mechanism, and the relationship between network dynamics and solution rheology, by measuring viscosity profiles and linear viscoelastic properties over a broad range of shear rates and frequencies. Once we understand the behavior of the model associative polymers in pure water, we can more easily understand the effect of surfactants, cosolvent, latex particles, and their complex interactions, on dispersion rheology.

Experimental Detail

To elucidate the self–association mechanism, we studied the rheological properties of aqueous solutions of model associative polymers, kindly supplied by Union Carbide Corporation (11). As shown in Table I, the model polymers' structure consists of linear water–soluble poly(oxyethylene) backbones of number average molecular weight 16600–100400 with hydroxyl, dodecyl, or hexadecyl endgroups. The hydroxyl terminated polymers serve as a control group by which we can measure the influence of the presence of linear alkyl hydrophobic endgroups on solution rheology. Intrinsic viscosity measurements of the model associative polymers corroborate the number average molecular weights listed in Table I, as we will report elsewhere. The notation C_{n-m}, where n represents the number of carbon atoms in the alkyl endgroup and m represents the polymer number average molecular weight in thousand Daltons, describes the structure of the polymer molecule. Stock solutions of 5% associative polymer by weight were prepared by adding a weighed amount of solid polymer to double distilled deionized (DDI) water which contained 5 ppm of hydroquinone inhibitor. The steady shear and oscillatory shear responses of solutions of

Table I: Structure and Nomenclature of Model Associative Polymers

Reference C_{n-m}	Structure§ R	Y	Calculated† Mole. Wt.
C_{12-17}	$C_{12}H_{25}$	2	17,400
C_{12-34}	$C_{12}H_{25}$	4	34,200
C_{12-51}	$C_{12}H_{25}$	6	50,700
C_{12-68}	$C_{12}H_{25}$	8	67,700
C_{12-85}	$C_{12}H_{25}$	10	84,500
C_{12-100}	$C_{12}H_{25}$	12	99,900
C_{16-18}	$C_{16}H_{33}$	2	17,500
C_{16-34}	$C_{16}H_{33}$	4	34,200
C_{16-51}	$C_{16}H_{33}$	6	51,000
C_{16-68}	$C_{16}H_{33}$	8	67,600
C_{16-84}	$C_{16}H_{33}$	10	84,300
C_{16-100}	$C_{16}H_{33}$	12	100,400
C_{0-17}	H	2	16,600
C_{0-33}	H	4	33,400
C_{0-50}	H	6	50,200
C_{0-67}	H	8	67,000
C_{0-84}	H	10	84,000
C_{0-100}	H	12	100,400

§ Structure is : R-O-(DI-PEO)$_Y$-DI-O-R, where DI is a diisocyanate, and PEO is CARBOWAX 8000 with a nominal number average molecular weight of 8200

†Number average molecular weight calculated from reaction stoichiometry

1% to 5% model polymer by weight were measured with a Bohlin VOR rheometer equipped with a set of Mooney–Couette concentric cylinders and a 1 degree angle cone and plate. Comparison of model solutions' rheological properties will help to elucidate the self–association mechanism.

Steady Shear Viscosity Profiles

Figure 1 displays viscosity profiles that are typical of aqueous solutions of model polymers with hexadecyl endgroups. In contrast to solutions of the polymers without hydrophobic endgroups, which have viscosities which are independent of shear rate for the polymer concentrations and molecular weights used in this study, the solutions of the model associative polymers with hexadecyl hydrophobic endgroups are Newtonian at low concentrations and become non–Newtonian as polymer concentration increases. The non–Newtonian viscosity profiles consist of a limiting low shear viscosity, a shear–thickening region at moderate shear rates, and a shear–thinning region at high shear rates, which is similar to the data of Bock et al. and of Maerker and Sinton for their model associating systems. Shear–thinning begins at the shear rate at which shear forces disrupt the association network. At the highest shear rates observed in this study, the high shear viscosities of solutions of model polymers with hydrophobic endgroups are still greater than those of solutions of the model polymers without hydrophobic endgroups, which indicates that the shear forces have yet to completely disrupt the hydrophobic intermolecular associations.

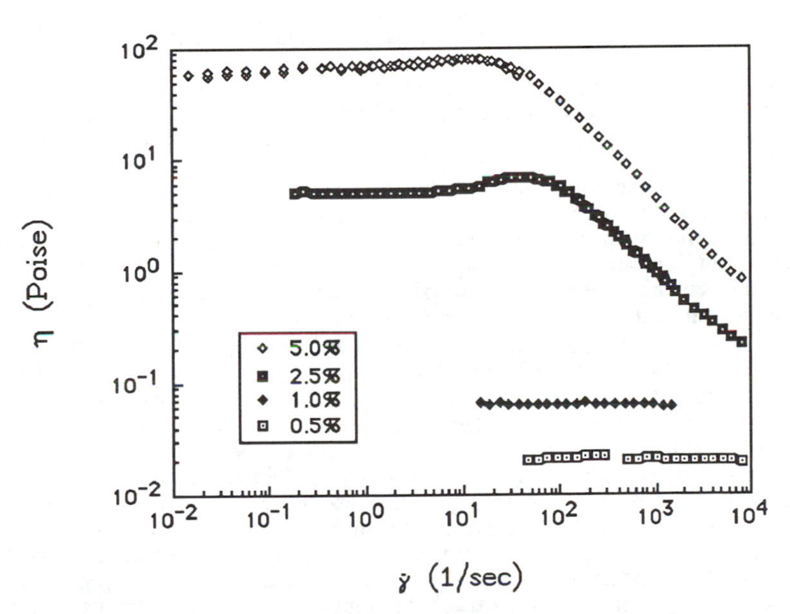

Figure 1: Steady shear viscosity profiles of C_{16}-100 model associative polymer in water at 24.5°C, which are representative of all solutions of model associative polymers with hexadecyl hydrophobic endgroups. Polymer concentrations are by weight.

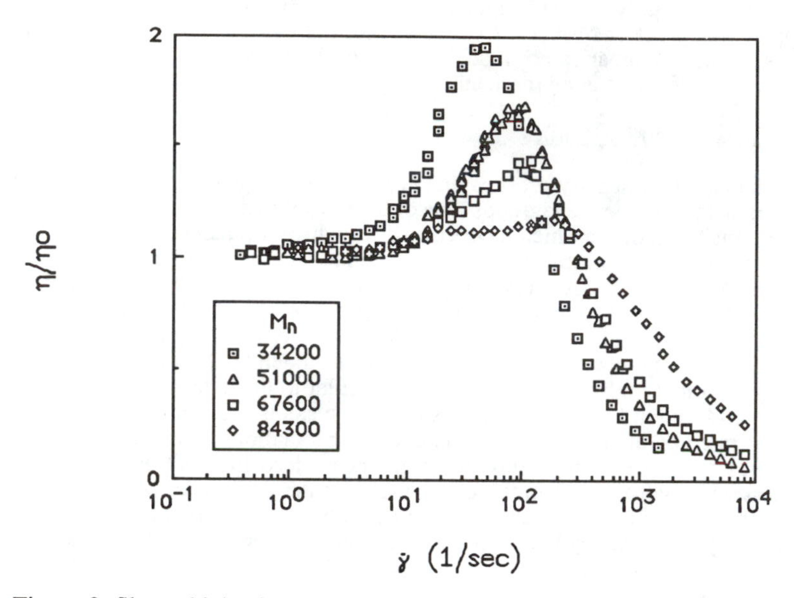

Figure 2: Shear-thickening steady shear viscosity profiles of 1% solutions of model associative polymers with hexadecyl hydrophobic endgroups, normalized against the low shear viscosity. Data taken at 24.5°C.

The lack of an apparent yield stress, which would cause the viscosity to become infinite as shear rate is decreased to zero, indicates that either the network is of finite extent or that the hydrophobic junctions which make up the network are constantly forming, breaking, and reforming by Brownian motion, or both. Otherwise, if the network junctions were permanent, and if the network was large enough to consider it a single cluster in solution, the solution would exhibit a yield stress. As in Witten and Cohen's theory on the shear-thickening behavior of ionomers, (12) a dynamic network can support stress and appear to have properties similar to those of permanent networks when the lifetime of an association junction is nearly as long as the hydrodynamic relaxation time of the association unit.

The shear–thickening region that occurs at a shear rate of nearly 10 sec^{-1} in the 2.5% and 5% curves of Figure 2 is not surprising, because shear–thickening occurs at nearly the same shear rate as in the data for model associating systems presented by Bock et al. and by Maerker and Sinton. The molecular network theory used by Vrahopoulou and McHugh (13) qualitatively predicts such a shear thickening region in the steady shear viscosity profile, and also predicts that the magnitude of viscosity enhancement in the shear–thickening region increases as the polymer molecular weight decreases, as seen in Figure 2. According to this theory, the shear–thickening region of the viscosity curve results from the extension of network chains under shear, which provides an additional energy dissipation mechanism. Since smaller chains reach the limit of their extensibility before longer chains, networks with smaller chains should exhibit larger increases in viscosity by this mechanism. The magnitude of the absolute increase in viscosity in the shear–thickening region shown in Figure 1 decreases with increasing polymer concentration, and, as Maerker and Sinton have also observed in their solutions, the shear rate at which shear–thickening begins decreases with increasing polymer concentration. Witten and Cohen attributed the shear–thickening viscosity, and the decrease in the shear rate required to induce it as polymer concentration increases, to a change from intramolecular to intermolecular association as molecules elongate under shear. According to this theory, many of the association junctions in a network at rest are intramolecular. During flow, shear forces stretch the polymer chains and break some of the intramolecular associations, allowing intermolecular junctions to form at the expense of intramolecular junctions, and enhancing viscosity without increasing the total number of association junctions. Since the shear rate required to produce a given degree of elongation decreases as solution viscosity increases, the shear rate at which shear thickening begins decreases as solution viscosity increases. Likewise, the magnitude of shear thickening decreases as solution viscosity increases because the relative number of intermolecular to intramolecular associations increases.

Figures 3 and 4 illustrate how the low shear viscosities of the model polymer solutions depend on polymer concentration, molecular weight, and the length of the alkyl endgroups. The low shear viscosities, obtained by using least squares regression to extrapolate the viscosity profiles to zero shear rate, depend upon the amount of network structure in the solution at rest. The inverse viscosity–molecular weight relationship shown by solutions of the model polymers with hydrophobic endgroups in Figures 3 and 4 becomes more pronounced as polymer concentration increases: the low shear viscosity depends upon polymer concentration as c^{β}, where β is greater than 1 and increases with increasing ratio of alkyl chain length to polymer molecular weight. Although the diisocyanates that connect the poly(oxyethylene) units in the model polymers' backbone are slightly hydrophobic, the primary hydrophobicity of the model polymer comes from its alkane endgroups. Thus, the viscosity behavior depicted in Figures 3 and 4 indicates that the driving force for intermolecular association increases with increasing length of the alkane endgroup relative to the hydrophilic polymer backbone, and that interaction between the hydrophobic endgroups of the polymer molecules increases the apparent molecular weight of the polymer in solution.

Viscoelastic Material Properties

Figure 5 presents storage $G'(\omega)$ and loss $G''(\omega)$ moduli typical of concentrated solutions of our model polymers with hexadecyl hydrophobes. All of the solutions were tested in the region of linear viscoelasticity, where the measured properties did not depend upon the amplitude of the oscillating shear strain. Hence, the measurements did not significantly perturb the network from its equilibrium rest state, and did not change the number of effective physical crosslinks in the association network. At low frequencies, $G'(\omega)$ has a slope of 2 and $G''(\omega)$ has a slope of 1 on a doubly logarithmic plot, which indicates that the response is viscous since the complex viscosity tends toward the zero shear limiting viscosity in this region. At higher frequencies, $G'(\omega)$ crosses $G''(\omega)$ to reach an entanglement plateau, and $G''(\omega)$ passes through a maximum. Maerker and Sinton observed a similar plateau in $G'(\omega)$ between 6 and 300 rad/s for their solutions due to entanglement coupling. Thus, the range of frequencies used in our experiments encompasses the longest relaxation times in the relaxation spectrum of our solutions. This viscoelasticity results from the association between the hydrophobes, since solutions of model

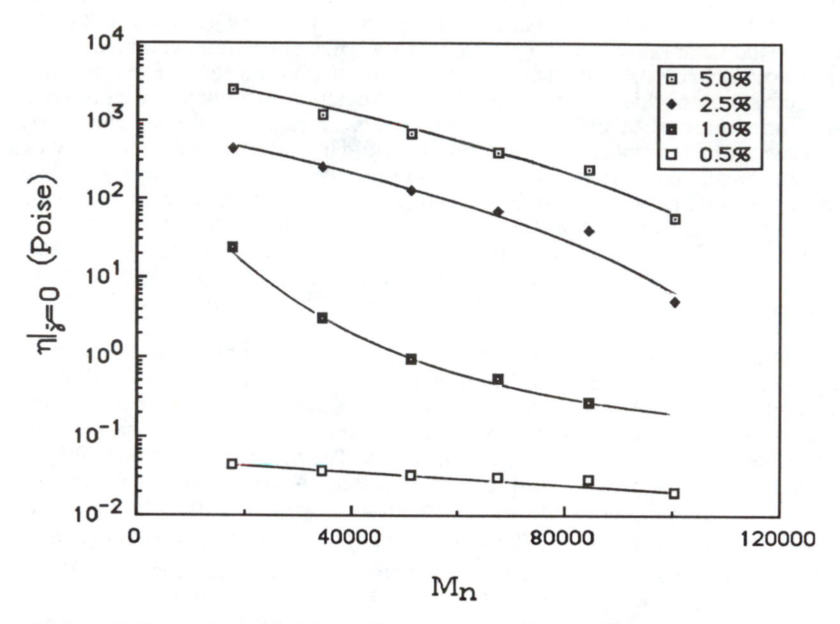

Figure 3: Low shear limiting viscosities of solutions of model polymers with hexadecyl hydrophobic endgroups.

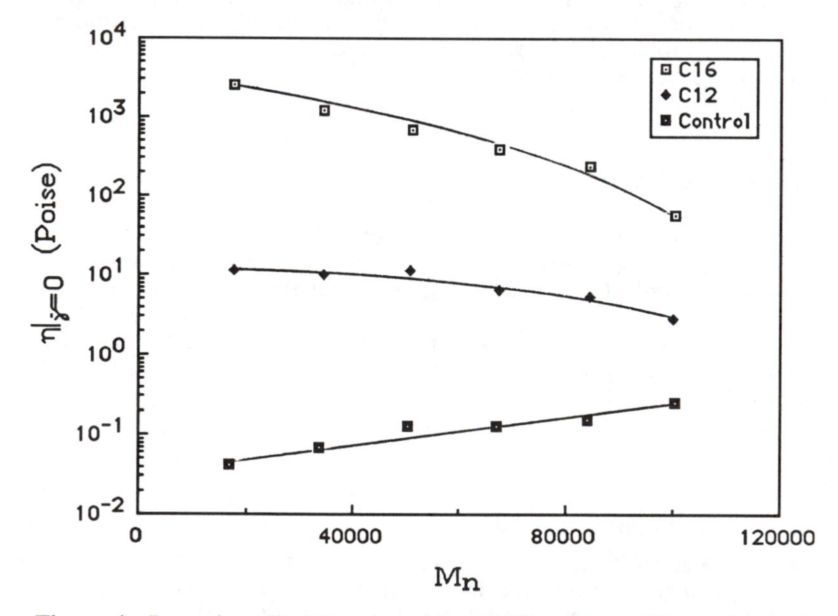

Figure 4: Low shear limiting viscosities of 5% solutions (by weight) of model associative polymers with alkyl endgroups (C16 or C12). The control polymers do not have alkyl endgroups.

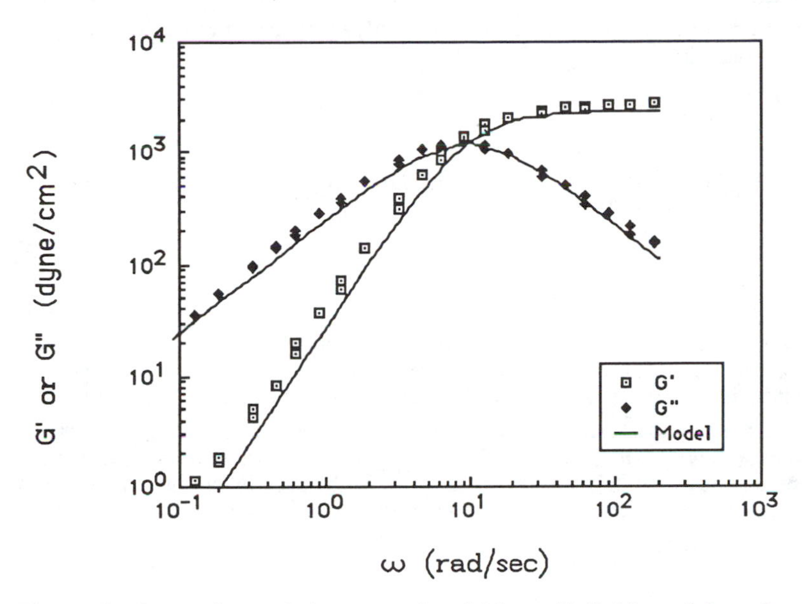

Figure 5: Comparison of the corotational Maxwell fluid model to the complex shear moduli of 2.5% solutions of the C_{16-34} model associative polymer in water at 24.5°C.

polymers without hydrophobic endgroups of the similar molecular weight and concentrations lack viscoelasticity. As indicated by the theory of Witten and Cohen, the hydrodynamic relaxation time measured by this method is characteristic of the lifetime of the association junction because the association network acts as a single rheological unit.

Since, as a first approximation the relaxation spectrum $H(\tau)$ equals $\frac{2}{\pi}G''(\omega)|_{\tau=1/\omega}$, the maximum in $G''(\omega)$ results from the longest relaxation times of the relaxation spectrum associated with the network, which we can use to calculate ν, the average molar density of network junctions by association. (14) In analogy to rubber elasticity, the pseudo–equilibrium modulus (G^o_{en}) is:

$$G^o_{en} = \nu RT = \frac{2}{\pi}\int_{-\infty}^{a'} G''\, d\ln(\omega) \tag{1}$$

where R is the gas constant, T is the temperature in Kelvin, and a', the frequency at the upper limit of integration, is chosen to include the maximum in longest times of the relaxation spectrum within the range of integration.

To determine the pseudo–equilibrium modulus, we can integrate Equation 1 graphically from the data, or analytically by either fitting the data with a empirical function or a constitutive equation. In this work, we use the corotational Maxwell constitutive equation (15), which appro· nates the relaxation spectrum by a relaxation time constant λ, so that we can physically interpret the result of the integration, viz.:

$$G^{o}_{en} = \frac{2\eta_o}{\lambda\pi}\int_{o}^{a}\frac{d(\lambda\omega)}{1+(\lambda\omega)^2} = \frac{2\eta_o}{\lambda\pi}\tan^{-1}(\lambda\omega)\Big|_{o}^{a} = \frac{2\eta_o}{\lambda\pi}\tan^{-1}(\lambda a) = \frac{\eta_o}{\lambda} = \frac{1}{J'} \qquad (2)$$

In Equation 2, $a'=\ln(a)$, η_o is the low shear limiting viscosity, and the relaxation time constant λ is the frequency at which $G'(\omega)$ equals $G''(\omega)$. As shown in Figure 5, the corotational Maxwell fluid model describes the oscillatory shear response well. The last simplification in Equation 2 results because the upper limit of the frequency range that encompasses the maximum in $G''(\omega)$ in the data is so large that the inverse tangent is always within 6% of $\pi/2$. The physical interpretation of this result is that the plateau value of the storage modulus equals the pseudo–equilibrium modulus for these solutions. In general, both the zero shear limiting viscosity and the relaxation time constant depend upon ν.

Discussion

The viscoelastic response and the enhanced viscosity of solutions of polymers with hydrophobic endgroups, as compared to solutions of polymers without hydrophobic endgroups, suggest that the contribution to network junctions by chain entanglement is small relative to that by association. Hence, the pseudo–equilibrium modulus of an associative polymer solution essentially measures the molar density of network junctions formed by association. As shown in Figure 6, when the viscosities presented in Figure 3 for all six model polymers with hexadecyl hydrophobes at various concentrations are plotted against the molar density of association junctions, a single line with a slope of 4/3 results. This correlation between rheological parameters and the molar association density indicates that the morphology, size, and strength of the association junctions, rather than traditional volume exclusion mechanisms, determine the rheological properties of associative polymer solutions.

The pseudo–equilibrium modulus is usually interpreted in terms of an apparent molecular weight between entanglement points: $\nu = G^{o}_{en}/RT = c/M_e$, where M_e is the apparent molecular weight between entanglement points and c is the associative polymer mass concentration. (16) Hence, the ratio $\bar{c}/\nu$, where $\bar{c}$ is the molar concentration of model associative polymer, is the dimensionless ratio of the apparent molecular weight between junctions to the polymer molecular weight, M_e/M_n. Thus M_e/M_n is 1 or greater for an association network. As shown in the Figure 7 for solutions of model polymers with hexadecyl hydrophobes, M_e/M_n is greater than one and decreases as concentration increases. This implies that several associative polymers must cooperate to span the distance between hydrophobic junctions.

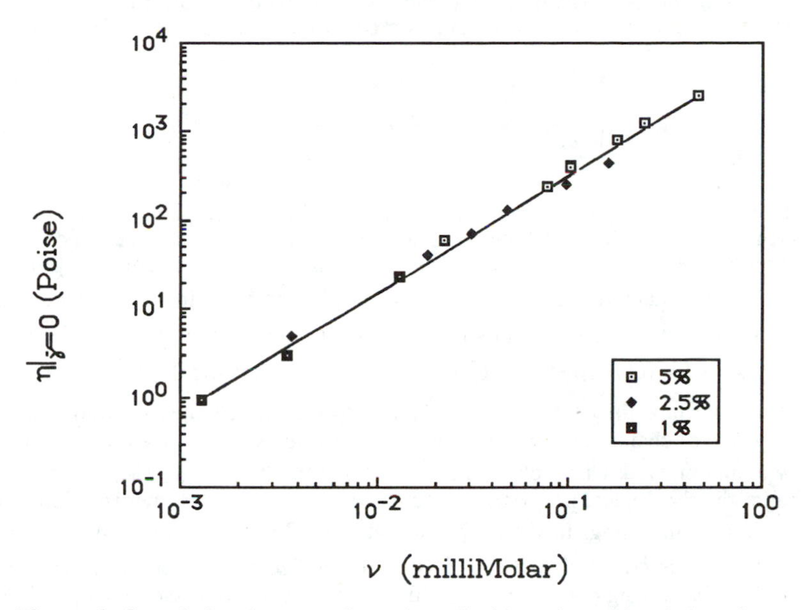

Figure 6: Correlation between low shear limiting viscosities (taken from Figure 3) for solutions of model associative polymer with hexadecyl hydrophobic endgroups and the molar density of association.

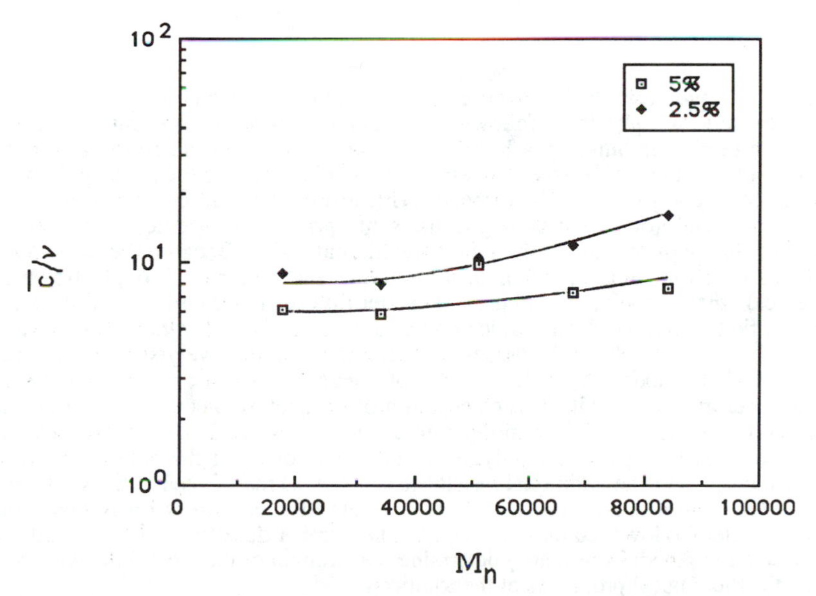

Figure 7: Number of model associative polymers between network junctions for solutions of model associative polymers with hexadecyl hydrophobic endgroups.

The rheological data of many polymers often exhibit a power law dependence on concentration and molecular weight, as predicated by the de Gennes c* theorem: (17)

$$\eta \sim [\bar{c}/c^*]^X \;,\; \lambda \sim [\bar{c}/c^*]^Y \qquad (3)$$

where c* is the overlap concentration. Experimental evidence demonstrates that the exponent X can be as high as 4, but is usually near 3.4 for most polymers. Following Equation 3, we regressed the rheological properties of the concentrated polymer solutions to an equation of the form: $\eta,\nu,\lambda = c^a M^b$, where c is in wt%. Selecting the results with best regression statistics yielded: $\eta = c^{3.36}/M^{1.67} = (c/\sqrt{M})^{3.36}$; $\nu = c^{2.47}/M^{1.24} = (c/\sqrt{M})^{2.47}$; and $\lambda = c^{0.88}/M^{0.44} = (c/\sqrt{M})^{.88}$, which are self–consistent because they satisfy $\eta = \nu^{4/3}$, and $\eta = \nu\lambda$. de Gennes showed that in a dense system of chains, such as found in a concentrated solution, the polymer chains follow a random walk where $c^* \sim M^{-1/2}$, so that $[\bar{c}/c^*] \sim (c/\sqrt{M})$. Hence, the scaling of the rheological properties of the model associative polymer solutions with $(c/\sqrt{M})$ is not surprising. In Figure 7, the ratio M_e/M scales as $(c/\sqrt{M})^{-.6}$, which indicates that the correlation length ξ (i.e., mesh size of the network), decreases rapidly with increasing concentration. This is comparable to de Gennes's prediction that the correlation length $\xi \sim [\bar{c}/c^*]^{-3/4}$. Therefore, the strength and completeness of the network decreases as molecular weight increases.

Conclusions

Physical interpretation of the rheological properties of aqueous model associative polymer solutions suggests the following qualitative association mechanism. At rest, Brownian processes build up a highly entangled association network, where the primary source of entanglement comes from associations among the model polymers' hydrophobic endgroups. Nevertheless, intramolecular associations between hydrophobic endgroups that belong to the same polymer molecule, which do not contribute to the overall strength of the network, can exist. Because the association junctions are dynamic (i.e., continuously building and rupturing through Brownian processes), the network is not an infinite cluster that spans the entire solution, but is instead a finitely sized hydrodynamic unit whose lifetime is longer than the relaxation time of the solution. Since the association junctions are relatively strong and long-lived, small to moderate small amounts of shear stress can extend the network structure before rupturing it. This extension promotes intermolecular associations to form at the expense of intramolecular associations, and also decreases the configurational entropy of the polymer chains that construct the network system. Both effects provide an additional resistance to flow, as manifested in the solutions' shear-thickening viscosity profiles. Larger levels of shear stress break down the network system to lower solution viscosity. The number density and functionality of the association junctions primarily determine the strength of the overall network, and hence, the rheological properties of the solutions.

Literature Cited

1. Schwab,F.G., In Water Soluble Polymer: Beauty with Performance; Glass, J.E., Ed.; Advances in Chemistry Series No. 213; American Chemical Society: Washington, DC, 1986; p. 369.
2. Croll,S.G.; Kleinlein,R.L., In Water Soluble Polymer: Beauty with Performance; Glass, J.E., Ed.; Advances in Chemistry Series No. 213; American Chemical Society: Washington, DC, 1986; p. 333.
3. Glass,J.E., In Water Soluble Polymer: Beauty with Performance; Glass, J.E., Ed.; Advances in Chemistry Series No. 213; American Chemical Society: Washington, DC, 1986; p. 391.
4. Glass,J.E.; Fernando,R.H.; Egland–Jongewaard,S.K.; Brown,R.G., J. Oil Colour Chem. Assoc. 1984, 67 No. 10, p.256.
5. Fernando,R.H.; Glass,J.E., J. Oil Colour Chem. Assoc. 1984, 67 No. 11, p.279
6. Fernando,R.H.; McDonald,W.F.; Glass,J.E., J. Oil Colour Chem. Assoc. 1986, 69 No. 10, p.263.
7. Thibeault,J.C.; Sperry,P.R.; and Schaller,E.J., In Water Soluble Polymer: Beauty with Performance; Glass, J.E., Ed.; Advances in Chemistry Series No. 213; American Chemical Society: Washington, DC, 1986; p. 375.
8. Bock,J.; Siano,D.B.; Valint,P.L.; Pace,S.J., Proc. of the ACS Div. of Polymeric Mater: Sci. and Eng., 1987, 57, p.487.
9. Maerker,J.M.; Sinton,S.W., J. Rheol. 1986, 30(1), p.77.
10. Lundberg, D.J.,Glass,J.E.,Eley,R.R., Proc. of the ACS Div. of Polymeric Mater: Sci. and Eng., 1989, 61, p.533.
11. D. Bassett, Union Carbide Corp., private communication
12. Witten,T.A.; Cohen,M.H., Macromol., 1985, 18,p. 1915.
13. Vrahopoulou,E.P., and McHugh,A.J., J. Rheol., 1987, 31(5),p. 371.
14. Ferry,J.D., Viscoelastic Properties of Polymers ; Wiley: New York, 1970; p.404.
15. Bird,R.B.; Armstrong,R.C.; Hassager,O., Dynamics of Polymeric Liquids, Volume 1: Fluid Mechanics; Wiley and Sons: New York, 1977; p.355.
16. Graessley, W.W., The Entanglement Concept in Polymer Rheology; Springer-Verlag: Berlin, 1974; p.58.
17. de Gennes, P.G., Scaling Concepts in Polymer Physics, Cornell University Press: Ithaca , 1979; p.54.

RECEIVED August 22, 1990

Chapter 14

Surfactant Influences on Hydrophobically Modified Thickener Rheology

David J. Lundberg[1], Zeying Ma, Karu Alahapperuna[2], and J. Edward Glass

Polymers and Coatings Department, North Dakota State University, Fargo, ND 58105

The commercial acceptance of most hydrophobically modified, water-soluble polymers has been in formulations containing soluble surfactants and latex dispersions. The influence of surfactants and, to a limited extent, latices are examined in this chapter. The role of excess "free surfactant" in the continuous phase is considered in the achievement of lower viscosities at low shear rates in dispersions containing latices less than 150 nm in diameter. The role of excess free surfactant is also considered in the achievement of higher viscosities at high shear rates and in the viscoelastic behavior of different hydrophobically modified, water-soluble polymers.

A variety of hydrophobe-modified, water-soluble polymers have been described previously (*1*). Only four of these have found commercial acceptance: hydrophobically modified hydroxyethyl cellulose (HMHEC; Chapters 18, 19, and 27 in ref. 1); hydrophobically modified ethoxylated urethanes (HEUR; Chapters 26 and 27 in ref. 1 and the preceding chapters in this volume); hydrophobically modified, alkali-swellable emulsions (HASE; Chapters 25, 27, and 28 in ref. 1 and Chapter 7 in this volume); and hydrophobically modified poly(acrylic acid) gels (Chapter 7 in ref. 1 and Chapter 6 in this volume). The hydrophobically modified gel is used in sun-screen products, and the others have found application in water-borne coatings. In coatings, HMHEC, HEUR, and HASE rheology modifiers function in the presence of several distinct phases and in

---

[1]Current address: Consumer Video and Audio Division, 3M Company, 236–2D–88, 3M Center, Saint Paul, MN 55144

[2]Current address: 3M Wahpeton Plant, 3M Company, 1300 3M Drive North, Wahpeton, ND 58075

0097–6156/91/0462–0234$06.00/0

the presence of one or more soluble surfactants. Surfactant influences on the viscosity of associative thickeners have not been studied extensively (*2, 3,* and Chapters 18 and 26 in ref. 1) and is the subject of this chapter. Surfactants are discussed in the context of their role in the performance of coatings containing small (100-nm) latices, in the viscoelastic response of HEUR solutions alone and in the presence of model latices, and in the context of surfactant influences on the viscosity response of coatings at high shear rates. All three influences have contributed to the commercial acceptance of associative thickeners in water-borne latex coatings.

Experimental Details

The HEUR associative thickeners discussed in this chapter have been previously described (*3, 4*), as have the procedures and materials used in model latex syntheses (*5*) and the rheological measurements (*6*).

Surfactant Influences

I. 100-Nm Latex Dispersions

Low-Shear-Rate Viscosities. Cellulose ethers became universal thickeners for water-borne latex coatings when the latter achieved commercial realization in the early 1960s. At that time, the use of small-particle-size latices was a general goal of formulators, because small-particle binders provide greater film integrity by increasing the pigment binding properties of the coatings (*7, 8*). This advantage has been outweighed by a number of disadvantages. The undesirable rheology of small-particle (i.e., less than 150 nm) latex coatings thickened with cellulose thickeners (Chapter 26 in ref. 1) has been one of the factors resulting in the use of larger particle binders over the past two decades. For example, with latex particles less than 150 nm, lower viscosities at high deformation rates result in thinner films that effect both poor substrate coverage and poor flow and leveling. Higher viscosities at low shear rates also are observed with small-particle latices, compounding their poor application performance. The commercial success of larger particle latices is reflected in a recent contribution by the Montreal society in which two large-particle (>500-nm) commercial latices, representative of large-sale-volume binders, were studied. Associative thickeners of the hydrophobe-modified ethoxylated urethane (HEUR) type were concluded to be cost ineffective relative to cellulose ether thickeners (*9*). This is consistent with previous studies from our laboratories (*10*) when large-particle binders are used. Associative thickeners provide desirable rheology to small-particle latex coatings (*10* and Chapter 26 in ref. 1), and the mechanism is addressed below.

Characterization of Model Latices. In the production of small and intermediate-size, all-acrylic latices, it is a common practice to incorporate a small percentage of acrylic or methacrylic acid in the synthesis of the latex (*11*); this improves the stability of the latex and is reflected in mechanical and freeze–thaw stability of the dispersion. The presence of chemically attached oligomeric acid (resulting from the incorporation of a small amount of methacrylic acid in the emulsion polymerization synthesis) on the latex surface will result in displacement of some of the synthesis surfactant, which will reside in the bulk phase (*5*) as "free surfactant." The free-surfactant concentrations of model MMA latices, except with the 220-nm, acid-monomer-free MMA latex (Table I), are higher than the critical micelle concentration (CMC) of the surfactant, indicating saturated (*12*) latices surfaces. The 220-nm, acid-monomer-free latex adsorbed additional nonionic surfactant (*5*) (i.e., 1.24×10^{-5} mole of $OP\text{-}(OCH_2CH_2)_{10}OH$ per gram of polymer), indicating that the synthesis surfactant covers only 70% of the surface. The free-surfactant concentration for the model latices and a commercial latex, typical of that used in the early 1960s, are given in Table I. The model latices were monodisperse. The surfactant used in both syntheses contained aromatic units, and their concentrations in the aqueous phase were determined by UV spectroscopy, as was the level of free surfactant in the commercial latex.

Table I. Characteristics of Model Latices

Latex	*D (nm)*	*Surface Acids*[a] *(meq/m²)*	*Surfactant (wt %) in aqueous phase (32% NVV)*
MMA/MAA (98/02)[b]	103	0.00077	0.068
MMA/MAA (96/04)[b]	97	0.00222	0.278
MMA/MAA (100/00)[b]	220	0.00028[c]	ND[d]
MMA/MAA (96/04)[b]	220	0.00655	0.145
MMA/EA/MAA[e] (39/60/01)[f]	117	0.00112	0.45

[a]Conductometric titration.

[b]Synthetic method previously described (*5*). A nonylphenol with an average of 20 oxyethylene units and a terminal sulfate group, $NP(OCH_2CH_2)_{20}SO_4^-NH_4^+$, was employed. A more hydrophobic surfactant analog, $NP(OCH_2CH_2)_9SO_4^-NH_4^+$ (3 wt % based on monomer), was used to achieve a 100-nm-particle-size latex (*13*).

[c]Acid group arises from catalyst fragments.

[d]ND: not detectable (i.e., <0.01%). The surface is not saturated with surfactant, and there is no free surfactant.

[e]Rhoplex AC234, a commercial latex.

[f]Determined by NMR.

Model Latex/Hydroxyethyl Cellulose Thickener Dispersions. The model latices were thickened at three concentrations with hydroxyethyl cellulose (HEC, M_v = 700,000). Cellulose-ether-thickened, 100-nm-particle latices, at a 0.32 volume fraction, are higher in viscosity at low shear rates than the 220-nm latices, and the difference increases with increasing HEC concentration

(Figure 1). The effective volume fraction of dispersions varies with particle size due to surface hydration. This has been defined in elegant dimensional analysis studies (*14*) of latex dispersions. The surface areas of 1 cm^3 of the monodisperse 100- and 220-nm latices are 60 m^2 and 27.3 m^2, respectively. An increase in the effective radii of latex particles due to the adsorption of surfactant and subsequent hydration has been estimated (*14, 15*) to be 5 nm. If a 5-nm-thick hydration layer is assumed, the effective volume fractions of a formulated 0.32 volume fraction of 100- and 220-nm latices are 0.42 and 0.36, respectively. This is a conservative estimate, considering the oligomeric surface acid segments on the latex's surface. The low-shear-rate viscosity of cellulose-thickened (1.0 wt %), 0.32 volume fraction, 100-nm MA/MMA/MAA(72/26/02) latex dispersion was not sensitive to salinity (0.17 M NaCl), indicating that the low-shear viscosities are dependent primarily on hydration, not electroviscous effects. Hydroxyethyl cellulose (HEC) thickens through its hydrodynamic volume, related to both the amount used and the polymer's molecular weight. This can be estimated through the root-mean-square, end-to-end distance of the thickener (*16, 17*). These estimates, combined with the effective volume fraction of the different size latices, produce very crowded dispersions (Table II), accounting for the notable viscosity increases observed in Figure 1.

Table II. Contribution of Hydroxyethyl Cellulose Diameter to 0.32 Dispersion Volume Fraction

Latex Size Fraction (nm)	*Effective Volume Fraction*	*Wt % HEC in Aqueous Phase*	*Estimated HEC* $[\langle r^2 \rangle^{0.5} = 23\ nm]$ *Volume Fraction*	*Total Effective Volume*
100	0.42	0.50	0.15	0.57
		0.90	0.27	0.69
		1.41	0.42	0.84
220	0.36	0.51	0.15	0.51
		0.89	0.26	0.62
		1.33	0.39	0.75

Associative Thickener/Model Latice Dispersions. In considering the influence of HEUR thickeners on the viscosity of small-particle latices, the uniformity of the dispersions must be considered. The two commercial HEURs that have achieved the greatest success (708 and 270) often exhibit syneresis in aqueous solutions, even in the absence of a dispersed phase. In model HEURs, synthesized in these laboratories, this type of behavior is noted in HEUR structures with limited solubility and with those exhibiting intramolecular hydrophobic bonding.

The second factor to consider in the uniformity of the dispersion is the free-surfactant level present in the continuous phase after synthesis of the latex. For example, the amount of surfactant used in the acid-monomer-free, 220-nm latex (i.e., the one synthesized without methacrylic acid) is insufficient to cover the surface of the final latex. The hydrophobes of the HEUR thickener would be expected to be adsorbed on the surface of such latices, perhaps interbridging particles, and in equilibria, these hydrophobes may interact also with the latices in which only small amounts of free surfactant are present. The addition of surfactant effects free-flowing dispersions in any of the thickened dispersions exhibiting phase separation or gel tendencies.

The viscosities at low shear rate of the 100- and 220-nm latices with increasing associative thickener concentration follow the same trend noted in the HEC dispersions; the small-particle dispersions are higher in viscosity. With the HEUR thickener, the low-shear-rate viscosity is influenced also by the amount of latex surface acid (*5*) (Figure 2). The differences in viscosity of associative thickener dispersions can arise also from other factors. For example, in this comparison, differences in the synthesis of the latices must be considered. A nonylphenol with an average of 20 oxyethylene units with a terminal sulfate group ($NP(OCH_2CH_2)_{20}SO_4^-NH_4^+$) was employed (*5*). Increasing the amount of this surfactant within a reasonable concentration range resulted in only a moderate decrease in latex particle size (170 to 190 nm) in the MMA/MAA(96/04) dispersions. It was necessary to increase the hydrophobicity (i.e., lower the degree of oxyethylene units to an average of nine $NP(OCH_2CH_2)_9SO_4^-NH_4^+$) and to increase surfactant concentration to 3 wt % based on monomer to achieve a 100-nm-particle-size latex (*13*). These differences result in a higher concentration of free surfactant with the 100-nm-particle MMA/MAA(96/04) latex, and this more-hydrophobic surfactant is more effective in increasing the viscosity of HEUR solutions. The greater efficiency of a more-hydrophobic nonionic surfactant (*2, 18*) in increasing the low-shear-rate viscosity of HEUR solutions also has been observed. Thus, the hydrophobicity of the surfactant and its concentration in the aqueous phase and the difference in surface area, with associated hydration layers effecting a greater effective volume fraction, are important parameters in defining the difference in low-shear-rate viscosities between the 100- and 220-nm latices.

Effect of Surfactant Addition. The viscosity of an associative thickener solution is very sensitive to free-surfactant concentration (*3, 19*). The low-shear-rate viscosity of thickener solutions exhibits a maximum that is dependent on surfactant type and concentration. With a model HEUR thickener, $NPIP(EtO)_{595}IPNP$ (Scheme 1 in Chapter 26 of ref. 1), the maximum is observed at approximately 1 critical micelle concentration (CMC) of a classical anionic surfactant, sodium dodecyl sulfate (SDS), and at approximately 130 CMC of a nonionic, branched surfactant, β-$C_{13}H_{27}(OCH_2CH_2)_9OH$. At first consideration, this difference appears to be related to the same number of micelles (*3*); this, however, may be superficial since the associative thickener may effect different micelle shapes with different surfactants. Viscosity maxima with commercial HEUR thickeners and the nonionic surfactant (β-$C_{13}H_{27}(OCH_2CH_2)_9OH$) are observed at surfactant concentrations of 0.03% (HEUR-200 4%), 0.3% (HEUR-270 2%), and 0.8% (HEUR-708 4%) (*19*).

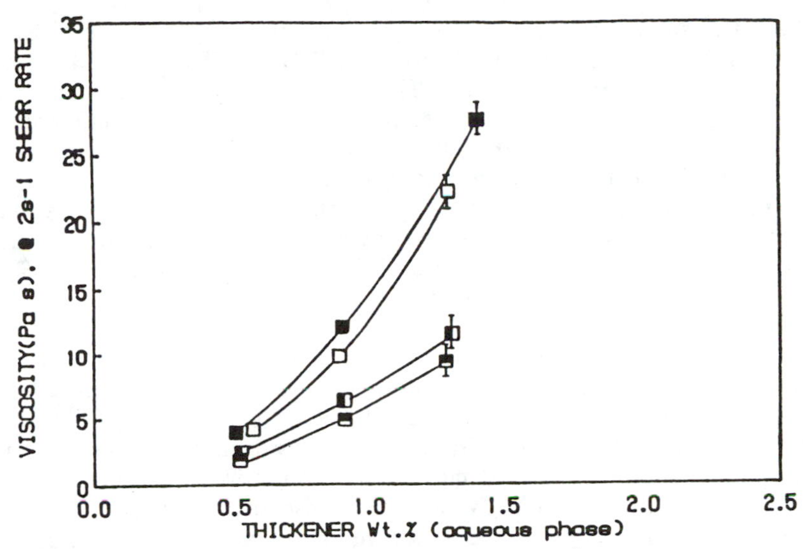

Figure 1. Dependence of low-shear-rate (2 s^{-1}) viscosity (Pa s) on hydroxyethyl cellulose (M_v = 700,000) concentration in the aqueous phase of 0.32 volume fraction latex dispersion. Latices are defined as follows:

■, 97 nm, MMA/MAA(96/04) with 0.00222 meq/m^2 surface acid
□, 103 nm, MMA/MAA(98/02) with 0.00077 meq/m^2 surface acid
◧, 220 nm, MMA/MAA(96/04) with 0.00655 meq/m^2 surface acid
⬒, 220 nm, MMA/MAA(100/00) with 0.00028 meq/m^2 surface acid

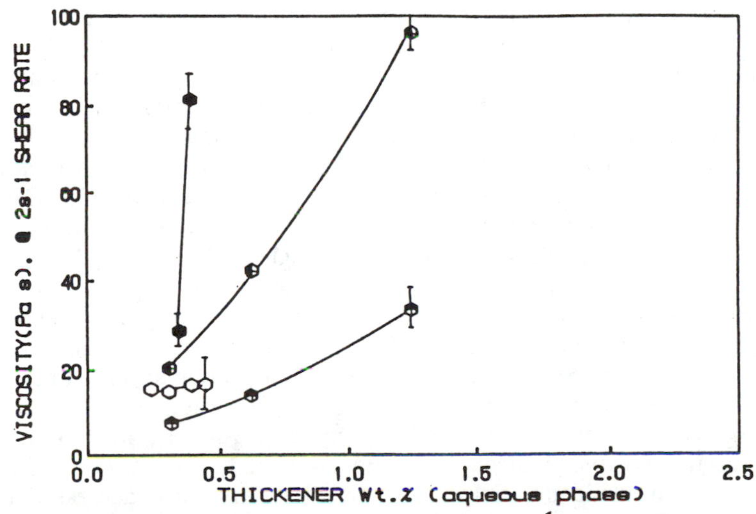

Figure 2. Dependence of low-shear-rate (2 s^{-1}) viscosity (Pa s) on HEUR-708 concentration in the aqueous phase in 0.32 volume fraction latex dispersions. Latices are described in the caption for Figure 1.

The free-surfactant concentrations of the 100- and 220-nm MMA/MAA(96/04) latices are 0.278 wt % $NP(OCH_2CH_2)_9SO_4^-NH_4^+$ and 0.145 wt % $NP(OCH_2CH_2)_{20}SO_4^-NH_4^+$, respectively. These values are much higher than the CMC of the surfactants (0.01 wt %, a difference in CMC due to a difference in the amount of oxyethylene units is not indicated by the supplier), indicating surface saturation of the latices (*12*). At the thickener concentration in the disperse-phase studies (<1.0%) a maximum associative thickener/surfactant viscosity would be expected at a free-surfactant concentration below 0.3% for HEUR-270 and below 0.8 wt % in a mixed, ethoxylated anionic/nonionic surfactant combination for HEUR-708.

The addition of 0.6 wt % surfactant (the concentration of nonionic surfactant used in a fully formulated coating) increases the surfactant concentration in the aqueous phase to a level of 0.9 wt % for the 100-nm latex. This is beyond that of the viscosity maxima of associative thickeners. Thus, in the presence of excess surfactant, the associative thickener network effecting a large hydrodynamic volume and viscosity would be disrupted, explaining the achievement of the lower viscosities at low shear rates with the 100-nm latex in a fully formulated coating.

II. Viscoelastic Responses

The most advertised "asset" of associative thickeners is their ability to reduce spatter when coatings are applied by roll. This behavior has been related to dynamic uniaxial extensional viscosities (DUEVs) (*6, 20*). In cellulose-ether-thickened paints, the DUEVs are greater than can be attributed to the thickener alone and are due to flocculent structures presumably with high aspect ratios. Although different deformation processes are involved, the elastic functions that reflect elasticity (i.e., DUEV, storage modulus, first normal stress difference, and die swell) under steady-state conditions have been qualitatively related. In this section some of these parameters are measured for HEUR–surfactant solutions.

The viscosity maxima in SDS/HEUR-270 solutions occur over a broad SDS concentration and at surfactant concentrations below 0.75 CMC of SDS (*21*). The addition of SDS to HEUR-270 solutions at 1.5 wt % effects a significant elasticity, as denoted by the higher DUEVs and die swells observed in the extensional measurements; these increases below the c^*_h (the concentration at which a notable increase in viscosity is observed due to hydrophobic associations) of HEUR-270 arise from intermicellar networks facilitated by SDS micelles. From these observations, one would expect a shear-thickening behavior in HEUR-270 solutions at 1.5% with increasing SDS concentration; this is observed (Figure 3), for reasons best defined in the poly(vinyl alcohol) borate study (*22*). At 0.75 CMC, both the shear and dynamic uniaxial extensional viscosities have decreased to the value of HEUR-270 without SDS. The low-frequency oscillatory data also reflect this transition and are discussed below for HEUR-200 dispersions.

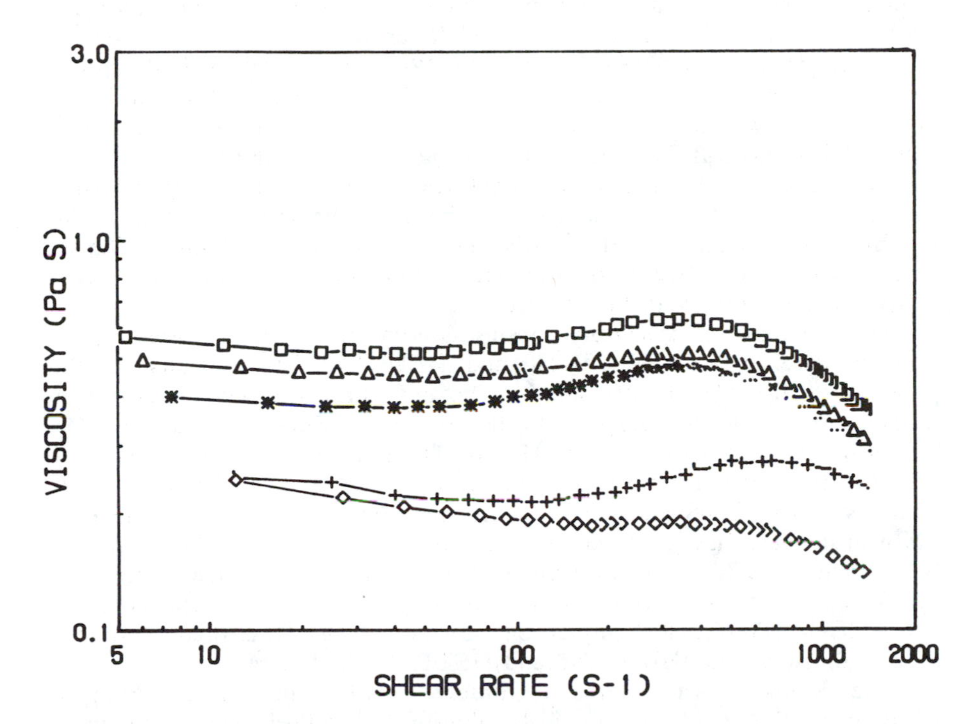

Figure 3. Dependence of viscosity on shear rate of a 1.5 wt % HEUR-270 solution with different critical micelle concentrations (CMCs) of sodium dodecyl sulfate (SDS). Symbols: ◇, no SDS added; Δ, 0.15 CMC SDS; □, 0.35 CMC SDS; *, 0.50 CMC SDS; and +, 0.75 CMC SDS. (Reproduced with permission from reference 21. Copyright 1991 Wiley.)

Latex Influences. The nature of HEUR thickener response in oscillatory studies in the presence of model latices can be similar to that observed in surfactant solutions or vary dramatically. For example, the responses are similar in the oscillatory behavior of HEUR-708 below its c^*_h; the dominant response in both surfactant solutions (*21*) and in acid-stabilized latex dispersions is the loss modulus, G″. The model, di-end-capped NPIP(EtO)$_{595}$IPNP HEUR (Figure 4a), of molecular weight approximately equal to that of HEUR-708, exhibits the opposite behavior: the dominant response in both SDS and nonionic surfactant solutions (*3*) and in dispersions containing the acid-stabilized model latex (*4*) is the storage modulus, G′, which denotes a high elastic character. With this model HEUR, both a moderate extensional response and a die swell are observed also despite its low molecular weight, while neither is observed with HEUR-708. A second model, (NP(EtO)$_{100}$IPDI)$_3$ HEUR (Figure 4e), in which the terminal hydrophobes are separated by only 100 oxyethylene units from a large internal hydrophobe, exhibits a dominant loss modulus in the presence of the acid-stabilized latex, similar to HEUR-708. Due to syneresis and other interactions of this model HEUR in SDS and nonionic-surfactant solutions, oscillatory measurements were not undertaken.

The difference in oscillatory response among surfactant and latex dispersions is observed with HEUR-200 and HEUR-270. For example, in the presence of SDS, HEUR-270 solutions do not exhibit the dominant storage response noted in the presence of the oligomeric-acid-stabilized latex (*5*). This disparity is observed also in HEUR-200/surfactant solutions (Figure 5). Without the latex, the dominant response at low frequencies is the loss modulus in both SDS and nonionic-surfactant solutions. In the presence of the acid-stabilized latex, the dominant response is the storage modulus (i.e., the dispersion is highly elastic) unless there is an excess of free surfactant (*5*). When a moderate level of surfactant is present with the acid-stabilized latex, HEUR-200-thickened dispersions are nonviscous. This dramatic difference is illustrated for HEUR-270 and HEUR-708 in Figure 6.

The strong elastic response observed with the linear dicapped NPIP(EtO)$_{595}$IPNP HEUR in SDS solutions is reflective of a network structure in the solution or dispersion; any extensive structure would reflect itself in a yield stress behavior, which is observed (*21*). The commercial HEUR polymers (with internal hydrophobes) do not exhibit yield stress behavior.

III. Viscosity at High (10^4 s^{-1}) Shear Rates

As noted earlier, the achievement of higher viscosities at high (10^4 s^{-1}) shear rates is an important contribution of associative thickeners to improved coatings performance. It contributes directly to hiding through a greater film thickness and indirectly to the flowout of the applied film through an inverse dependence of flow and leveling behavior on film thickness (*23*).

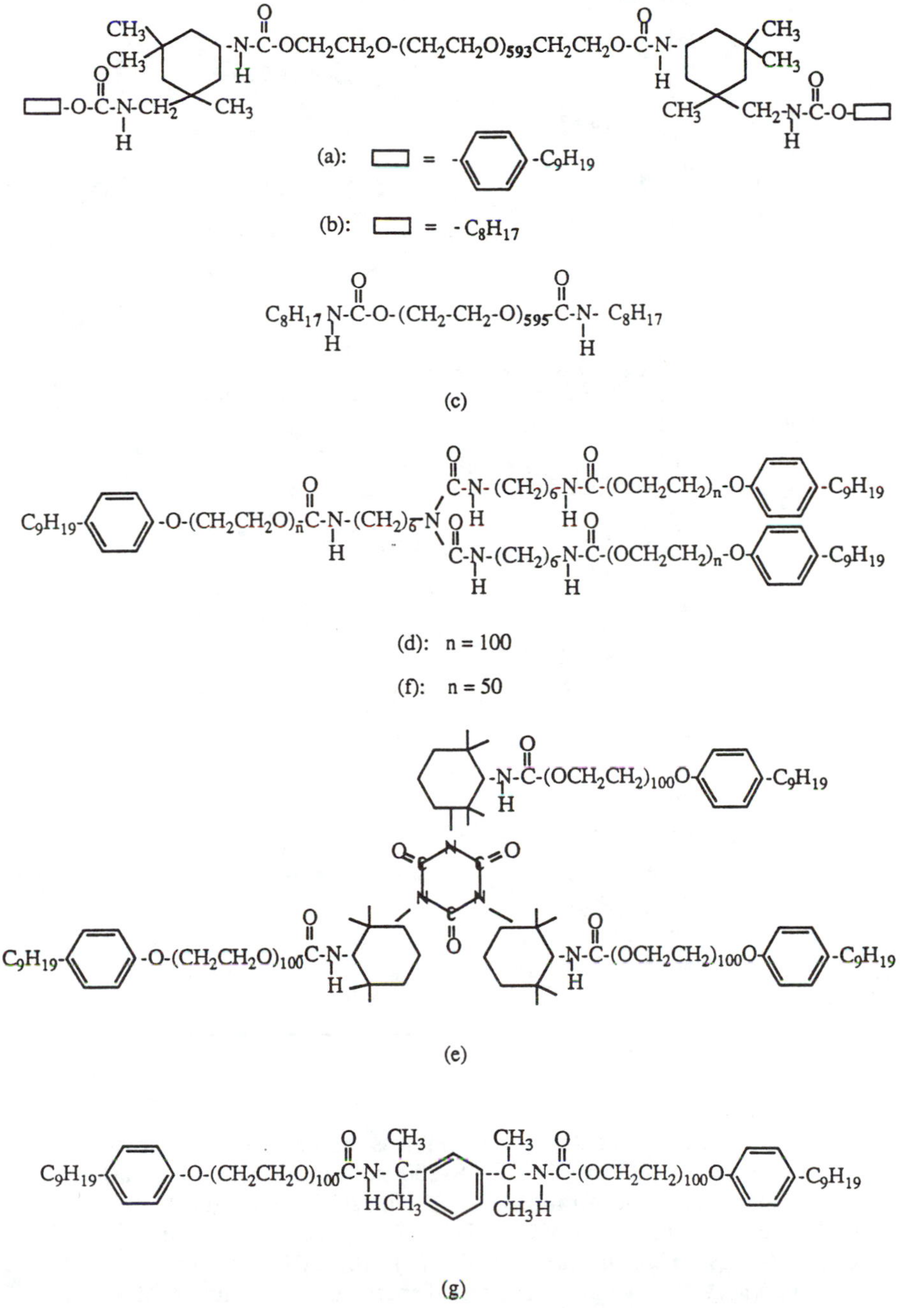

(a): ☐ = -C_6H_4-C_9H_{19}

(b): ☐ = -C_8H_{17}

C_8H_{17}-NH-C(=O)-O-$(CH_2$-CH_2-$O)_{595}$-C(=O)-NH-C_8H_{17}

(c)

(d): n = 100

(f): n = 50

(e)

(g)

Figure 4. Selected models of hydrophobically modified ethoxylated urethane structures used in high-shear-rate-viscosity studies (*see* Figure 7).

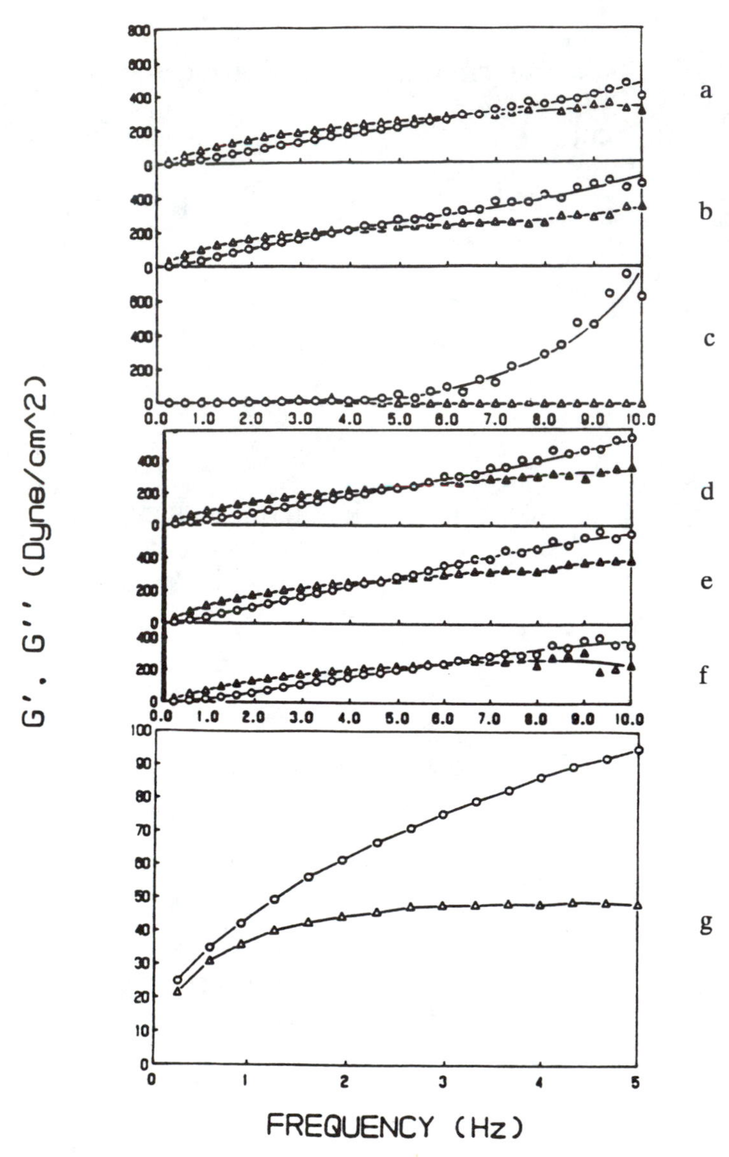

Figure 5. Storage modulus (G′) and loss modulus (G″) dependence on frequency of 4.0 wt % HEUR-200 solutions containing different surfactants and concentrations. (a) No surfactant added; (b) 0.50 CMC SDS; (c) 1 CMC SDS; (d–f) concentrations of β-$C_{13}H_{27}(OCH_2CH_2)_9OH$ are 1 CMC (d), 10 CMC (e), and 100 CMC (f); and (g) a 0.32 volume fraction dispersion of 220 nm MMA/MAA (98/2) latex containing 0.00084 meq/m^2 surface acid at pH = 9 with 0.138 wt % free surfactant in the aqueous phase. HEUR-200 concentration in the aqueous phase is 1.6 wt %. Symbols: ○, storage modulus; Δ, loss modulus.

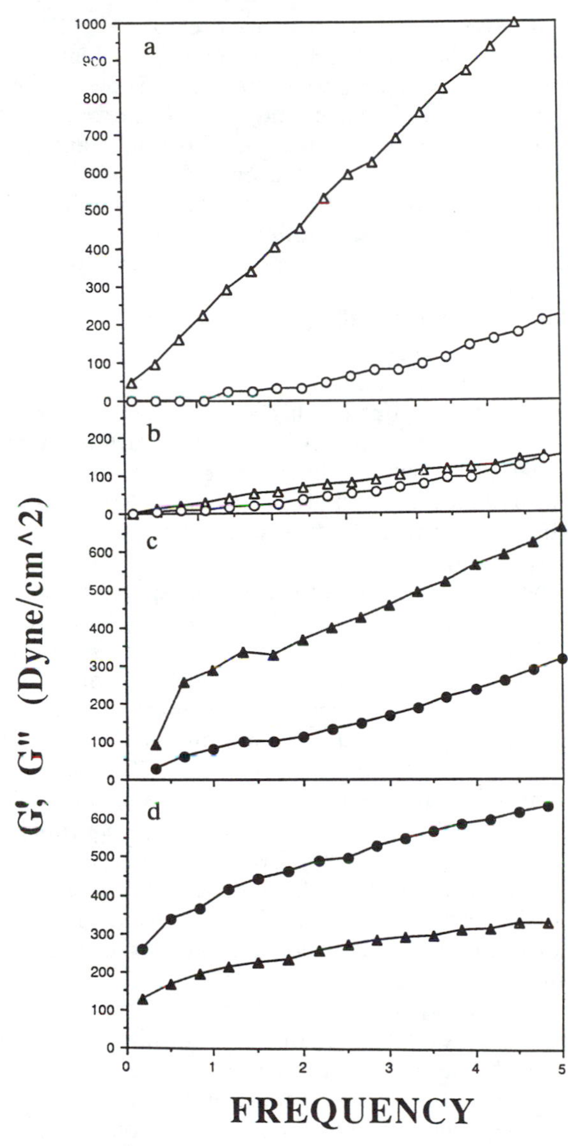

Figure 6. Dependence of storage modulus (G′, ○) and loss modulus (G″, Δ) on frequency of commercial HEUR polymers in the presence of sodium dodecyl sulfate (SDS, open symbols) and a 0.32 volume fraction dispersion of 220-nm MMA/MAA (96/4) latex containing 0.00648 meq/m^2 surface acid at pH = 9 with 0.138 wt % free surfactant in the aqueous phase (closed symbols). (a) HEUR-708, 4 wt %; SDS, 5.0 CMC (1.15 wt %). (b) HEUR-270, 1.5 wt %; SDS, 0.35 CMC (0.080 wt %). (c) HEUR-708, 0.75 wt %, 0.138 wt %; $NP(OCH_2CH_2)_{20}SO_4^-NH_4^+$. (d) HEUR-270, 0.5 wt %, 0.138 wt %; $NP(OCH_2CH_2)_{20}SO_4^-NH_4^+$.

Influence of HEUR Architecture. The importance of HEUR-hydrophobe molar concentrations in obtaining higher viscosities at high deformation rates was emphasized earlier (Chapter 26 in ref. 1). Three model HEURs were examined, the two discussed above and one synthesized by the reaction of *m*-tetramethylxylene diisocyanate with a 100-mole ethoxylated hydrophobe (structure g in Figure 4). This synthesis bypassed the insolubility obtained with hydrophobe capping of lower molecular weight polyethylene glycol diisophorone precursors and also produced a diblock secondary model of the trimer obtained by reaction of the same ethoxylated hydrophobe with the isocyanurate of isophorone (structure e in Figure 4).

The absence of a center hydrophobe resulted in very high dispersion viscosities at low shear rates (Figure 12 in Chapter 26 in ref. 1), higher than a commercial hydroxyethyl cellulose with M_v = 700,000. The essential feature of the HEUR polymers effecting higher viscosities at high deformation rates (Figure 13 in Chapter 26 in ref. 1) in the previous study was the presence of an internal hydrophobe. In chapter 12 of this volume, high viscosities at high shear rates are achieved in model HEURs containing multiple internal toluene–diurethane-linked hydrophobes by decreasing the size of the terminal hydrophobes.

To extend the above studies, the influence of an additional series of model diblock and trimer HEUR thickeners on the high-shear-rate viscosities were examined. When examined as a function of HEUR concentration, it is evident (Figure 7) that decreasing the terminal hydrophobe size in model HEURs with molecular weights approximately equal to those described in the previous chapter, but without internal hydrophobes, does not produce higher viscosities at high shear rates. The small, terminal-hydrophobe-modified poly(oxyethylene)s are in fact very poor viscosifiers, and significant amounts are required, approximately equivalent to the molecular weight of unmodified poly(oxyethylene) (*23*). All of the above studies reflect the influence of hydrophobe contact residence times on the fluid's response at high deformation rates. This, however, is not the only factor; the conventional low-molecular-weight surfactant contributes to the high-shear-rate viscosity.

Influence of Surfactant. The extent of multiple associative thickener hydrophobe participation in the formulation of surfactant micelles should be a function of the ratio of these two components. The phenomenon is apparent in the following study. The addition of a model latex, a methyl methacrylate/methacrylic acid (96/04) dispersion thickened with a hydrophobe-modified styrene/maleic acid poly(electrolyte) terpolymer, decreases the low-shear-rate viscosities at low thickener concentrations, but increases in low-shear-rate viscosity are observed at these higher surfactant levels when the thickener concentration is increased (Figure 8). There is a critical ratio of surfactant to associative thickener that provides the highest viscosity and a change in the concentration of either component results in a lower viscosity at low shear rates. The same phenomenon is observed in HEUR-708 dispersions (*13*).

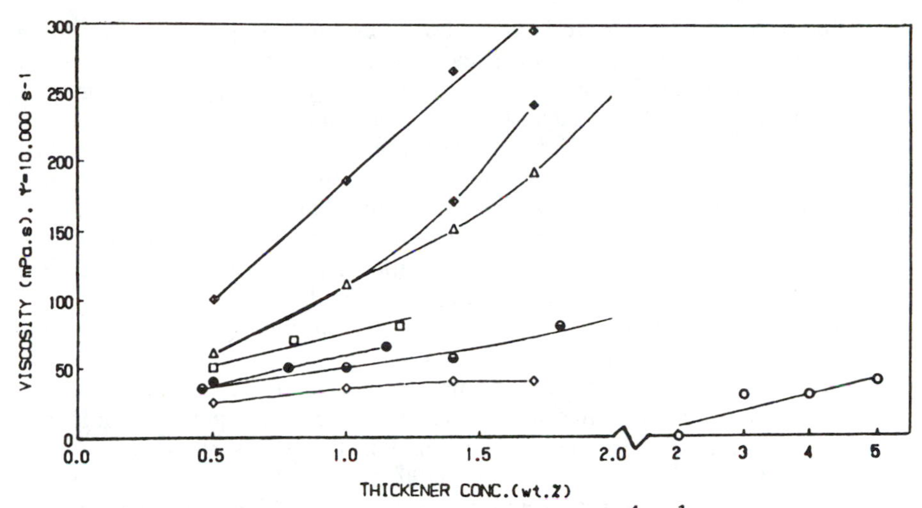

Figure 7. Dependence of viscosity (mPa s) at 10^4 s^{-1} on weight percent of model HEUR compositions in aqueous solutions. Thickeners are utilized in a latex coating with 117-nm acrylic latex at 0.32 volume fraction in a 21 PVC formulation. Thickeners: □, hydroxyethyl cellulose; ●, 4a; ◒, 4b; ○, 4c; ◆, 4d; ◆, 4e; ◇, 4f; Δ, 4g.

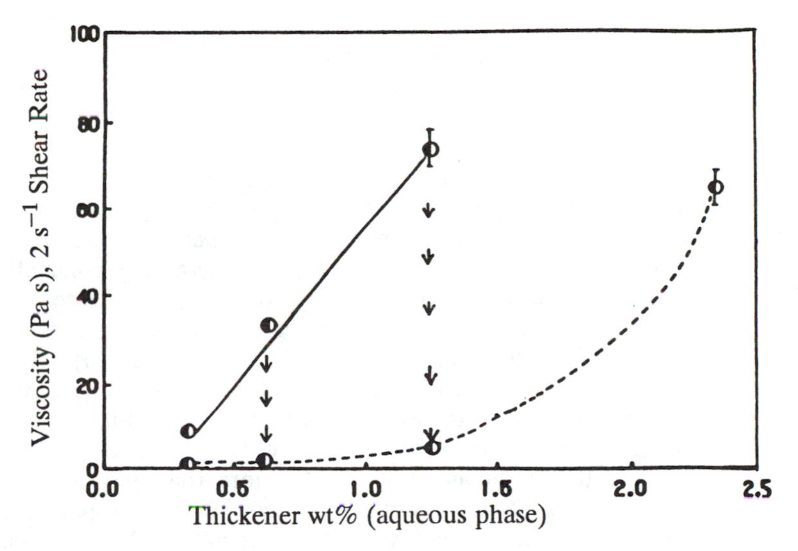

Figure 8. Effect of 0.6% nonionic surfactant, β-$C_{13}H_{27}(OCH_2CH_2)_9OH$, on low-shear-rate (2 s^{-1}) viscosity (Pa s) of thickened 0.32 volume fraction dispersions is indicated by arrow. Solid line, no nonionic surfactant addition; broken line, 0.6% nonionic surfactant, β-$C_{13}H_{27}(OCH_2CH_2)_9OH$ is added. Latex: 220 nm, MMA/MAA(96/04) with 0.00655 meq/m^2 surface acid. Thickener: Hydrophobically modified styrene/maleic acid poly(electrolyte).

What is not so apparent is that the surfactant concentration optimum for a maximum viscosity at low shear rate is not the optimum amount required for a maximum in the viscosity at high shear rates; a greater amount of conventional surfactant is required for the latter. This distinctive difference is apparent in the response of the hydrophobe-modified poly(electrolyte)/nonionic-surfactant mixtures in fresh water (Figure 9) and saline (Figure 10) solutions. The maxima in high-shear-rate viscosity for the hydrophobe-modified poly(electrolyte) requires a higher nonionic-surfactant concentration in both fresh and saline solutions, and there is an additional shift of the maxima to higher surfactant levels in the saline solutions.

The hydrophobe-modified, poly(electrolyte) solutions (i.e., with the surfactants levels noted to give viscosity maxima at 2 and 10^4 s^{-1}) were examined by forced-oscillation measurements (Bohlin VOR rheometer) to estimate the dynamic moduli in the plateau region, but the plateau could not be achieved within the frequency range of constant-stress rheometer. Shear wave propagation (at approximately 1200 rad s^{-1}) was then measured (Chapter 19 in ref. 1) by Goodwin's group at Bristol University, and in this high-frequency range, the SMAT solution with the higher surfactant concentration was observed to have the higher wave rigidity modulus.

In the presence of latices, the hydrophobe-modified, poly(electrolyte)-thickened, 220-nm latex dispersions containing high surface acid are higher in high-shear-rate viscosity than the thickened dispersions with the acid-monomer-free, 220-nm latex. Surfactant addition decreases low-shear-rate viscosities, but an increase in the high-shear-rate viscosity is observed (Figure 11). Similar behavior is observed in HEUR-708 dispersions. These studies suggest that the hydrophobes of the associative thickener participate in what are probably nonspherical surfactant micelles at high deformation rates, to yield higher viscosities.

Surfactants, in general, undergo changes in micellar shape with increasing concentration (*24*). Transitions in surfactant geometry from individual entities to spherical and then to rodlike micelles and finally to liquid crystals have been described for cationic surfactants with increasing concentrations. Cationic surfactants are not used in coating formulations. These transitions are less detailed in rheological response for classical anionic surfactants such as SDS, but they have been qualitatively defined (*25, 26*). With cationic surfactants, shear can induce structure and viscosity in surfactant solutions (*27*). The phenomena has been recently proposed for nonionic surfactants (*28*) of the type used in this study in the presence of salts. Thus, the influence of the surfactant can provide more than just a viscosity increase by inducement of the associative thickener's hydrophobe in spherical micelle networks. The geometry of the associative thickener can be expected to influence this behavior. Our studies in this area are continuing.

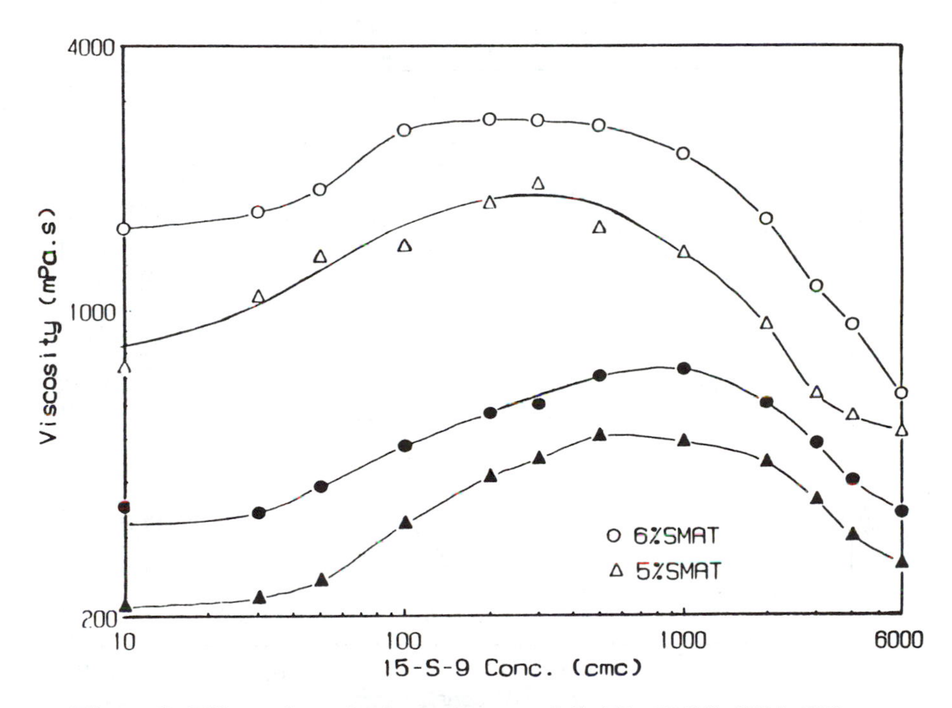

Figure 9. Effect of nonionic surfactant, β-$C_{13}H_{27}(OCH_2CH_2)_9OH$, on viscosity (mPa s) at high (10,000 s^{-1}) and low (2 s^{-1}) shear rates of 5 and 6 wt % hydrophobically modified styrene/maleic acid poly(electrolyte). Closed symbols, viscosity at a shear rate of 10,000 s^{-1}; open symbols, viscosity at a shear rate of 2 s^{-1}.

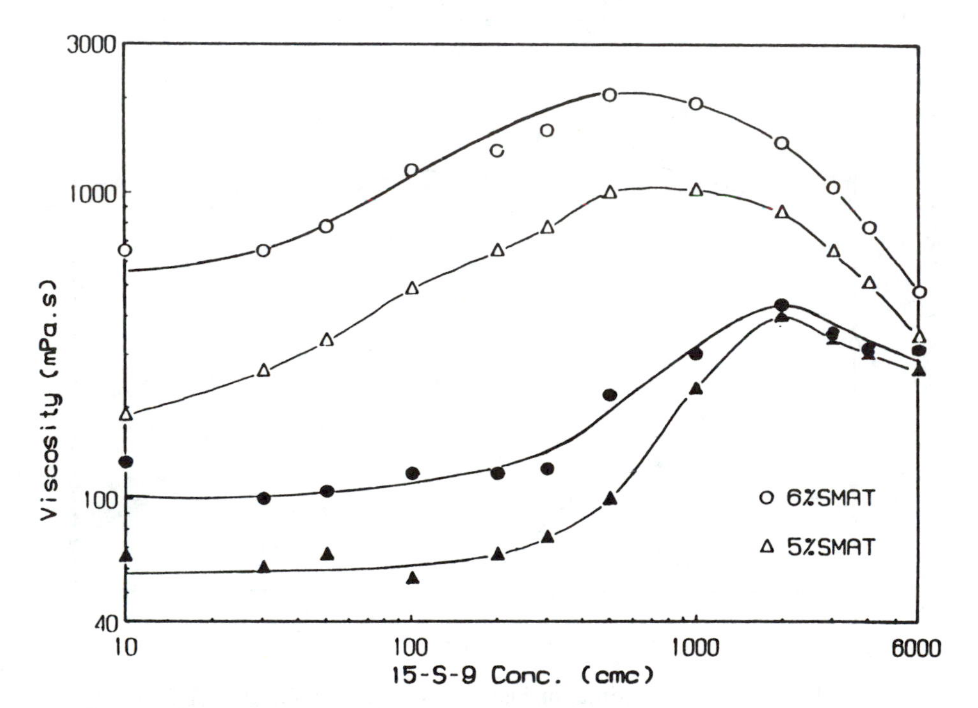

Figure 10. Dependence of nonionic surfactant in 5 wt % NaCl aqueous solutions on viscosity of hydrophobically modified styrene/maleic acid poly(electrolyte). Symbols are described in the caption of Figure 9.

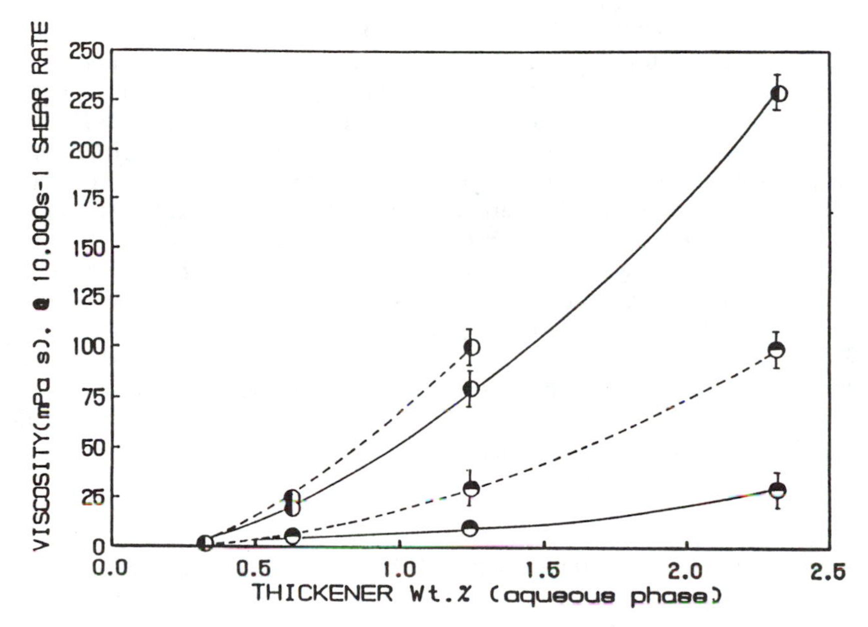

Figure 11. Effect of 0.6% nonionic surfactant, β-$C_{13}H_{27}(OCH_2CH_2)_9OH$, on high-shear-rate (10,000 s^{-1}) viscosity (mPa s) of 0.32 volume fraction dispersions thickened with a hydrophobically modified styrene/maleic acid poly(electrolyte). Solid line, no nonionic surfactant addition; broken line, with 0.6% nonionic surfactant, β-$C_{13}H_{27}(OCH_2CH_2)_9OH$ added. Latices are defined as follows:

◐, 220 nm, MMA/MAA(96/04) with 0.00655 meq/m^2 surface acid

◓, 220 nm, MMA/MAA(100/00) with 0.00028 meq/m^2 of residual catalyst acid (*5*, *13*).

Conclusions

Viscosities of hydrophobically modified ethoxylated urethane (HEUR) associative thickeners are sensitive to conventional surfactant structure and concentration, and this sensitivity is important to the function of HEUR thickeners as rheology modifiers in water-borne latex coatings. Certain HEURs have the potential of building elastic networks in the presence of surfactant. In coatings containing large-particle-size latices with chemically grafted hydroxyethyl cellulose surface fragments, this may be an important characteristic for cost-effective thickening, for there is little contribution of latex surface hydration or latex interaction with the associative thickener to the overall viscosity of the dispersion. With smaller latices traditionally stabilized with oligomeric surface acids, there is a significant contribution of surface hydration and latex interactions to dispersion viscosities. Thickening of small-particle latices with hydroxyethyl cellulose (i.e., via hydrodynamic volume) effects viscosities that are too high at low shear rates (LSR) and that are too low at high shear rates (HSR). HEUR thickeners appear to affect desirable rheology in small-particle latices because of the excess surfactant required to synthesize such latices. This excess free surfactant, with the free surfactant added with titanium dioxide affects desirable rheology in small-particle coatings by disrupting HEUR networks at low shear rates and by promoting micelle shape transitions and higher viscosities at HSR. The influence of HEUR geometry is important also in these transition, and studies in this area are continuing.

Acknowledgments

The financial support of this study by E.I. DuPont and the FMC Corporation is gratefully acknowledged.

Literature Cited

1. *Polymers in Aqueous Media: Performance through Association;* Glass, J. E., Ed.,; Advances in Chemistry 223; American Chemical Society: Washington, DC, 1989; Chapters 13–28.

2. Murakami, T.; Fernando, R. H.; Glass, J. E. *J. Oil Colour Chemists Assoc.* **1988,** *71*(10), 315–323.

3. Lundberg, David J.; Glass, J. Edward; Eley, R. R. *Proc. ACS Div. Polym. Materials: Sci. & Engin.* **1989,** *61,* 533–538.

4. Lundberg, David J.; Fossum, Eric; Glass, J. Edward *Proc. ACS Div. Polym. Materials: Sci. & Engin.* **1990,** *62,* 663–667.

5. Karunasena, A.; Glass, J. E. *Prog. Org. Coatings* **1989,** *17*(3), 301–321.

6. Soules, D. A.; Fernando, R. H.; Glass, J. E. *J. Rheol.* **1988,** *32,* 181.

7. Schaller, E. J. *J. Paint Technology* **1968,** *40*(525), 433.

8. Bowell, S. T. In *Applied Polymer Science;* Craver, J. K.; Tess, R. W., Eds.; American Chemical Society: Washington, DC, 1975; Chapter 41.

9. Hall, J. E.; Hodgson, P.; Krivanek, L.; Malizia, P. *J. Coating Technol.* **1986,** *58*(738), 65.

10. Fernando, R. H.; Glass, J. E. *J. Oil Colour Chemists Assoc.* **1984,** *67*(11), 279–282.

11. Hoy, K. L. *J. Coatings Technol.* **1979,** *51*(651), 27.

12. Kronberg, B.; Stenius, P. *J. Colloid Interface Sci.* **1984,** *102*(2), 418.

13. Karunasena, A.; Glass, J. E. *Prog. Org. Coatings,* in press.

14. Krieger, I. M. *Advances in Colloid and Interface Science* **1972,** *3,* 111.

15. Nielsen, L. E. *Polymer Rheology;* Marcel Dekker: New York, 1977.

16. Sperry, P. *J. Colloid Interface Sci.* **1981,** *82,* 62.

17. Glass, J. E. *Soc. Petrol. Engin. Publ. No. 7872,* International Symposium on Oilfield and Geothermal Chemistry Proceedings; Houston, TX, 1979.

18. Murakami, T., M.S. Thesis, North Dakota State University, 1987.

19. Bergh, J. S.; Lundberg, D. J.; Glass, J. E.; *Prog. Org. Coatings* **1989,** *17,* 155–173.

20. Glass, J. E. *J. Coatings Technol.* **1978,** *50*(641), 56–71.

21. Lundberg, D. J.; Glass, J. E.; Eley, R. R. *J. Rheol.* in press.

22. Maerker, J. M.; Sinton, S. W. *J. Rheol.* **1986,** *30*(1), 77–99.

23. Glass, J. E. In *Water-Soluble Polymers: Beauty and Performance;* Advances in Chemistry Series 213; Glass, J. E., Ed.; American Chemical Society: Washington, DC, 1986; Chapter 21.

24. Hoffmann, H.; Ebert, G. *Angew. Chem. Int. Ed. Engl.* **1988,** *27,* 902.

25. Missel, P. J.; Mazer, N. A.; Carey, M. C.; Benedek, G. B. In *Solution Behaviour of Surfactants;* Mittal, K. L.; Fendler, E. J., Eds.; Plenum Press, 1980, 373–388.

26. Kekicheff, P.; Grabielle-Madelmont, C.; Ollivon, M. J. *Colloid Interf. Sci.* **1989,** *131,* 112.

27. Rehage, H.; Wunderlich, I.; Hoffmann, H. *Prog. Colloid Polym. Sci.* **1986,** *72,* 51.

28. Penfold, P.; Staples, E.; Cummins, P. G. *Polym. Preprints* **1990,** *31*(2) 98.

RECEIVED March 11, 1990

POLYMER–POLYMER AND POLYMER–SOLVENT INTERACTIONS

Chapter 15

Polymers as Lubricating-Oil Viscosity Modifiers

G. Ver Strate and M. J. Struglinski

Polymers Group and Paramins Technology Division, Exxon Chemical Company, Linden, NJ 07036

The basic principles of "viscosity modification" of lubricating oils by high molecular weight polymers is discussed. The Newtonian viscosity-temperature characteristics of the oil are not strongly affected by the polymer. The use of polymers simply permits low viscosity basestocks with good temperature characteristics to be employed.

A variety of polymer types are used commercially, each having a different balance in cost-performance characteristics.

A semiquantitative discussion of the polymer solution physics is presented which explains the general performance of the different polymer types as a function of temperature and strain rate. Certain details of non-Newtonian behavior at low temperature are not completely understood.

It is economically and technically useful to employ polymer-containing "mineral" oils as lubricating fluids for internal combustion engines. There are many interesting scientific issues to be understood in the performance of such fluids as lubricants. The "original" technology which employed simple oil soluble polymers such as polyisobutylene, to create fluids with desirable viscosity and viscosity-temperature characteristics, has matured to see block, long chain branched, narrow molecular weight distribution and intramolecularly tapered composition materials yield specific tailored properties. The contribution of viscoelastic effects to the balance of fuel economy and wear is not yet completely resolved. Some polymers also contain polar functional groups so that they can act as dispersants for sludge etc. as well as "viscosity modifiers."

0097–6156/91/0462–0256$06.00/0

The basis for this technology has been reviewed to varying degrees in a number of articles (1-13). These references provide a sampling of the technology as it exists today. In certain instances there appears to be some misconception as to what is going on. We present a brief review of what we perceive to be the facts.

Multigrade Oil Formulation. Petroleum refineries produce a range of hydrocarbon "cuts" for use as lube oil basestocks, which contain molecules ranging in molecular weight from a few hundred to a few thousand. These hydrocarbon fluids act as useful lubricants via a hydrodynamic mechanism, although in typical gasoline or diesel fired engines some boundary lubrication also occurs. It should be noted that mineral and synthetic basestocks are Newtonian liquids above their cloud point, viscosity is independent of shear stress.

The glass transition temperature of these oils is in the -70°C ± 30°C range and they exhibit a very significant viscosity change in the temperature range from -40°C to 170°C where typical gasoline engines operate. A mineral oil which has high enough molecular weight to have adequate viscosity at high temperature, e.g. 2.5-5.0 cP at 150°C, will have too high a viscosity, >30,000 cP, at -25°C. This is an order of magnitude too high to permit engine cranking with batteries and starter motors typically found in today's engines. Some mineral oil basestocks have better viscosity-temperature characteristics than others, and synthetic basestocks (usually mixtures of polyalphaolefins and di- or triesters of a dicarboxylic acid and a polyalkylene glycol) are particularly good, but expensive to make. All natural paraffinic basestocks used as crankcase lubricants have methylene sequence-based waxes in them (linear and 2-methyl alkanes having carbon numbers between 18 and 30), which can also increase low stress viscosity at low temperatures if anisotropic wax crystals form.

The lower molecular weight basestocks (e.g. SAE 5W vs. SAE 30) generally have a less steep temperature dependence of viscosity. This is basically a free volume effect, with T_g decreasing as molecular weight decreases, since they have similar compositions apart from molecular weight. As can be seen in Figure 1, the difference between an SAE 5W and an SAE 30 oil at 100°C is about a factor of 2.5 in viscosity, whereas at 10°C it is about a factor of 10. If the SAE 5W basestock is thickened with enough polymer (in this case, a random ethene-propene copolymer) to give it the same 100°C viscosity as the SAE 30 oil, it generally maintains the viscosity-temperature slope of the SAE 5W oil (Figure 1). This is referred to as a "multigrade" oil, and given the designation SAE 5W-30, for reasons to be discussed below. Polymer is added to a concentration, c, such that $c[\eta]$ is about 1, corresponding to the region where the polymer coils begin to overlap as the solution moves with increasing concentration from the dilute to the semi-dilute regime. The intrinsic viscosity $[\eta]$ is 0.5 to 2 dl/gm for polymers used as VM's so the polymer is present at about 0.5 to 2 weight %. This polymer concentration has only a small effect on the overall η-T relation, unless some strong association or phase change behavior occurs. The Newtonian viscosity of the SAE 5W-30 oil is four times lower than that of the SAE 30 at 10°C, and the margin grows further as the temperature is lowered. This is why any oil soluble polymer that undergoes no unusual "phase" change behavior can be a so-called "viscosity modifier." We do not show data here below about 10°C, because of complications due to wax crystallization at lower temperatures.

This also does not take into account the non-Newtonian shear thinning that occurs at high shear stresses for the polymer-containing oil, resulting in still lower viscosities. It turns out that at high shear stress, the SAE 5W-30 oil described above has the same viscosity at -25°C as a Newtonian SAE 5W oil. Thus, multigrade oils can be produced using viscosity modifiers, due to a more favorable viscosity-temperature relationship from the use of a lower molecular weight mineral oil, and shear thinning of the polymer at high shear rates.

There are important perturbations around this basic principle. If there is polymer coil expansion or contraction with temperature due to (1) rotational isomeric backbone effects, (2) excluded volume effects, (3) micellization, (4) crystallization, or some other phenomenon, the viscosity temperature characteristics will be further modified. Generally these effects are small, on the order of factors of 2, compared to factors of >10 changes due to differences in basestock type, over the same temperature range.

Factors of 2 are not to be scoffed at, however. If one is trying to meet a low temperature oil viscosity specification (e.g., SAE J300 (14)) of 30,000 cP max at -25°C, there is a big difference between 35,000 and 17,500 cP. (This is related to oil pumpability as measured by the MRV and TP-1 mini-rotary viscometer) (15).

Viscosity Modifier Characteristics. As Flory said long ago (16), in cases where solvent character is not changing greatly with temperature, as is typical with lube oils, $d\eta rel/dT$ is controlled mostly by $d\langle S^2\rangle/dT$, where ηrel is the ratio of the polymer solution viscosity to the base oil viscosity and $\langle S^2\rangle$ is the mean square radius of gyration of the polymer. There are currently five polymer types used as commercial viscosity modifiers: poly(ethene-co-propene), polyalkyl methacrylates (where the alkyl esters are typically made from C_6 to C_{18} alcohols), hydrogenated poly(isoprene-b-styrene), hydrogenated poly(styrene-b-(butadiene-r-styrene)), and hydrogenated polyisoprene "stars". Of these, only the methacrylates have positive $d\langle S^2\rangle/dT$. Coil dimensions of the other polymer backbones actually shrink as T increases (17-19), as shown in Figure 2. On this basis, the alkyl methacrylates should be the best polymers for obtaining low viscosity at low temperature. Notice that the change in coil dimensions is on the order of 50-70% over this temperature range, compared to more than an order of magnitude change in viscosity of an SAE 5W mineral oil (Figure 1).

But there are other performance and cost criteria. Since polymers cost more than oil, one would like to use as little polymer as possible. Thickening power increases with increasing molecular weight, and therefore high molecular weight systems would be preferred. When the molecules get long enough they can be mechanically broken by the stresses in an engine (which can be at

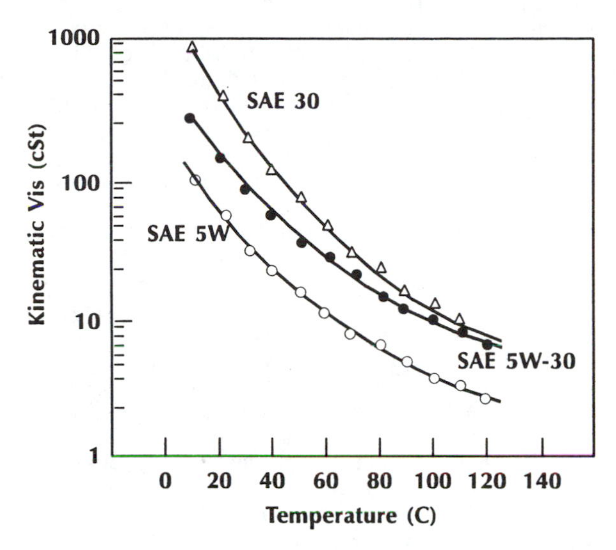

Figure 1. Temperature Dependence of Kinematic Viscosity for SAE 5W and 30 Single Grade Oils, SAE 5W-30 Multigrade Oil.

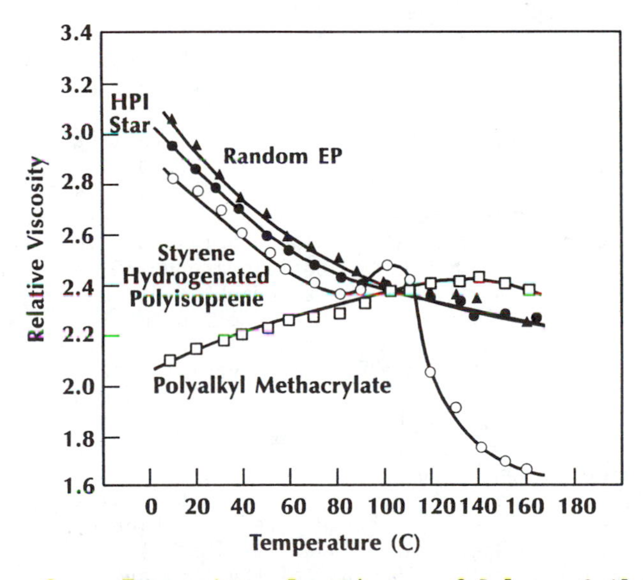

Figure 2. Temperature Dependence of Polymer Coil Dimensions, Shown as Solution Relative Viscosity. All Solutions Correspond to SAE 5W-30 Multigrade Oils.

least on the order of (10^7 sec^{-1})(.03 poise) =$3x10^5$ dynes/cm^2 shear stress, in flows that can also include complex squeezing or extensional components). Two basic shear stability grades are sold with either 20% or 10% maximum viscosity loss in a bench test (the "Kurt Orbahn" diesel fuel injector) (20). In this test, a formulated oil is circulated repeatedly through a diesel fuel injector, for a total of 30 passes. The polymer molecules undergo some complicated extensional flow the solution enters a nozzle and upon exit as oil droplets are formed, and break near the center of each chain.

Some polymers are inherently more stable than others for a given thickening power. For all carbon-carbon bond type polymers, inherent bond strengths are not too different from one polymer to the next (21,22). Thickening goes as $[\eta] \sim \langle S^2\rangle^{3/2}/M$ (Zimm model), while shear stress per polymer molecule goes as $[\eta]M\eta_s$ at constant shear rate. The intrinsic viscosity, $[\eta]$, for a given M, increases the fewer the number and size of side groups and the stiffer the chain (i.e., large characteristic ratio, C_∞) as $\langle r^2\rangle = C_\infty nl^2$, where n is the number of backbone bonds and l the bond length. By this measure poly(ethene-co-propene) is best, followed by styrene-b-HPI or styrene-b-(HPB-r-styrene) followed by polyalkyl methacrylates. The data in Figure 3 agree reasonably well with these predictions, with the exception of the HPI stars and the poly(styrene-b-HPI), which are more stable than expected for their $[\eta]$. They will be discussed in greater detail below. Polyisobutylene is now essentially off the market due to poor mechanical and oxidative stability. Polymethacrylates, while not terribly cost-effective, have a desirable viscosity-temperature relationship and therefore low zero-shear viscosity at low temperatures, and survive mainly in markets where outstanding low temperature performance at low shear stresses is desired.

Recent Viscosity Modifier Developments. Against this basic set of concepts, there have emerged "high tech" viscosity modifiers. Examples are shown in Figure 4. Although the Shell Kraton-type S-b-HPI diblock polymers have been around for many years they present an interesting case (23). These polymers have molecular weight from 80,000 to 120,000, and contain around 30 wt% styrene. They have extremely narrow molecular weight distributions, with $M_w/M_n<1.20$, as shown in Figure 5. Gel permeation chromatography data were obtained at 135°C in TCB. A typical poly (ethene-co-propene) VM and polyalkyl methacrylate are also shown, both of which have essentially most probable molecular weight distributions ($M_w/M_n=2$, $M_z/M_w=1.5$). The styrene blocks of S-b-HPI polymers have low solubility in hydrocarbon oils. Near 100°C the "solution" undergoes a reversible micellization phenomenon (see (4) for example) wherein 5° to 15° chains aggregate causing an abrupt viscosity change (refer to Figure 2). Below 80°C the diblock polymers are in a star-like associated state, with the styrene blocks at the core of the star and the HPI arms dangling out into the solution solubilizing the micelle. As the temperature approaches 100°C, the styrene core swells, increasing the viscosity somewhat. Above 100°C, the micelles dissociate and the viscosity drops until when dissociated the viscosity is that of the unassociated diblock polymer. Recent SANS measurements

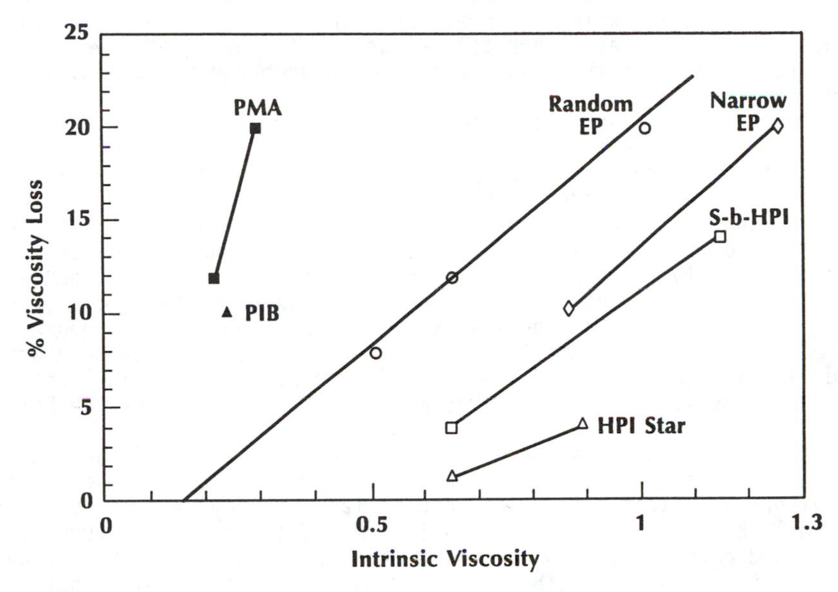

Figure 3. Mechanical Stability of VM Polymers in the Diesel Fuel Injector Test (ASTM D3945A).

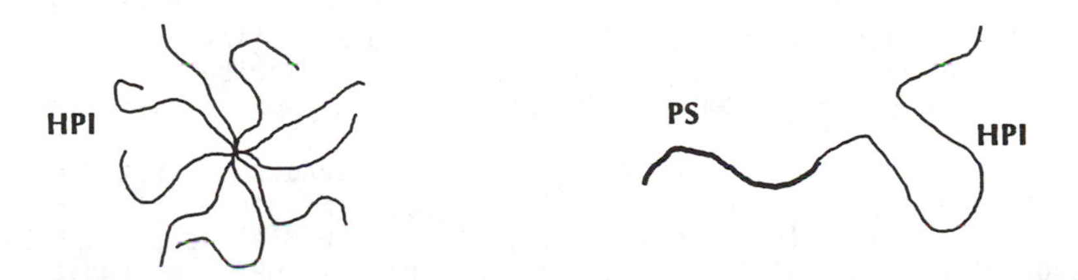

Styrene-b-Hydrogenated(Butadiene-r-Styrene)

Figure 4. Anionic VM Polymer Structures.

indicate that at high temperatures, the micelle aggregation number decreases as the increased styrene solubility shifts the equilibrium to free chains, but the number of micelles does not change (24). There are also indications of intermicellar interactions at low temperatures which become less important as the aggregation number decreases at higher temperatures. At low temperatures, the authors note that there is negligible interchange between micelles and free chains.

Laboratory shear stability tests are normally run below 100°C, while engines typically operate above 100°C. The bench test shear stability of the block polymer is quite good for its thickening ability. The fragments break near the end of the styrene block, rather than in the middle of the chain, indicating that the micelles do not dissociate as they pass through the diesel injector. Polymer fragments of the shorter styrene-rich block and the longer HPI block each contain reasonable quantities of styrene presenting interesting questions about how the micelles degrade. Fired engine shear stability tests, run at elevated temperatures and involving thermooxidative as well as purely mechanical degradation, result in greater breakdown than predicted by the bench test.

Multiarm star-branched hydrogenated polyisoprenes are also available from Shell(25). They are prepared by linking living polyisoprene arm anions of appropriate molecular weight with divinyl benzene, forming a 10-16 arm nodular star structure, and then hydrogenating. Typical arm molecular weights are on the order of 30,000 to 50,000. The resulting molecular weight distribution, while not as narrow as linear anionic block polymers, is still narrower than most other types of commercial polymers, with M_w/M_n=1.2-1.4. These structures are very shear stable for their thickening power, as anticipated from the low $d\eta rel/d(break)$ and the internal stress distribution (stored energy) in the molecule under stress (26,27). They have a similar configuration in solution to associated styrene-b-HPI described above, but because they are chemically crosslinked they do not dissociate at high temperatures.

Linear poly(styrene-b-(styrene-r-hydrogenated butadiene)), S-b-HPB, polymers are also available (28). Simple poly(styrene-r-butadiene) have also been used (29). The S-b-HPB structure exhibits association in solution similar to the S-b-HPI. An additional consideration is the mode of butadiene incorporation. 1,4-Addition, when hydrogenated, gives essentially polymethylene sequences, which if not controlled can crystallize at low temperatures and interact with basestock wax. Typically, polymerization modifiers are used to produce essentially equimolar amounts of 1,2- and 1,4-addition, resulting in ethyl branching in the HPB section. This breaks up the methylene sequences and reduces or eliminates the crystallinity.

Most recently there has emerged Ziegler Natta polymerization technology which is capable of producing narrow MWD poly(ethene-co-propene) with controlled intramolecular compositional distributions (30). Figure 5 shows the MWD of a copolymer produced with this technology, where M_w/M_n=1.35 and M_z/M_w=1.15. Narrow MWD per se leads to good mechanical shear stability at a given $[\eta]$

(see Figure 3), since viscosity depends on a somewhat lower power of the molecular weight than susceptibility to breakdown. However, by incorporating methylene sequences into the polymer chain, which are capable of associating due to crystallization, it is possible to exert an influence on coil size and viscosity beyond the normal $d\langle S^2\rangle/dT$ effects (11,31).

A variety of EP's with segments of high enough average ethene content to crystallize near or below 23°C have been prepared (11) using conventional soluble Ziegler catalysts in a "mix free" reactor, either a batch or a plug flow reactor. Conditions were chosen such that initiation was rapid compared to propagation, and chain transfer and termination steps were minimized. By varying the time at which monomers were added to the reactor, blocky sequences rich in ethene or propene were formed. Depending on the arrangement of the sequences, relative viscosity may increase, with even the appearance of weak gels, or it may pass through a maximum such that the low temperature coil dimensions are smaller than when dissociated at high temperatures (Figure 6). In this case the association and coil collapse makes up for the positive $d\langle S^2\rangle/dT$ deficiency of EP's compared to polyalkyl methacrylates and produces nearly equivalent low shear stress properties at low temperature. In the EP case wax crystal modifiers (LOFI's) must be added. Similar behavior has been noted by others (12).

"Good" and "bad" polymer structures for use in lube oils are shown in Figure 7. Here, the incremental ethene content is shown as a function of position along the polymer chain. Sections of the chain with ethene contents above approximately 65 wt% can crystallize in these mineral oil solutions at temperatures below around 30°C. In a good structure, a methylene-rich section is near the center of each chain, surrounded by propene-rich sections on either end. Poor structures have the methylene-rich sections on the ends and the low ethene section in the middle. Total intensity light scattering in a variety of solvents including lube oils reveals that the apparent MW of the agglomerates increases monotonically as the solution temperature is lowered, even for "good" structures. Unlike the poly(styrene-b-HPI) system (4), there is not a finite micellization occurring in the blocky EPs. Complex high molecular weight aggregates which have a lower total effective contribution to viscosity are formed in the case where the aggregating segments, consisting of high ethene content sequences, are sterically hindered in the center of the chain contours. Figure 8 presents some speculative reasoning for the steric effects.

Polymers with crystallizable or associating segments present some interesting complications in the standard handling and testing procedures used for lube oils. A polymer designed to associate and reduce viscosity at 0.7 wt% in a finished formulated oil can become a solid, gelled mass at 7 wt% in a concentrated solution, if the concentrated solution is allowed to cool from its normal handling temperature above ~40°C, as shown in Figure 9. Here viscosity data at 25°C are shown as a function of the product $c[\eta]$. Data for linear polymers should fall along a common curve in the absence of associative effects. Notice that compared to random, amorphous EP copolymers, the associating systems all exhibit higher viscosities

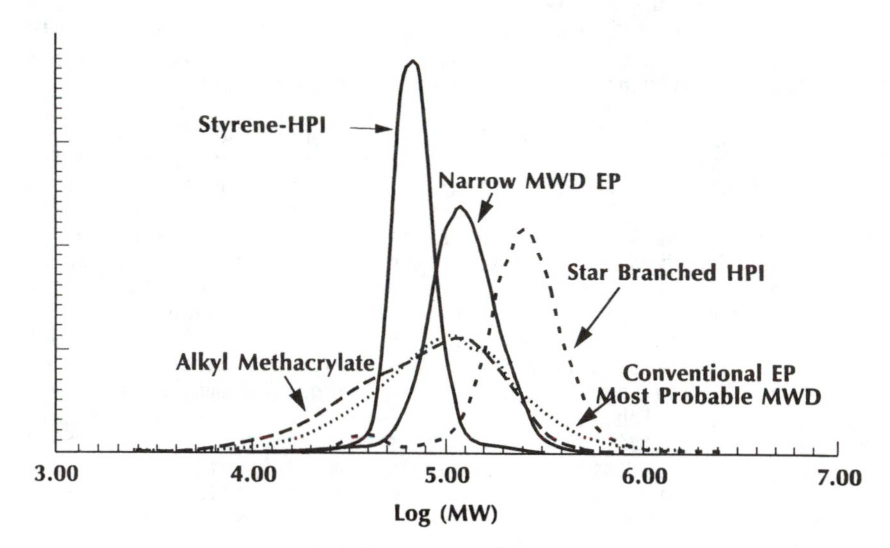

Figure 5. Typical VM Molecular Weight Distributions. All Data are on an EP Elution Volume Calibration Basis.

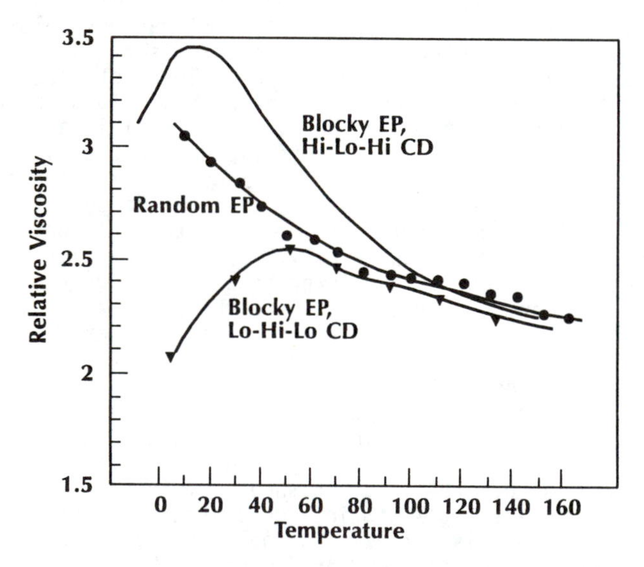

Figure 6. Temperature Dependence of Narrow MWD EP Copolymers Having Different Intramolecular Composition Distributions. The Lo-Hi-Lo CD Refers to the "Good" Structure in Figure 7, and the Hi-Lo-Hi to the "Bad" Structure in Figure 7.

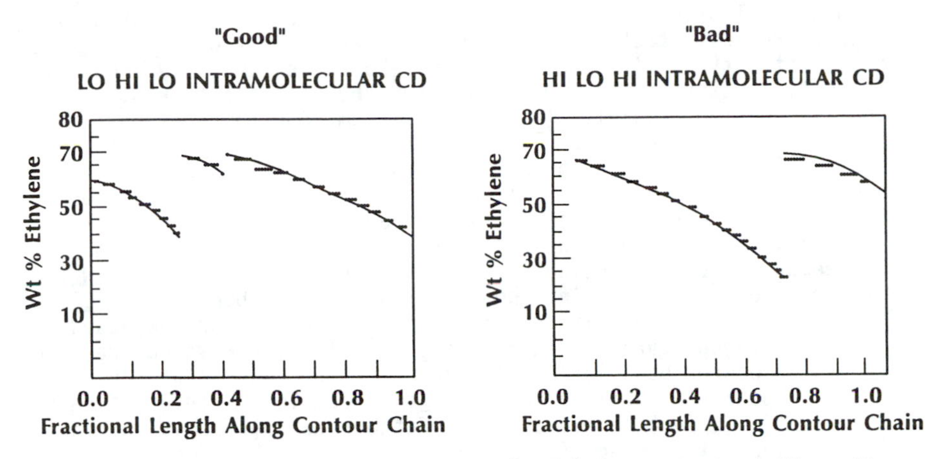

Figure 7. Calculated Incremental Ethylene Content Along the Polymer Chain for "Good" and "Bad" Narrow MWD EP Structures.

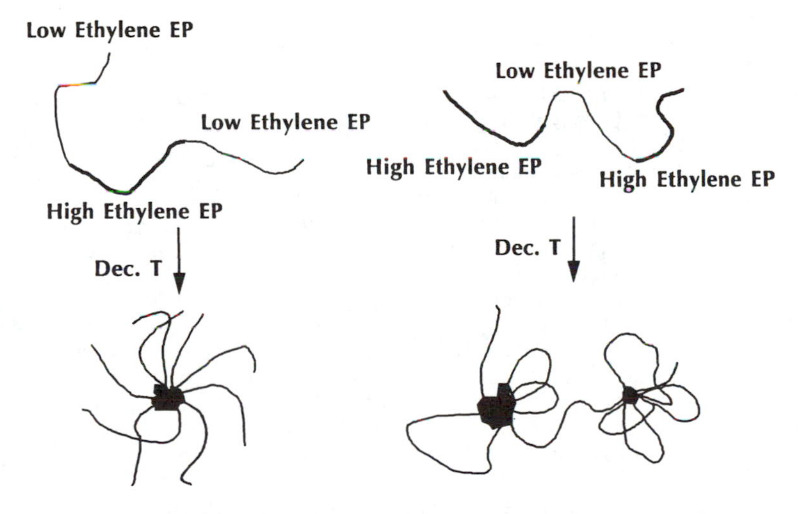

Figure 8. Change in Coil Dimensions for "Good" and "Bad" Structures. Chains with High Ethene Section in Center of Chain, Left, Associate at Low Temperature With Little Intermolecular Connectivity, Compared to Structure With High Ethene Sections on Ends of Chain, Right.

for $c[\eta]>4$. At higher concentrations, the associative systems first exhibit a yield stress, then form soft gels. The gels are thermally reversible, and finished oil viscometric properties are unaffected below $c[\eta]\sim2$.

Other elastic effects can occur in one of the routine low temperature oil tests, the cold cranking simulator or CCS (14). This test, typically run at -15 to -25°C, is basically a Couette viscosity measured at $\sim10^7$ dynes/cm^2 shear stress. An oil solution is cooled rapidly (45-60 sec.) from room temperature to the desired test temperature and the stress applied to the rotor. It takes 2-3 sec. to reach the ultimate rotational rate. As the oil solution is sheared, the well known rod climbing or "Weissenberg" effect may occur. Interestingly, it can take 7-15 seconds for the elasticity to develop, which is on the order of $>10^5$ shear units. It seems to be a nucleation related effect and may show the influence of stresses in the solution on phase behavior (32,33). The extent to which it occurs can be decreased or increased by adjusting temperature. In general, for structures which are good at -20°C "poorer behavior" is observed at ~-5°C, where at the start of the apparent "coil collapse" is less complete. Interestingly, the elastic effect can be avoided if the shear stress is increased slowly, over a time span of 1-2 min.

Upon full formulation of the oil by addition of polyisobutylene dispersants etc., which are essentially low MW polymeric non-solvents, the effect can become suppressed, similar to lowering temperature and improving coil collapse and association.

Performance Comparison. With this introduction to the concepts of viscosity modification let us make a comparison of the several types in a specific formulation. Representative data are shown in Table I. Here, several different VMs were dissolved in a paraffinic mineral oil which has a kinematic viscosity of 5.2 cSt at 100°C. Enough polymer was added to achieve a solution viscosity at 100°C of 13.5 cSt, which is representative of an SAE 10W-40 oil. Enough lube oil flow improver was added to minimize basestock wax-polymer interactions. The CCS viscosity is shown in the second column. Differences of 300 cP are real and useful. The question arises, why is the CCS lower for the narrow MWD EP-3 and 4 and the anionic block polymers than for PMA or amorphous EP-1? We have already discussed why there are differences in MRV values in Column 3 but we restate some of that here. The fourth column compares high and low strain rate data at -20°C. All but the PMA are significantly "shear thinned" at the CCS test conditions. We give the following qualitative interpretation of the low temperature properties.

As temperature is lowered from 100°C, "normal" polymer solution behavior results in increasing viscosity due mainly to the change in solvent viscosity. Superposed on that trend is the polymer contribution which can be tracked by η_{rel}, as described previously. With the exception of PMA, the rest of the polymers basically have poly(ethene-co-propene) as the soluble, viscosity producing structure in solution. $d\langle S^2\rangle/dT$ is positive for PMA, and negative for EP, so as the temperature drops the relative viscosities for PMA and EP diverge. At -20°C η_{rel} has dropped to

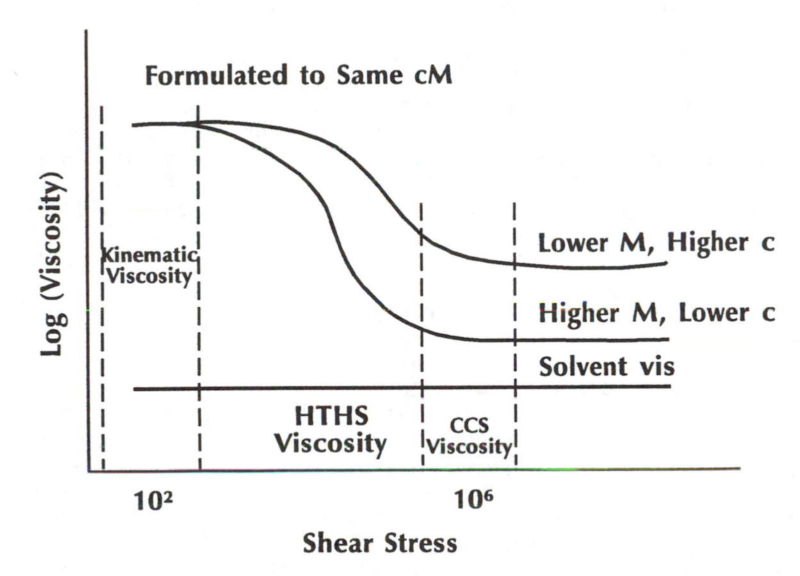

Figure 10. Schematic Representation of the Shear Rate Dependence of Viscosity for Oils Formulated to the Same Zero-Shear Viscosity, Containing Polymers With Different Molecular Weights. The "Solvent Viscosity" Could be Perturbed by the Polymer Thus Confusing an "Upper Newtonian Limit."

Table I: Polymer Structure-Performance Comparison

Polymer	c	CCS, -20°C	η_o, -20°C*	CCS/η_o
PMA	2.8 wt%	3300 cP	4000 cP	0.82
Random EP-1	1.0	3200	12000	0.27
Random EP-2	0.9	2600	6000	0.45
Narrow EP-3	0.63	2700	5000	0.54
Narrow EP-4	0.90	2700	5500	0.49
HPI Star 200	1.56	2700	9500	0.28
HPI Star 250	0.90	2700	9500	0.28
S-b-HPI 40	0.88	2450	9000	0.27
S-b-HPI 50	1.56	2500	9000	0.28
S-b-(S-r-HPB)	1.65	2700	5500	0.50

* η_o at -20°C measured with Mini-Rotary Viscometer.

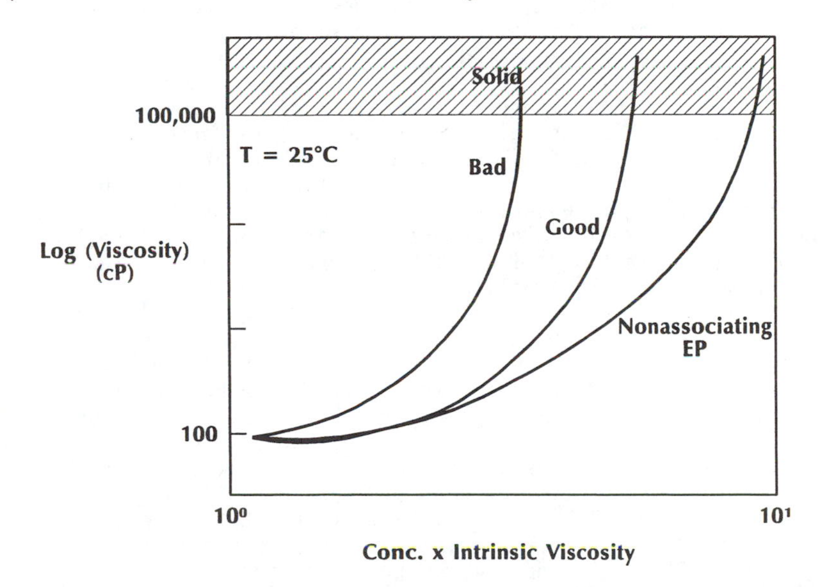

Figure 9. Viscosity as a Function of Coil Overlap Parameter, c[η].

about 1.5 for PMA and has risen to about 3.9 for EP. Thus the low strain rate (essentially zero-shear) MRV viscosities differ by a factor of ~3 at -20°C. The 12,000 cP value for EP-1 may not be completely free of polymer association or wax crystal interaction effects. This value is independent of cooling rate and oil de-waxing as long as an effective wax crystal modifier is used, but it could be as much as 2000-3000 cP too high due to polymer association or wax effects. Extrapolation of η_{rel} from temperatures above 10°C is not precisely linear.

In spite of the large difference in MRV viscosity between PMA and EP-1, the CCS values at -20°C differ by only a few percent. The is due to the fact that the EP, a much higher intrinsic viscosity polymer, exhibits more "shear thinning" than PMA at the high strain rate and shear stress of the CCS measurement. PMA cannot be used at high intrinsic viscosity (where it would "shear thin" more) due to its low mechanical shear stability. CCS viscosity is controlled much more by the absolute concentration of polymer chains in the oil than by the low strain rate thickening of the chains. CCS deformations are such that only short sections of chains are able to contribute fully to the viscous dissipation. Thus the lower polymer concentration for EP-1 provides a CCS advantage.

EP-2 presents an additional complication. This is a high ethylene content polymer that can exhibit some crystallinity. Similar polymers have recently been described elsewhere (12), although we do not necessarily agree with their interpretation of the molecular mechanism. Depending on the exact composition and heterogeneity of compositional distribution, the polymer associates in oil at temperatures below its pure, undiluted polymer melting point. This association can increase or decrease the viscosity compared to the unassociated η_{rel}. The solutions can gel and yield stresses may be observed. Thermal history also becomes important. If the polymer is prepared correctly for the basestock being used, it is possible to have the polymer associate in a way that reduces η_{rel} and MRV to values only half or less of those of the low ethylene content EP-1. This low η_{rel} when combined with the shear thinning observed at CCS conditions yields low CCS viscosities as well. The apparent shear thinning, η(CCS)/η(MRV), is only half that of low ethylene EP-1. This is reasonable, considering that about half of the viscosity contribution for EP-2 was already lost thru polymer association to some smaller effective hydrodynamic volume, even at low strain rate.

The two specialty EPs, EP-3 and EP-4, have the low-high-low ethylene structure described previously. These narrow MWD polymers can be manufactured at high [η] yet still retain good mechanical stability. Thus, oils can be formulated at still lower polymer content which should help reduce CCS viscosity. These polymers also undergo association effects and show deviations from "normal" η_{rel} vs. T behavior. Association effects here produce MRV viscosity almost equal to PMA. This is a remarkable result. As with the random high ethylene polymer EP-2, the apparent shear thinning is less than for EP-1, but that is again probably due to the fact that the MRV viscosity is so low. Structures that are "hydrodynamically small" do not shear thin as effectively as free chains

with their full dimensions. EP-3 and EP-4 end up with lower CC viscosity than EP-1, showing a combined shear thinning and associ ation effect. We do not yet understand why EP-3 and EP-4 which ar present at significantly different concentration yield essentiall the same CCS. One might have expected EP-3 to have given a lowe value. When one is dealing with significant agglomeration, howeve it may not be easy to predict results such as this. Furthermore as seen below, when "identical" polymer types of differi molecular weights are formulated to the same c[η], the CCS vi cosities are very similar. c[η] seems to normalize shear thinni effects at CCS conditions. This would not be true at some "upp Newtonian" viscosity limit for strain rate dependence.

The many-arm HPI 200 and 250 star samples should be somewh easier to understand. Here one expects no perturbations in t η_{rel} vs. T behavior since the hydrogenated polyisoprene structu should not associate. In fact, the η_{rel} behavior is not unusua but the MRV values at -20°C are somewhat lower (~20%) than the l ethylene EP-1. This probably indicates that the EP-1 data are n completely free of interpolymer or wax interaction. One mig anticipate that the branched structures in these dilute to sem dilute solutions might shear thin less than linear EP-1, but fact they are quite non-Newtonian and end up with only slight higher CCS than EP-2, -3, and -4. Like EP-3 and -4, CCS doesn seem to depend on the molecular weight, since the higher polym concentration for HPI 200 does not produce an increased CCS. She thinning to 25-30% of the zero-shear value apparently does n cause as much differentiation as one might expect from t schematic shown in Figure 10. The onset of non-Newtonian behavi is expected to correlate with β_0cRT/(η-η_0)M, i.e. as c/M f our solutions (34). β_0 is a dimensionless shear rate locati the onset of shear rate dependence of viscosity, and is typical near 1.5 in this concentration range. The slope of the viscosit shear rate curve in the non-Newtonian region should be similar constant c[η]. The fact that the CCS is actually run at consta shear <u>stress</u> should make the differences even larger.

The S-HPI-1 and -2 polymers again have EP-like "arms" solution extending from the styrene micelle centers. Thus, η_r vs. T is similar at high temperatures to EP-1, EP-3, or HPI st 250. The MRV of 9,000 cP is nearly the same as 200 or 25 significantly higher than EP-3 or -4. The CCS viscosity measurably below all other polymer types. η(CCS)/η(MRV) = 0.27 close to HPI 200 or EP-3, but the absolute viscosity is 300 lower. This value could result from one or both of the followi reasons.

Perhaps the polystyrene micelle core becomes somewhat small at low T and helps keep η_{rel} slightly smaller than for free chains. Thus the 9,000 cP MRV value is really below the star H at 9,500 and the EP-1 at 12,000 cP. Perhaps the real star does n shear thin as much as EP-1 or the S-b-HPI micelle. Alternativel it seems reasonable that stress could pull the micelles apart some extent. Viscosity falls off dramatically for S-b-HPI polyme above 110°C, where the micelles appear to dissociate. Why couldn this happen to some extent under stress at -20°C? Perhaps sca tering experiments on sheared solutions, the addition of sma

amounts of solvent to "soften" the styrene domains, or even "hysteresis" in strain rate cycle history could help to answer this question.

The S-b-(S-r-HPB) has a good combination of CCS and MRV. Perhaps there is enough styrene in the S-r-HPB block to cause those segments to come out of solution as compared to the HPI segments in the S-b-HPI polymer. The S-r-HPB segment may further collapse into a center micelle as T drops giving good $d\eta_{rel}/dT$, due to a real solvent quality effect and not rotational isomeric effects. The shear thinning of this structure is not as good as the S-b-HPI structure, however. Furthermore, we do not know $d\langle S^2\rangle/dT$ for this polymer.

Conclusions

Lube oil technology provides an interesting opportunity to work creative science from molecule synthesis to semidilute solution behavior of polymers. There are many opportunities remaining to produce an optimum structure that is best for mechanical stability, low temperature viscometric properties, fuel economy and dispersancy.

Literature Cited

(1) Muller, H.G. Tribology International 1978, 11, 189.

(2) Bartz, W.J. SAE Technical Paper, 890728, February 1989.

(3) Fein, R.S. Ind. Eng. Chem. Fundam 1986, 25, 518.

(4) Candau, F.; Heatley, F.; Price, C.; Stubbersfield, R. Eur. Pol. J. 1984, 20, 685.

(5) Spiess, G.; Johnston, J.; Ver Strate, G. in Additives for Lubricants and Operational Fluids; Bartz, K., Ed.; Technische Akademie Esslingen 1987; p. 8.10-1.

(6) Ramesh, K.; Clifton, R. J. Tribology 1987, 107, 215.

(7) Marui, E. et al. Trans ASME 1987, 109, 696.

(8) Klamann, D. Lubricants and Related Products, Verlag Chemie Weinheim (1984) p. 185.

(9) Spearot, J.; Murphy, C. SAE Technical Paper, 880680, 1988.

(10) Watson, R.; McDonnell, T. SAE Technical Paper, 841208, 1984.

(11) Ver Strate, G. Rolduc Meeting, April 22, 1989 in press.

(12) Kapuscinski, M.; Sen, A.; Rubin, I. SAE Technical Paper, 892152, September 1989. Sen, A., Rubin, I. Macromolecules, 23, 2519 (1990).

(13) Rozeanu, L.; Shavit, A.; Maayan, M. SAE Technical Paper 850545, 1985.

(14) Society of Automotive Engineers Specification J300.

(15) ASTM D3829 and D4684 test methods.

(16) Flory, P. J. Statistical Mechanics of Chain Molecules; Wiley, 1969; p. 37.

(17) Kurata, M.; Tsunashima, Y.; Iwama, M.; Kamada, K. In Polymer Handbook; Brandrup, J.; Immergut, E., Eds.; John Wiley, NY, 1975.

(18) Mark, J. Rubber Chem. Tech. 1973, 46, 593.

(19) Mark, J. J. Chem. Phys. 1972, 57, 2541.

(20) DIN 51382.

(21) Rooney, J. G.; Ver Strate, G. Liquid Chromatography of Polymers and Related Materials III; J. Cazes ed., Marcel Dekker.

(22) Odell, J.; Keller, A. J. Poly. Sci. B Physics 1986, 1899.

(23) St. Clair, D.; Evans, D. U.S. Patent 3 772 196, 1973.

(24) Schouten, M.; Dorrepaal, J.; Stassen, W.; Vlak, W.; Mortensen, K. Polymer 1989, 30, 2038.

(25) Eckert, R., U.S. Patent 4 116 917, 1978.

(26) Ver Strate, G.; Graessley, W. Polymer Preprints 1979 20 149.

(27) Ver Strate, G.; Graessley, W.; Kresge, E. U.S. Patent 4 620 048, 1986.

(28) Trepka, W., U.S. Patent 4 412 087, 1983.

(29) Schiff, S.; Johnson, M.; Streets, W. U.S. Patent 3 554 911, 1971.

(30) Ver Strate, G.; Ju, S.; Cozewith, C. Macromolecules 1988, 21, 3360.

(31) Ver Strate, G.; Bloch, R.; Struglinski, M.; Johnston, J.E.; West, R.; U.S. Patent 4 804 794, 1988.

(32) Rangel-Nafaile, C.; Metzner, A.; Wissbrun, K. Macromolecules 1984 17, 1187.

(33) Katsaros, J.D.; Malone, M.; Winter, H. Polymer Eng. Sci. 1989, 20, 1434.

(34) Graessley, W.W. Adv. Polymer Sci. 1974,16, 1.

RECEIVED July 18, 1990

Chapter 16

Solution Rheology of Ethylene–Propylene Copolymers in Hydrocarbon Solvents

Isaac D. Rubin and Ashish Sen[1]

Texaco Research Center, P.O. Box 509, Beacon, NY 12508

Dilute solution viscosities were determined for ethylene-propylene copolymers with 60-80 mole % ethylene in a series of hydrocarbon solvents at -10 to 50 C. Highest viscosities were obtained in methyl cyclohexane and lowest in toluene and tetrahydronaphthalene. In poor solvents, the slightly crystalline copolymers with 80 % ethylene had considerably lower intrinsic viscosities at low temperatures than the amorphous ones with 60-70 % ethylene. In better solvents, the two groups behaved identically. The behavior of the partially crystalline copolymers in poor solvents at low temperature is ascribed to ordering of the longer ethylene sequences into partially ordered domains.

Ethylene-propylene copolymers (EP) with 60 or more mole % ethylene are widely utilized as viscosity index improvers. VI improvers fill a unique need in our motorized society and are used by almost all of us. They are the materials which provide all-season properties to motor oils; typically, most of today's oils contain 0.5-3.0 % polymer.

Addition of polymer to an oil increases its viscosity. Ideally, this increase should be very small at low temperature where the oil itself is sufficiently viscous, and become progressively larger as the temperature is raised, resulting in a reasonably

[1]Current address: Dow Chemical Company, Midland, MI 48640

0097–6156/91/0462–0273$06.00/0

constant viscosity over the entire temperature range encountered in service. This flattening of the viscosity-temperature curve makes the lubricant useful over a wider temperature range and eliminates the need for seasonal oil changes. Under ideal conditions, at low temperature, interactions between the oil and VI improver are weak; polymer molecules are coiled up and small in size, and interfere little with the flow of oil. As the temperature is raised, interaction improves and the polymer expands significantly, resulting in appreciable oil thickening and improved lubrication.

Work in our laboratory on EPs of widely differing molecular weights and compositions dissolved in oils showed that high ethylene content and small amounts of crystallinity led to lower viscosity at low temperatures, e.g. -30°C, than in comparable solutions with totally amorphous copolymers and greater viscosity at 100°C (1). The purpose of this work is to gain a better understanding of the effect of structure, composition and morphology of EP copolymers used as VI improvers on their viscosity in oils over the temperature range encountered by motor oils in service. With this in mind, we investigated the influence of model hydrocarbon solvents, representative of different base oil components, on the solution viscosity of five EP copolymers in the temperature range of -10°C to 50°C. The model hydrocarbon solvents were: hexane, isooctane (2,2,4-trimethylpentane), methyl cyclohexane, toluene and tetralin (tetrahydronaphthalene). The first two represented paraffinic base oil components; methyl cyclohexane, a naphthenic component; and toluene and tetralin, simple and complex aromatic components. All were purchased materials of high purity and were used as received. The copolymers were prepared by conventional means with a soluble Ziegler-Natta catalyst composed of an alkylaluminum halide and a vanadium salt.

COPOLYMER PROPERTIES

This study was carried out with linear copolymers incorporating from 58 to 80 mole % ethylene. The increase in ethylene contents was accompanied by a rise in the mean number of ethylene units in the chain in sequences of 3 or more, $\bar{N}$, from 3.9 for EP-2 with 58 % ethylene to 6.5 and 6.1 for EP-4 and EP-5 with 80 % ethylene; the fraction of ethylene sequences containing 3 or more ethylenes, $En{\geqslant}3$, rose from 0.30 to 0.62 and 0.65. These data are shown in Table I. They were obtained from C-13 NMR runs in o-dichlorobenzene on a Varian VXR-300 spectrometer using the method of Randall (2) and Johnston et al (3). Table I also gives IR results obtained on films pressed from the copolymers (4). They are consistent with the NMR results, showing an increase in the CH2/CH3 ratio from 2.0 to 7.5-7.8 as the ethylene concentration in the EPs changed from 58 to 80 mole %.

Weight average molecular weights, $\bar{M}w$, and molecular weight distributions, $\bar{M}w/\bar{M}n$, for the copolymers are given below.

EP Copolymer	EP-1	EP-2	EP-3	EP-4	EP-5
$\bar{M}w \times 10^{-3}$	148	252	192	207	322
$\bar{M}w$ / $\bar{M}n$	2.9	3.5	3.4	3.6	4.1

All results were obtained by gel permeation chromatography (GPC) in 1,2,4-trichlorobenzene at 135° C utilizing a Waters 150° C GPC unit equipped with five columns ($1x10^6$, $1x10^5$, $1x10^4$, $1x10^3$, and 500 Å). Eighteen polystyrene samples ranging in $\bar{M}w$ from 3,000 to 3,300,000 were used for calibrating the equipment.

Melting points, crystallinities and second order transitions were obtained using differential scanning calorimetry (DSC). Samples EP-1 through EP-3 were completely amorphous while EP-4 and EP-5 contained about 5-9 % crystalline material. The data were obtained on a Perkin Elmer DSC-7 Differential Scanning Calorimeter at a heating rate of 10° C/min. calibrated as recommended by the manufacturer. Crystallinity was also determined from wide-angle x-ray scattering measurements (WAXS) on a Scintag Pad V diffraction system. As shown below, considering the difficulty of measuring such small amounts of crystallinity, agreement between the two methods was quite good.

EP Copolymers	EP-1	EP-2	EP-3	EP-4	EP-5
Tg, °C, DSC	-68	-59	-63	-50	-46
Tm, °C, DSC	--	--	--	43	42
Crystallinity, %, DSC	0	0	0	9.3	7.4
Crystallinity, %, x-ray	--	--	--	8.4	5.2

RELATIVE AND SPECIFIC VISCOSITIES

All viscosity measurements were made in Ubbelohde viscometers in a bath controlled to within ±0.01°C on solutions with concentration of 0.4 to 1.0 gm/dl, depending on copolymer molecular weight. Since solvent flow times were at least 100 sec. and intrinsic viscosities of most samples were less than 3.00 dl/gm, no corrections were made for kinetic energy or rate of shear (5,6). Densities of the solvents at the different temperatures were measured with a Mettler/Paar DMA 45 Digital Density Meter and copolymer concentrations were corrected for the change of solvent density with temperature.

The relative viscosity of a polymer solution is its kinematic viscosity divided by that of the solvent; at any temperature, the relative viscosities of a polymer in a series of solvents thus provide a ready means of ranking their solvating power. Figure 1 compares the relative viscosities of EP-5, one of the partially crystalline samples, in all five solvents at 20° C. As can be seen, methyl cyclohexane gave the highest viscosity and tetralin, the lowest. Methyl cyclohexane was thus thermodynamically the best solvent, followed by hexane and isooctane, while toluene and tetralin were poorest. Results at other temperatures and for other copolymers were analogous.

Specific viscosity is obtained by subtracting the kinematic viscosity of the solvent from that of the solution and dividing the result by the kinematic viscosity of the solvent. It thus shows at any temperature the solution thickening attributable to the solute. Specific viscosities of all copolymers as a function of polymer concentrations in methyl cyclohexane and toluene at -10° C are given in Figures 2 and 3, respectively. In methyl cyclohexane, the samples were arranged in the expected order and their viscosities increased with molecular weight and concentration. In toluene, however, as depicted in Figure 3, the situation was appreciably different. The viscosities of the partially crystalline EP-4 and EP-5 were lower than those of the completely amorphous EP-1 to EP-3. This occured in spite of the fact that EP-5 had the highest molecular weight and the molecular weight of EP-4 was higher than those of EP-1 and EP-3. The behavior in the other three solvents at -10° C was similar; in all cases the specific viscosities of EP-4 and EP-5 were lower than expected on the basis of concentration and molecular weight. At 20° C and above, the specific viscosities in all five solvents were arranged in the same order as in Figure 2.

These data demonstrate that in poor solvents, as temperature is lowered, parameters related to copolymer composition and structure such as ethylene content, size and number of ethylene sequences, and crystallinity can have more influence on solution viscosity than molecular weight.

INTRINSIC VISCOSITIES

The concentration dependence of viscosity can be represented by the following equation (6):

$$n_{sp}/c = [n] + K1[n]^2 c + K2[n]^3 c^2 + \ldots.. \quad (1)$$

where n_{sp} is specific viscosity; n_{sp}/c, reduced viscosity; [n], intrinsic viscosity; and c, polymer concentration. Without the third and higher terms, this is the familiar

TABLE I. Molecular characteristics of ethylene-propylene copolymers

EP copolymer	Ethylene (mole %) (NMR)	$\bar{N}$ (NMR)	En≥3 (NMR)	(CH2)/(CH3) (FTIR)
EP-1	60	4.2	0.35	2.0
EP-2	58	3.9	0.30	2.1
EP-3	70	4.8	0.56	5.1
EP-4	80	6.5	0.62	7.8
EP-5	80	6.1	0.65	7.5

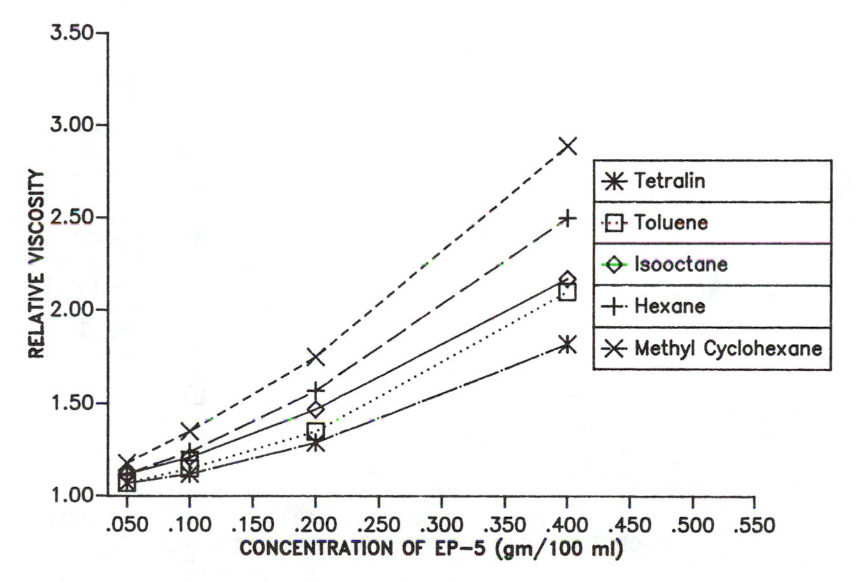

Figure 1. Relative viscosities of EP-5 in different solvents at 20 °C.

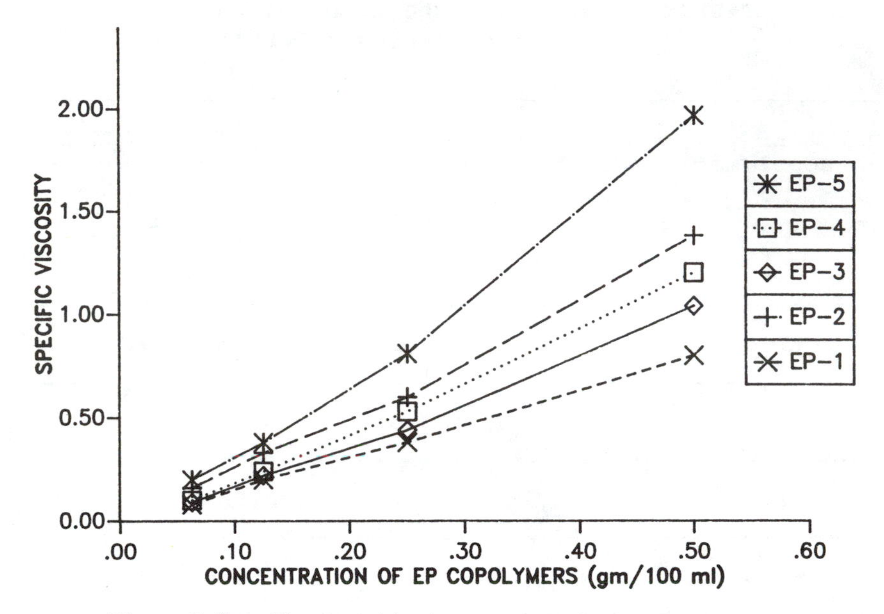

Figure 2. Specific viscosities in methyl cyclohexane at −10 °C.

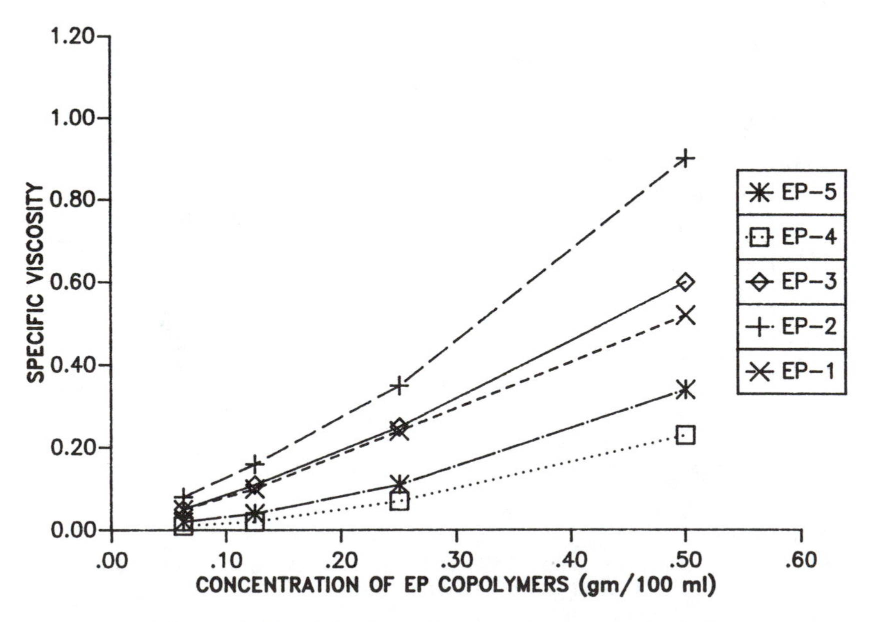

Figure 3. Specific viscosities in toluene at −10 °C.

Huggins equation (7), and the dimensionless parameter K1 is the Huggins constant. Intrinsic viscosities were obtained from this equation by extrapolation of plots of η_{sp}/c vs. c to infinite dilution while Huggins constants were calculated from the slopes of the plots.

Figure 4 shows the temperature dependence of intrinsic viscosities in methyl cyclohexane. For all cases, intrinsic viscosities decreased modestly as the temperature rose; the biggest decrease in [η], about 20 %, was obtained for EP-5. This indicates some deterioration in polymer-solvent interaction and solubility with increasing temperature. Analogous curves for toluene are plotted in Figure 5. The intrinsic viscosities for the amorphous samples changed only modestly with temperature, first increasing as the temperature rose from -10° C to 20° C, and then remaining reasonably constant or decreasing slightly as it rose further to 50° C. This was not the case for the partially crystalline samples EP-4 and EP-5. For these samples, [η] increased by over 500 % between -10° C and 10° C, after which it increased much more moderately when the solutions were warmed further to 50° C. Such a rapid rise in [η] with temperature has been ascribed to an endothermal heat of mixing and increasing solvent power with rising temperature (6,8,9).

Plots of the ratio of intrinsic viscosities for the copolymers in methyl cyclohexane and toluene at the two extreme temperatures, -10° C and 50° C, against mole % ethylene are shown in Figure 6. This ratio exceeded one for all copolymers in methyl cyclohexane, indicating a decrease in [η] with temperature but was essentially independent of molecular weight and composition. It was less than one in toluene, showing the improved polymer-solvent interaction as the temperature was raised and decreased enormously for the two copolymers with 80 mole % ethylene and small amounts of crystallinity.

Intrinsic viscosities and Huggins constants between -10° C and 50° C for EP-1, EP-2, and EP-5 in hexane, isooctane and tetralin are summarized in Tables II-IV. In tetralin, intrinsic viscosities for EP-1 and EP-2 did not change significantly with temperature. However, the viscosity for EP-5 increased by 200% between -10° C and 20° C and remained essentially constant between 20° C and 50° C. Tetralin thus exhibited the same general features as toluene, behaving as a poor solvent for partially crystalline EP copolymers below 20° C and a satisfactory one for the amorphous samples. The viscosity-temperature relationships for EP-1 and EP-2 in hexane and isooctane were essentially similar to those discussed above for toluene and tetralin. For instance, for EP-1 and EP-2, [η] values in hexane decreased from 1.05 and 1.70 dl/gm at -10 C to 0.74 and 1.25 dl/gm at 50° C. The corresponding [η] decreases in isooctane were

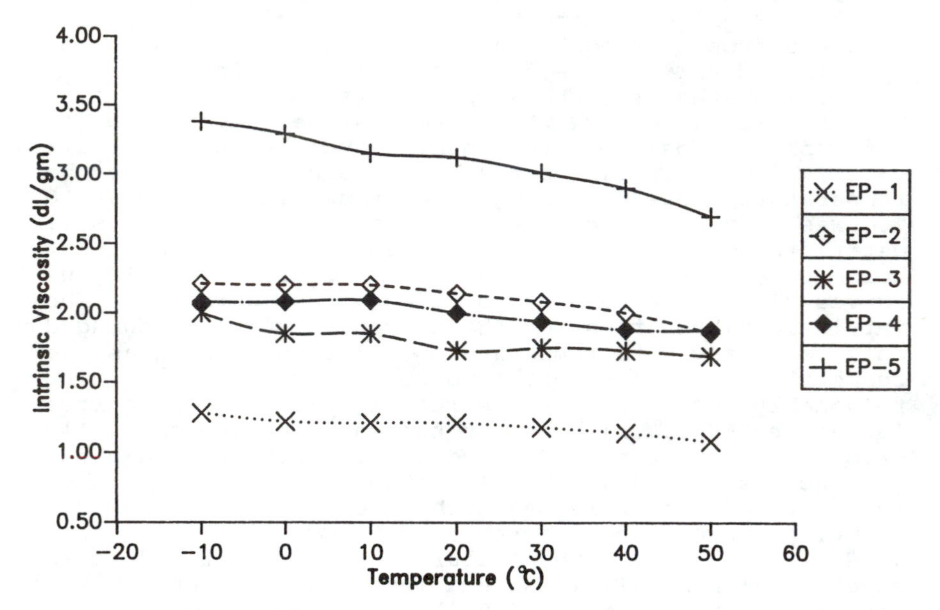

Figure 4. Intrinsic viscosities in methyl cyclohexane.

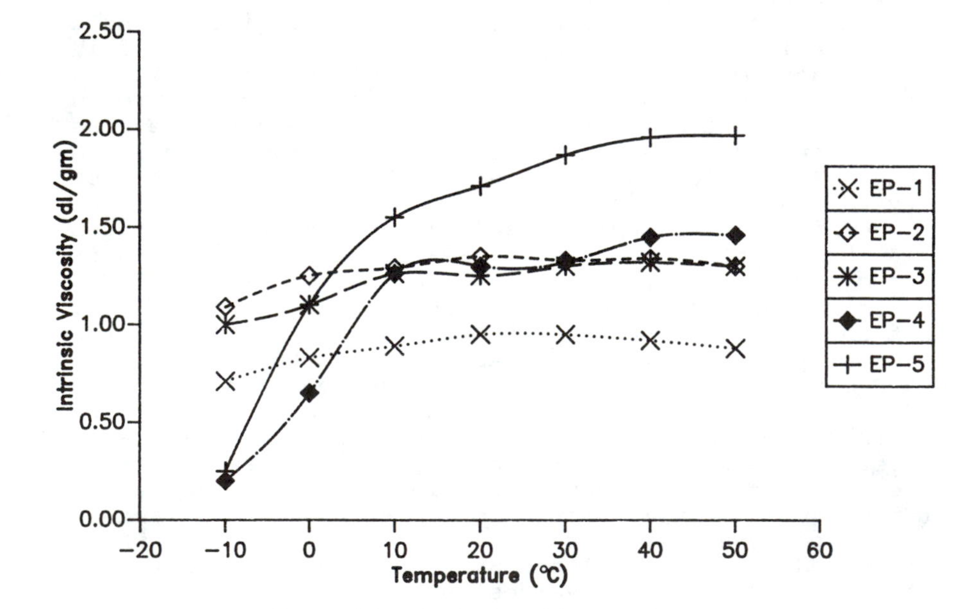

Figure 5. Intrinsic viscosities in toluene.

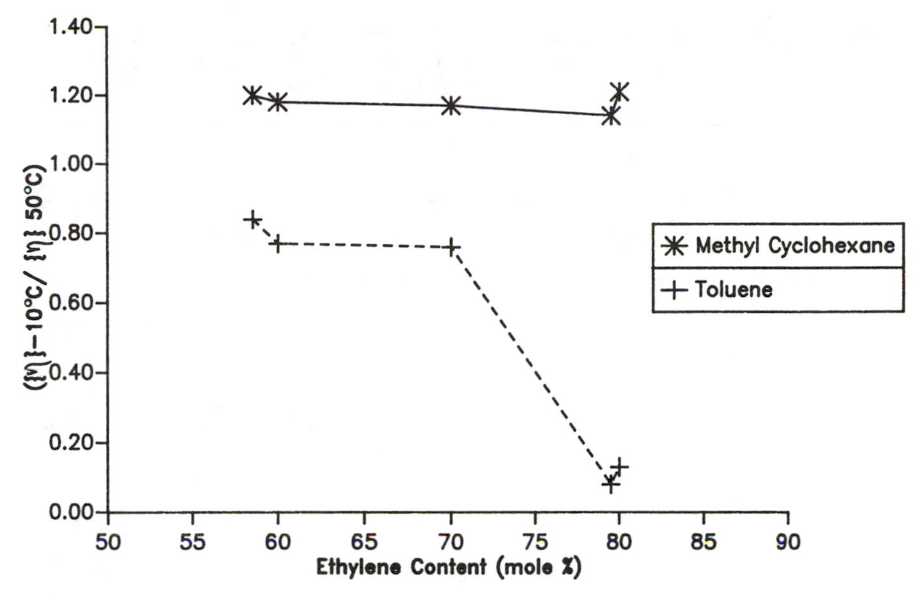

Figure 6. Ratio of intrinsic viscosities at two extreme temperatures, −10 and 50 °C.

TABLE II. Intrinsic viscosities, [n], and Huggins constants, K1, of EP-1, EP-2 and EP-5 in hexane

Temperature (°C)	EP-1 [n] (dl/gm)	EP-1 K1	EP-2 [n] (dl/gm)	EP-2 K1	EP-5 [n] (dl/gm)	EP-5 K1
-10	1.05	0.52	1.70	1.54	1.20	5.12
0	1.05	0.53	1.65	1.52	2.20	4.85
10	0.90	0.56	1.55	1.54	2.30	4.53
20	0.93	0.63	1.50	1.55	2.20	4.32
30	0.75	0.69	1.35	1.60	2.10	4.26
40	0.73	0.65	1.30	1.67	2.05	4.30
50	0.74	0.62	1.25	1.66	2.00	4.28

TABLE III. Intrinsic viscosities, [n], and Huggins constants, K1, of EP-1, EP-2, and EP-5 in isooctane

Temperature (°C)	EP-1		EP-2		EP-5	
	[n] (dl/gm)	K1	[n] (dl/gm)	K1	[n] (dl/gm)	K1
-10	1.10	0.62	1.70	0.83	0.45	3.06
0	0.98	0.64	1.60	0.82	1.10	3.10
10	0.95	0.67	1.50	1.33	1.85	2.67
20	0.95	0.67	1.50	1.37	1.90	2.64
30	0.90	1.02	1.40	1.56	1.80	2.47
40	0.85	1.04	1.35	1.86	1.80	2.82
50	0.80	1.02	1.30	1.82	1.75	2.84

TABLE IV. Intrinsic viscosities, [n], and Huggins constants, K1, of EP-1, EP-2, and EP-5 in tetralin

Temperature (°C)	EP-1		EP-2		EP-5	
	[n] (dl/gm)	K1	[n] (dl/gm)	K1	[n] (dl/gm)	K1
-10	0.90	0.35	1.10	1.33	0.70	2.55
0	0.95	0.32	1.10	1.17	1.15	2.67
10	1.00	0.38	1.15	1.25	1.90	1.50
20	0.95	0.38	1.15	1.33	2.10	1.58
30	1.00	0.42	1.15	1.38	2.10	1.60
40	1.00	0.50	1.15	1.35	2.05	1.63
50	0.95	0.46	1.10	1.34	2.05	1.62

from 1.10 and 1.70 dl/gm at -10°C to 0.80 and 1.30 dl/gm at 50°C. Intrinsic viscosity changes of EP-5 were likewise similar to those in toluene and tetralin. In hexane, the value increased rapidly between -10 and 10°C from 1.20 dl/gm to 2.30 dl/gm and then decreased slowly to 2.00 dl/gm as the temperature rose to 50°C. In isooctane, it increased from 0.45 dl/gm at -10°C to 1.90 dl/gm at 20°C and then gradually fell off to 1.75 dl/gm at 50°C. Hexane and isooctane thus also behaved as poor solvents for EP-5 below 10-20°C.

HUGGINS CONSTANTS

Huggins constants, K1, for all copolymers in methyl cyclohexane and toluene are plotted as a function of temperature in Figures 7 and 8. In methyl cyclohexane, all K1 values increased with temperature. In toluene, a much poorer solvent, they passed through a minimum and then rose. As expected from the [n] values, results for EP-5 showed the greatest change; K1 decreased from 2.94 at -10°C to 2.35 at 20°C and then increased to 2.58 at 50°C. The K1 values in the other solvents, shown in Tables II-IV, exhibited similar trends with respect to their intrinsic viscosities. In general, the trend for the K1 values was in the opposite direction from that observed for intrinsic viscosities, as has been reported in the literature (9,10).

INTERPRETATION OF RESULTS

In considering the results discussed above, it is necessary to take into account both the solvents and copolymers. It is clear that methyl cyclohexane was the best solvent as it gave the highest relative and intrinsic viscosities for all copolymers. Intrinsic viscosities in methyl cyclohexane decreased with rising temperature, indicating that heats of mixing were negative over the entire temperature range. Since [n] decreased, copolymer-solvent interaction diminished as temperature rose and, therefore, the EPs contributed less to solution viscosity at elevated temperatures, just the opposite of what is required for good VI improvers.

Intrinsic viscosity-temperature data in the other solvents showed that they were good solvents for the amorphous copolymers, EP-1, EP-2 and EP-3, though not quite as good as methyl cyclohexane. Figure 5, which is typical for these solvents, showed that [n] either increased slightly or did not change much with temperature. However, below 10-20°C these solvents had vastly reduced solvating power for the two copolymers

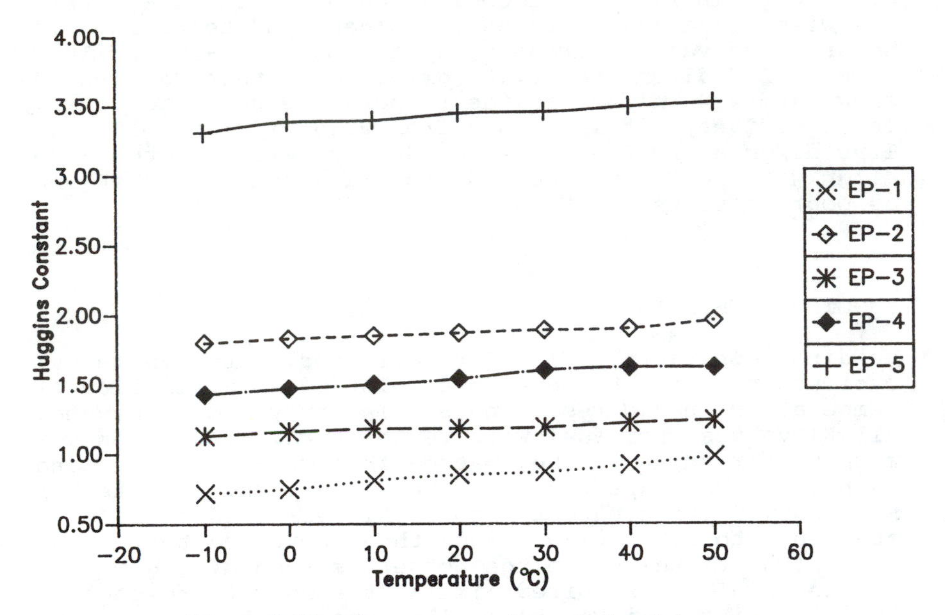

Figure 7. Huggins constants in methyl cyclohexane.

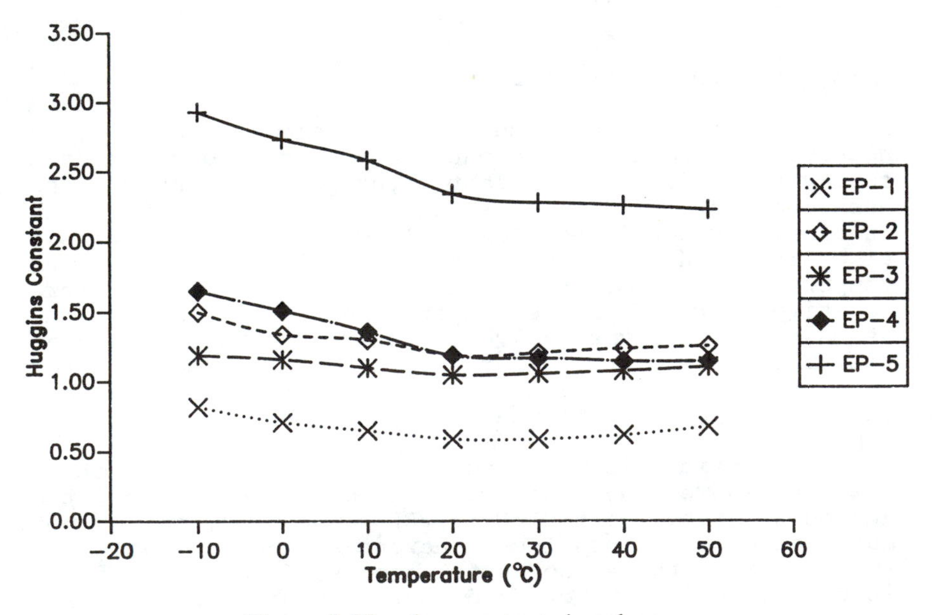

Figure 8. Huggins constants in toluene.

with high ethylene content and small amount of crystallinity. Our data showed that the naphthenic structure favored solubility and solvent-polymer interaction in comparison with paraffinic solvents and that the copolymers were least soluble in aromatics. The distinction between good and poor solvents was most pronounced for the partially crystalline copolymers EP-4 and EP-5 at low temperature.

The effect of the EP structure on solubility can be seen most clearly by comparing the behavior of the copolymers with 58-70 mole % ethylene to those with 80%, particularly at low temperatures. In the poorer solvents, the intrinsic viscosities of the two partially crystalline copolymers decreased rapidly as the temperature was lowered below about 10° C. No such decrease was observed for the three amorphous copolymers. The small [η] values for EP-4 and EP-5 at low temperatures in all solvents except methyl cyclohexane can be understood by examining the differences in structure between the two sets of copolymers. As shown in Table I, EP-4 and EP-5 had a larger fraction of longer ethylene sequences than the three totally amorphous copolymers, and it is presumably these longer ethylene sequences which formed the crystallites in the bulk copolymer. We believe that at low temperature in reasonably poor solvents these ethylene sequences can organize into partially ordered domains or aggregates which are held in solution by the more soluble mixed ethylene-propylene sequences and shorter ethylene and propylene segments incapable of crystallizing. This picture is similar to one proposed some time ago by Filiatrault and Delmas (11). These partially ordered domains would lead to contraction of the copolymer in solution and some reduction in viscosity. In addition, they could also give solutions with fewer chain entanglements and increased copolymer mobility, leading to further viscosity decrease (12).

It is thus clear from this study that the presence of a high amount of ethylene (80 mole %) and longer ethylene sequences in EP copolymers can significantly lower the solution viscosity of EP copolymers in relatively poor solvents at temperatures below 20° C. Dilute solution viscosities of amorphous EP copolymers show relatively little variation with temperature under similar conditions.

REFERENCES

1. Kapuscinski, M. M.; Sen, A.; Rubin, I. D. SAE Pub. No. 892152, 1989.
2. Randall, J. C. Polymer Sequence Determination, C-13 NMR Method, Academic Press, New York, 1977.
3. Johnston, J. E.; Bloch, R.; Ver Strate, G. W.; Song, W. R. US Patent 4,507,515, 1985.

4. Ham, G. E. High Polymers, Interscience, New York, 1964.
5. Fox, T. G., Jr.; Fox, J. C.; Flory, P. J. J. Amer. Chem. Society, 73, 1901, 1951.
6. Bohdanecky, M.; Kovar, J. Viscosity of Polymer Solutions, Elsevier Publishing, New York, 1982.
7. Huggins, M. L. J. Am. Chem. Soc. 64, 2716, 1942.
8. Maderek, E.; Wolf, B. A. Angew. Makromol. Chemie, 161, 157, 1988.
9. Schott, N.; Will, B.; Wolf, B. A. Makromol. Chem., 189, 2067, 1988.
10. Schmidt, J. R.; Wolf, B. A. Macromolecules, 15, 1192, 1982.
11. Filiatrault, D.; Delmas, G. Macromolecules, 12, 65, 69, 1979.
12. Rubin, I. D.; Stipanovic , A. J.; Sen, A., presented at 1990 meeting of Society of Tribologists and Lubrication Engineers (STLE), Denver, May 1990.

RECEIVED July 18, 1990

Chapter 17

Rheological Properties of High-Molecular-Weight Poly[(methyl methacrylate)-*co*-ethylacrylate-*co*-butylacrylate] Solutions

Influence of Polymer–Solvent Interactions

Wendel J. Shuely and Brian S. Ince

U.S. Army, SMCCR-RSC-P, Aberdeen Proving Ground, MD 21010–5423

The effects of polymer-solvent interactions on rheological viscoelastic properties is being investigated. Relatively small volume fractions, 0.02-0.06, of an ultra high molecular weight rheological processing aid, terpolymer poly(methylmethacrylate-co-ethylacrylate-co-butylacrylate), poly (MMA/EA/BA), form polymer solutions in the semidilute regime. Over 30 solutions were formulated to define several interaction categories. Polymer-solvent interactions were characterized by several methods: polymer cohesion phase diagram coordinates, limiting viscosity number, proton donating strength, and solubility with control homopolymers. Rheological measurements included steady shear first normal stress difference, apparent viscosity, hysteresis, transient, dynamic viscosity and storage and loss moduli. Relationships between degree or type of interaction and rheological properties have been formulated.

The fluid dynamics of liquids can be modified and controlled by the addition of polymer additives (1-3). Ultrahigh, megadalton (million gm/mole) molecular weight (MW) polymers are effective at low concentrations. Most specific industrial process applications involving the control of fluid dynamics are proprietary; examples of applications that involve free surface flow are roll splatter, bulk liquid spraying, aero-stripping, anti-misting of fuels, and spray droplet distribution control in crop pest management. The process phenomena of interest in applications often are characterized by large nonlinear deformations, simultaneous shear and extensional flows, complex geometries and free surface flows. The ideal geometries and flows imposed by rheological evaluation are a useful framework for understanding the effects of material properties on these complex processes.

The large deformations present in the process phenomena of interest suggest correlations with steady shear measurements instead of the small strains from dynamic mode measurements. However, oscillatory measurements were also recorded for potential correlation to steady shear rheological properties at low shear rates. Given this selection of steady shear measurements, two rheological properties were considered: first normal stress difference and apparent viscosity, including the power law coefficient for the shear rate dependence. The most reliable quantitative measure of solution elasticity under simple steady shear is first normal stress difference. Correlations of both of these rheological properties with solvent effects were investigated, although first normal stress difference had shown more promising prediction of process fluid dynamics and had allowed direct comparisons of polymer solutions over a wider variety of concentration, MW and solvent regimes.

Polymer-Solvent Interactions: The rheological and viscoelastic properties of polymer solutions are influenced by the polymer MW and MW distribution (MWD), the chemical structural features and configuration, the concentration, and polymer-solvent interactions (4). The quantitative range of influence of polymer-solvent interactions is relatively limited when compared with these other variables. Furthermore, concentration, MW and structural properties provide a continuum of solution properties, although the values of these variables might produce solutions in different coil density regimes. On the other hand, solvent effects are finite in that they are bounded between nominally theta solvents and maximally interacting solvents. Most studies only employ theta and 'good' solvents to bracket the extent of solvent effects. One purpose of this investigation was to determine the influence of the full spectrum of solvents between near theta conditions and highly expanded polymer coils on rheological properties and, in addition, determine the influence of several other qualitatively unique solvent sets on rheological behavior. Among these solvent sets are: (1) solvents that are insoluble with the major comonomer components and only 'dissolved' by preferential interaction with the minority comonomer components, (2) solvents with specific hydrogen bond interaction as proton donor, weak acceptor solvents, and (3) solvents that precipitate during shear.

Semidilute Regime: Variables held constant were polymer composition, MW, MWD, concentration and temperature. The concentration of 4.7 ± 0.1 g/dL resulted in coil density based on a concentration x limiting viscosity number (LVN) product of ca. 5-20, spanning the semidilute regime.

Characterization: Several physical, chemical and rheological characterization methods are being applied. Polymer Cohesion Phase Diagrams (PCPD) of the terpolymer (5) and homopolymer controls were employed to identify solvents with specific solution interactions. The phase diagrams consist of a bounded area on a plot of solubility or insolubility in terms of a self-consistent solvent parameter set (Hansen parameters, ASTM D3132 solvent parameters (6), etc.). The boundary coordinates approximate theta conditions or 'poor' solvents and the coordinates toward the center provide a selection of 'good' solvents. This semi-quantitative selection of solvents was then further quantified by measurement of LVN. LVN or intrinsic viscosity measurement provided an estimate of coil expansion and degree of interaction (7). Linear Solvation Energy Relationship (LSER) scales were used to identify and quantify proton donating, weak or nonacceptor solvents (8). The LSER system of scaling solvent parameters is more rigorous than the various cohesion parameter systems and includes specific terms for proton donating or acceptor strength. Unique solvents with relatively strong proton donating capability relative to proton acceptor strength (e.g. chloroform, methylene chloride) were identified employing LSER data (8). Straightforward solubility determinations (6) were employed to classify solvent/nonsolvent interactions with homopolymer controls for each of the terpolymer repeat units. Rheological measurements provided a characterization of viscous and viscoelastic properties: steady shear hysteresis, first normal stress difference (FNSD), apparent viscosity and critical shear rate for onset of FNSD are under evaluation. Dynamic and transient data have also been recorded for evaluation.

Experimentation

Materials. The specific polymeric additive investigated is poly(methylmethacrylate-co-ethylacrylate-co-butylacrylate), (poly(MMA/EA/BA)), Rohm and Haas Acryloid (now Paraloid) K125 (9). The $\overline{Mw}$ is 1.5 megadalton (million gm/mole) by light scattering (LS) and $\overline{Mw}/\overline{Mn}$ = 1.8 (10). Size exclusion chromatography (SEC) and Light Scattering (LS) show the tail extending from 1 to ca. 15 megadalton. The structure is that of a linear random copolymer produced by aqueous emulsion polymerization (recovered by spray drying). A lot (3-6326) of several hundred Kg has been set aside for detailed characterization and rheological process correlations.

The structure and molar monomeric ratios were determined by NMR with Mass Spectral (MS) confirmation and are (MMA:82, EA:12, BA:6) (11). All monomeric repeat units are dipolar, proton acceptors. The homopolymer control samples were employed only for solubility determinations and were as follows: poly(methylmethacrylate) (PMMA), DuPont Elvacite 2041, lot no. 2200, $\overline{Mw}$ ca. 0.4 megadalton; poly(ethylacrylate) (PEA), Rohm and Haas, lot no. PS-1, $\overline{Mw}$ ca. 2 megadalton; and poly(butylacrylate) (PBA), Scientific Polymer Products, lot no. 3, $\overline{Mw}$ ca. 0.06 megadalton. The commerical quality solvents were used as received.

Procedures. Solubility Determinations: The poly(MMA/EA/BA) solutions were prepared at a concentration of 4.7 $\pm$ 0.1 g/dL. The PEA was received as an aqueous emulsion and precipitated using methanol. The PBA was received as a 26% solution in toluene and dried to constant weight. All polymers were dried overnight in a vacuum oven at ca. 40 ^{o}C before solution preparation. All homopolymer PMMA, PEA, PBA solution concentrations ranged from ca. 5-10 g/dL and all solutions were prepared at room temperature (ca. 25 ^{o}C). All solutions were placed in a shaker ca. 1-5 weeks before final visual solubility determinations were made.

LVN: The LVN was employed as a measure of the coil expansion of the polymer coil in the solvent. A poor solvent, approaching a theta solvent, has a low LVN relative to a good solvent, which has a high LVN. Procedures for obtaining the LVN from dilute solution extrapolations have been published (7). LVN measurements were performed by Springborn Laboratories, Enfield, CT.

Rheology: Rheological measurements were performed using the Rheometrics Fluids Rheometer (RFR) Model 7800 with cone and plate geometry at a temperature of 25.0 $\pm$ 1.0 ^{o}C. The initial rheological data from the steady rate sweep experiment were further analyzed; viscosity data were reduced to obtain power law coefficients. FNSD versus shear rate squared was reduced to obtain FNSD coefficients and zeroshear FNSD. For many solvents, the onset of measurable FNSD occurred in or above the transition to the nonlinear region which has complicated data analysis. The linear region analyzed was defined as occurring above 1% full scale normal transducer output and including the linear data until the correlation coefficient dropped below 0.98.

Results and Discussion

Solvent Sets. Preferential polymer-solvent interactions can be viewed in complexity as a range from a matrix of a homopolymer with a single solvent through a copolymer with cosolvents. The system investigated consists of a terpolymer with single solvents. The investigation was structured to qualitatively determine solvent-nonsolvent classes for each solvent with the three comonomeric components by independent determination of solubility with homopolymer analogs of the comonomers. Each of the three homopolymers, PMMA, PEA and PBA, was individually tested for solubility in each solvent. For a terpolymer, this generates eight hypothetical classes. Based on the qualitative rheological results and the small number of solvents in several classes, these were collapsed into two classes as shown by the division in Table I. Note that in the nomenclature used herein, the "/" denotes the usual copolymerized polymer, poly(MMA/EA/BA). The "-" is used for homopolymer solubility classes based on combinations of independent homopolymer solubility experiments; for example, PEA-PBA represents a group of solvents in which both PEA and PBA are soluble, but not PMMA. Polymer-solvent interactions were further evaluated according to a matrix of nonpolar, polar, and hydrogen-bond interactions. The poly(MMA/EA/BA) is a completely aprotic dipolar structure. The solvent sets are defined below.

Table I. Formation of Solubility Classes by Permutation of Solubility (S) and Insolubility (I) of Component Homopolymers of the Terpolymer

Terpolymer Solubility Classes		Component Homopolymer Solubility PMMA	PEA	PBA
1. PMMA-PEA-PBA Soluble		S	S	S
2. PMMA-PEA Soluble	1)	S	S	I
3. PMMA-PBA Soluble		S	I	S
4. PMMA Soluble		S	I	I
5. PEA-PBA Soluble		I	S	S
6. PEA Soluble	2)	I	S	I
7. PBA Soluble		I	I	S
8. PMMA-PEA-PBA Insoluble (Not applicable)		I	I	I

1) Soluble with 82-100% of comonomer content, e.g. MMA, MMA-BA, MMA-EA or MMA-EA-BA.

2) Soluble with 6-18% of comonomer content, e.g. BA, EA or EA-BA, labelled "Nonsolvent w/0.82+".

Aprotic Solvent Set: The majority of solvents were aprotic dipolar solvents, further classified by a nonsolvent/solvent determination for each terpolymer component, as described above. Aprotic is used in the usual sense here as indicating absence of a proton capable of hydrogen bonding.

Theta Solvent Set: Solvents were identified that were theta or near-theta solvents and also belonged to the class demonstrating solubility with all three terpolymer components, that is, none of the homopolymer controls were insoluble with the terpolymer theta solvent.

Nonsolvent w/0.82+ Solvent Set: The term "Nonsolvent w/0.82+" defines an operational subset of solvents in which the solvent dissolves the terpolymer but was a nonsolvent for at least the majority 0.82 MMA fraction based on a solubility determination with PMMA homopolymer. A comparison of the rheological properties of "Nonsolvent w/0.82+" set with theta solvents was of interest since the low viscometric coil size estimates were similar but based on different physiochemical phenomena. The theta solvents appear to systematically extend the relationship of increasing FNSD and apparent viscosity with decreasing LVN. The solvents that are in the "Nonsolvent w/0.82+" set display extremely high viscoelasticity as evidenced by FNSD, apparent viscosity and dynamic properties. In general, hysteresis experiments do not demonstrate structure formation in these solvents although over 82% (PMMA) of the chain is nominally insoluble in the solvent.

Proton Donor Solvent Set: There are a select number of solvents with low to moderate proton donating strength, but even weaker acceptor strength (in most media); examples are chloroform, methylene chloride, pentachloroethane, trichloroethylene and pentachlorocyclopropane. These solvents should specifically interact with proton acceptor carbonyl and ester moieties of the polymer solute. All such solvents investigated showed enhanced coil expansion evidenced by LVN values clustered higher than any other 'good' solvents and correspondingly low values for rheological properties.

Shear Induced Precipitation Solvent Set: Polymer solutions made from four solvents were discovered to undergo shear induced precipitation during rheological measurements: the solvents were dimethyl methylphosphonate, dimethyl formamide, 1-methyl-2-pyrrolidone and trimethyl phosphate. Since only 17 instances of nonaqueous shear induced precipitation have been reported (12), the possibility of additions to this unique class of polymer-solvent systems was further investigated. The steady shear FNSD and apparent viscosity were recorded during the shear induced precipitation process as long as homogeneous solution was present but neither the FNSD nor the apparent viscosity from the precipitation experiments were used in the regression analyses. Clear viscous solutions had been introduced into the cone and plate fixture and low viscosity solvent with solid white polymer pieces were recovered. The precipitations were repeatable and could not be prevented by utilization of various sample preparation and measurement procedures. The water content of the solvents was analyzed and found to be several times higher in the precipitated solutions.

The polymer particles were found to redissolve in fresh, low water content solvent. The phase diagram boundaries of the polymer-solvent system with respect to sorption of water as a cosolvent were evaluated. The precipitating solvents have coordinates on or near the boundary between solubility/insolubility. Thermodynamically, an addition of a small volume fraction of sorbed water as a cosolvent would produce a nonsolvent mixture. The shear then may only initiate precipitation or accelerate the kinetics of precipitation. Previously published polymer-solvent pairs showing shear-induced precipitation will be evaluated for similar water cosolvent effects.

Correlations. Solvent Interaction Effect on FNSD: The range of solvent influence on FNSD can be viewed by inspection of Figure 1. FNSD vs shear rate squared is graphed for examples of each solvent set. The full range of FNSD data is presented including high shear rate data well into the nonlinear region. Descriptions of solvent abbreviations and solubility classes are listed in Table II.

Table II. Codes and Solubility Classes for Representative Solvents (Figure 1)

No.	Codes	Solvent	Solubility Class
1	$CHCL_3$	Chloroform	Proton donating PMMA-PEA-PBA*
2	ACP	Acetophenone	Good PMMA-PEA-PBA
3	DEM	Diethyl malonate	Moderate PMMA-PEA-PBA
4	2HP	2-heptanone	Near theta PMMA-PEA-PBA
5	3HP	3-heptanone	Near theta PMMA-PEA
6	4HP	4-heptanone	Preferential PEA-PBA
7	DPGMME	Dipropyleneglycol monomethylether	Intramolecular PMMA-PEA-PBA
8	TPPO	Tripropyl phosphate	Preferential PEA-PBA

* See Table I for Terpolymer Solubility class codes.

Figure 2 contains a plot of First Normal Stress Difference (FNSD) vs Limiting Viscosity Number (LVN) for a moderately high shear rate of 400/sec. The LVN values provide a quantitative indication of degree of solvent interaction; the LVN values range from about 1.2 to 4.1 representing a low to high degree of coil expansion. One can note that certain solvent sets fall into limited LVN ranges. Only the proton donating solvents ('+','★') have LVN values in the high range between 3.3 to 4.1. Even the lowest values for a proton donating solvent were higher than the best 'good' solvent based on non-specific interactions.

The LVN range from about 1.4 to 3.3 on the regression line in Figure 2 contains three solvent sets.

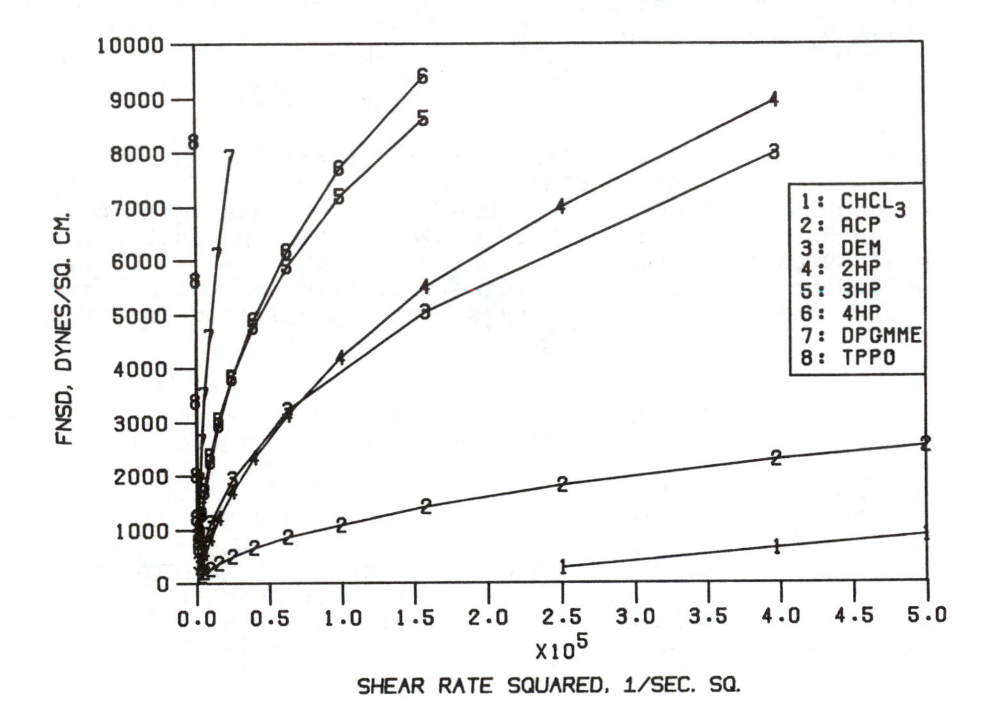

Figure 1. First Normal Stress Difference (FNSD) vs Shear Rate Squared for Examples of All Solvent Sets with Poly(MMA/EA/BA), for Solutions at 4.7 ± 0.1 g/dL and 25.0 ± 1.0° C: Proton Donating PMMA-PEA-PBA(1), PMMA-PEA-PBA (2,3,4), PMMA-PEA (5), PEA-PBA (6,8) and Intramolecular PMMA-PEA-PBA (7). See Tables I and II for abbreviations.

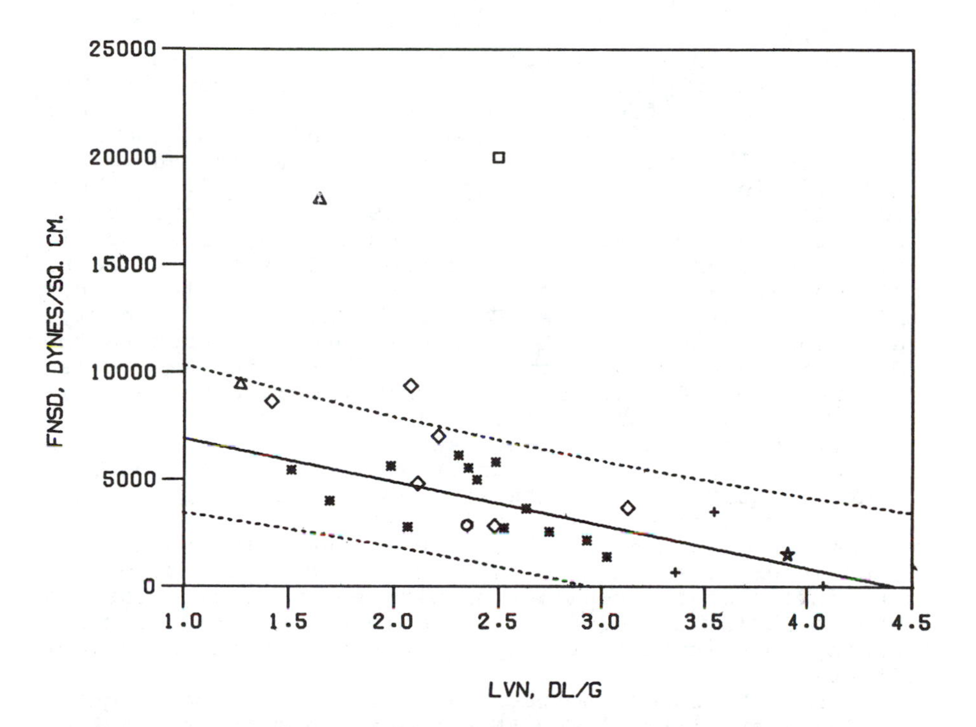

Figure 2. First Normal Stress Difference (FNSD) vs Limiting Viscosity Number (LVN) for All Solvent Sets with Poly(MMA/EA/BA) for Solutions at 4.7 ± 0.1 g/dL and 25.0 ± 1.0 °C: Regression Line = PMMA-PEA-PBA (*) and Proton Donating PMMA-PEA-PBA (+). Not included in the Regression fit: Intramolecular PMMA-PEA-PBA (o), PMMA-PEA (◇), Proton Donating PMMA-PBA (☆), PEA-PBA (Δ) and Proton Donating PEA-PBA (□). See Table I for abbreviations.

The majority of the solvents are contained in the set defined as soluble with all terpolymer components (Code = PMMA-PEA-PBA; '*'). The solvent set consisting of solvents soluble with both PMMA and PEA are contained within this range (Code = PMMA-PEA; '◇'). Qualitatively, the PMMA-PEA set, soluble with 94% of the terpolymer, appears to be following the general trend of the 100% soluble PMMA-PEA-PBA set, however, due to the limited data set, these have not been included in the regression in Figure 2.

The third solvent set in the 1.4 to 3.3 range has been defined as intramolecular hydrogen bonded solvents. These solvents form both cyclic (usually 5-7 member rings) intramolecular hydrogen bonds and acyclic intermolecular hydrogen bonds. Although their LVN values fall within the usual range of 1.4 to 3.3, their viscoelasticity, as measured by FNSD values, are quite scattered, do not fall on the regression line and, in some cases, are not within the Figure 2 axis. In fact, only one intramolecular hydrogen bond solvent falls into the plot axis (Code = PMMA-PEA-PBA; 'O'), methyl salicylate, with an ortho hydroxy proton bonded to the adjacent carbonyl moiety. The other intramolecular hydrogen bonded solvent, dipropyleneglycolmonomethylether, with a six-member intramolecular ring containing a hydroxy proton hydrogen bonded to the ether moiety, is off-scale in Figure 2, at an LVN value of 1.64 and FNSD value of 28,000. From the limited data in this set, one cannot yet determine if this set will form a unique regression equation or will be generally anomalous. Trace impurities, such as water or other protic solvents, can influence the equilibrium between intramolecular and intermolecular hydrogen bonding solvent conformations and thereby contribute to the appearance of anomalous solvent effects.

The LVN range between 1.5 to 2.0 appears quite complex, at first. The FNSD values seem to be rather randomly distributed between 3,000 and 25,000 dynes/sq. cm., although some solvents lie upon the regression line. Actually, FNSD values range off-scale up to 900,000 dynes/sq. cm. (not plotted) within this 1.5 to 2.0 LVN range. One can begin to unravel the apparent random nature of the viscoelastic FNSD measurements by noting the solubility classifications and solvent data sets associated with the data points. For the 1.5 to 2.0 LVN range, no instance of anomalously high (defined as: above the regression line 95% confidence interval) FNSD measurements were recorded for solvents soluble with the 82% to 94% majority comonomer content. All solvents that were soluble with the 18% comonomer fraction (Code = PEA-PBA; 'Δ') have FNSD values at or above the regression 95% confidence interval; this includes several PEA-PBA solvents that are off-scale at very high FNSD values.

FNSD coefficient can be independent of solvent effects for solutions of high MW-concentration products within the concentrated, network entangled regime (4). Within the semidilute regime for C x LVN = 5-20, and $\overline{C}$ x $\overline{Mw}$ = 7.1 g/dL megadalton, the FNSD increases with decreasing LVN as shown in Figure 2. The FNSD data has been analyzed as a function of specific shear rates of 100, 400, 1000/sec and as a function of FNSD coefficient at the onset of significant (>1% normal force transducer full scale) normal force response. The relationships at various shear rates are similar and data obtained to date can be summarized by the following equation.

$$\text{FNSD (@ 400/sec)} = 8900 - (2000 \times \text{LVN})$$

The equation represents all of the 17 PMMA-PEA-PBA soluble solvents including the proton donors ('*' and '+'). Excluded are the 11 of 28 solvents comprising the intramolecular PMMA-PEA-PBA (O), PMMA—PEA soluble (◇), proton donating PMMA-PBA (☆), PEA-PBA soluble (Δ), proton donating EB (□) and shear-induced precipitation solvent set (not shown on plot).

Solvent Influence on Apparent Viscosity: The solvent influence as estimated by LVN is summarized in the equation.

$$\text{Apparent Viscosity (@ 10/sec)} = 8.1 \times e^{-(0.02 \times \text{LVN})}$$

The results include over 16 solvents soluble in at least 82 mole percent of the copolymer content and cover all solubility classes. The slope shows a nominal decrease of apparent viscosity with increasing LVN but is not statistically different from a zero slope at a polymer concentration of 4.7 g/dL. Limited experiments at lower concentrations of 2-3 g/dL found a slightly increasing apparent viscosity with increasing LVN; at higher concentrations of 6.0 g/dL a slightly decreasing apparent viscosity with increasing LVN was found. Therefore, the incipient slope change from positive to negative occurs between C x $\overline{Mw}$ values of 4.5 and 7.1 g/dL megadalton. Overall, ten additional solvents were studied to better define the FNSD vs LVN relationship; additional apparent viscosity vs LVN data did not significantly influence the slope.

Conclusions

Range of Solvent Effects. It would be desirable to be able to summarize the magnitude of the solvent effect on rheological viscoelastic properties in the same manner that MW and concentration have been presented in the past. The effect of MW and concentration on rheological properties has been appropriately summarized in terms of the slope of the property-variable relationship, for example, change in FNSD per decade change in $\overline{Mw}$ (4).

Solvent effects are more complex in that the slope and even the direction of the relationship depend on the coil density regime; in addition, there is increased complexity due to the various special solvent effects considered here (preferential, proton donating). One approach to summarizing the solvent effect data takes advantage of the finite extent of solvent interaction between the two extremes of insoluble or theta conditions and maximal coil expansion. One can then ratio the rheological properties at or near these two limits in solvent interaction. One can also view the range or ratio of rheological properties with respect to the range in LVN or intrinsic viscosity over the solvent set.

These LVN, apparent viscosity and FNSD values, and the ratios for the limits of the solvent set are listed in Table III. Note that these ratios apply to a specific semi-dilute concentration-MW value and that solvent effects disappear at high values of FNSD at the high coil densities within the entangled regime. The ratio of high/low rheological values are listed in Table III for the example of a moderate coil density within the semidilute regime of C x $\overline{Mw}$ = 4.7 x 1.5 = 7.1 g/dL megadalton.

Table III. Solvent Effect Ranges and Ratios for Rheological Properties of Polymer Solutions at a Moderate Coil Density (Concentration x $\overline{Mw}$ = 7.1 g/dL megadalton) (See Table I for Abbreviations)

	Solvent Set			
Property	Within PMMA-PEA-PBA Soluble	Within Proton Donor	Within PEA-PBA Soluble	Between Proton Donor and PEA-PBA
LVN (dL/g)	(3.03/1.27) = 2.4	(4.08/3.36) = 1.2	(2.12/1.27) = 1.7	(4.08/1.27) = 3.2
Apparent Viscosity (poise) @ 10/sec	(11.7/4.05) = 2.9	(16.1/2.52) = 6.4	(29.2/5.18) = 5.6	(29.2/2.52) = 11.6
First Normal Stress Difference (dynes/sq. cm.) @ 400/sec	(6800/1420) = 4.8	(6660/100.) = 67.	(900K/9400) = 96.	(900K/100.) = 9000.

Two types of ranges are listed; 'within' a solvent class refers to the ratio between the highest/lowest values for a single class. 'Between' solvent classes refers to the ratio between the highest value of one class and the lowest of the other class. The properties listed in the first column are LVN, apparent viscosity at low shear (10/sec), and first normal stress difference at a moderate shear rate (400/sec). The solvent sets defined in adjacent columns include: within PMMA—PEA-PBA soluble, within proton donor, within PEA-PBA soluble, and between proton donor and PEA-PBA soluble.

The PMMA-PEA-PBA soluble solvent set provided a LVN ratio of 2.4, ranging from near theta solvents to maximal values from non-specific interactions. The range of apparent viscosity and first normal stress values measured from this range is about 3X and 5X, respectively. The PMMA-PEA-PBA soluble solvent set of nonspecific interactions is bracketed at higher LVN values by the proton donor set and at lower LVN values by the preferential PEA-PBA soluble set. Both of these solvent sets have LVN ranges of about half of the 'regular' solution range, although the preferentially soluble solvent set and near theta solvent LVN values overlap. Both of these solvent sets have wider ranges and larger ratios of apparent viscosity (about 6X) and first normal stress difference (67X-96X) than the PMMA-PEA-PBA soluble set.

The last column records the measurement ranges and their ratios between the extremely high LVN and low rheological property values of the proton donor set and the low LVN and extremely high rheological property values of the preferentially PEA-PBA soluble set. The LVN range has been extended to 3.2X, the apparent viscosity ratio to 11.6X, and first normal stress difference to 9000X. Therefore, although the normal range of solvent influence on rheological properties between 'good' and theta solvents is limited, the selection of polymer-solvent pairs can extend the viscous properties by about one decade and viscoelastic properties by over three decades.

Molecular Interactions. At the molecular level, the interactions can be interpreted in the usual manner, whereby the specific proton donor interactions further minimize polymer-polymer interactions and thereby minimize the friction factor influencing viscous flow and transient entanglements influencing viscoelasticity. The set of solvents soluble with only 6-18% of the comonomer can be interpreted in a general way as containing substantial sequence lengths of the insoluble PMMA; both intracoil and intercoil polymer-polymer contacts would occur and high frictional interaction and transient entanglement densities would be present. Rheo-optical studies might detect any non-statistical, ordered structure in these solutions.

Literature Cited

1. Joseph, D. D.; Beavers, G. S.; Cers, A.; Dewald, C.; Hoger, A.; Than, P.T.; J. Rheo. 1984, 28(4), 325.

2. Matta, J. E.; Harris, J. L.; J. Non-Newtonian Fluid Mech. 1983, 12, 225.

3. Joseph, D. D.; Matta, J. E.; Chen, K.; J. Non-Newtonian Fluid Mech. 1987, 24, 31.

4. Vinogradov, G. V.; Malkin, A. Ya.; In Rheology of Polymers; Springer-Verlag, Berlin, Heidelberg, New York, Ed.; 1980.

5. Shuely, W. J.; Proc. International Union of Pure and Applied Chemistry Symposium, 1986.

6. ANSI/ASTM D 3132-72.

7. ANSI/ASTM D 2857-70 (Reapproved 1977).

8. Kamlet, M. J.; Abboud, J. L. M.; Abraham, M. H.; Taft, R. W.; J. Org. Chem. 1983, 48, 2877.

9. Rohm and Haas Company, MR-2737 D/ce, (1975).

10. Chu, B.; Ying, Q.; Lee, D.; Wu, D. Macromolecules, 1985, 18, 1962.

11. Szafraniec, L. L. ARCSL-TR-79073, (1980), NTIS No. ADA082295.

12. Rangel-Nafaile, C.; Metzner, A. B.; Wissbrun, K. F. Macromolecules, 1984, 17, 1187.

RECEIVED November 26, 1990

Chapter 18

Thermal Characteristics of Waxy, High-Pour-Point Crudes That Respond to Rheology Modifiers

D. S. Schuster[1] and J. H. Magill[2]

[1]Department of Chemical Engineering, Bucknell University, Lewisburg, PA 17837

[2]Department of Chemical and Petroleum Engineering and Department of Materials Science and Engineering, University of Pittsburgh, Pittsburgh, PA 15261

Characteristics of waxy, high pour point crude oils that respond to chemical rheology modifying pour point depressants were determined using Differential Scanning Calorimetry. Waxy oils that show a reduced pour point after treatment crystallize over a broad temperature range, have a broader ΔH_c than do their saturate components, and show a reduced ΔH_c after treatment. After treatment these oils also show a phase transition as do their saturate components.

Waxy, high pour point crudes are low viscosity, Newtonian fluids at high temperatures but exhibit non-Newtonian behavior owing to the precipitation of waxes as the crude is cooled. At the pour point temperature a solid mass is formed.

Transporting these crudes presents technical and economic problems, the magnitude of which depends on the pour point of the oil, transportation method, and ambient temperature (1-5). For example, a 32.2°C (90°F) pour point crude will present congealing problems in most ambient conditions. It may congeal on the walls of tankers resulting in a stock loss, and if it congeals in a pipeline, the required restart pressure may exceed the burst pressure of the line. Whereas, a 10°C (50°F) pour point oil presents few congealing difficulties, but wax may precipitate out of solution and stick to pipe walls or form a sludge at the bottom of a storage tank even in warm climates. This wax deposition can block flow lines, reduce throughput, clog pumps, and inhibit the performance of metering devices that measure transferred crude oil (4).

The goal of successful treatment of high pour crudes with a polymeric pour point depressant chemical is to reduce the temperature of congealing, inhibit waxes from precipitating out of solution, reduce the yield strength of congealed crude, and decrease the crude's viscosity. Unfortunately, pour point depressants are not used to their fullest potential because either the treatment cost is prohibitive, or they are only marginally effective on certain crudes regardless of the treatment level (5).

0097–6156/91/0462–0301$06.00/0

In order to improve their effectiveness and economics, research has been conducted to improve pour point depressant formulations and to study the mechanism of pour point depression (4,6-10). Irani suggests that the interactions of the various precipitating high molecular weight species has resulted in specific additives being effective only with very specific types of crude. A U.S. patent claims that polars interfere with the propagation of the wax, and thus improve the pour point of the oil (8). Additionally, the presence of asphaltenes in the waxes appears to correlate with the maximum extent of pour point depression that can be obtained with a crude oil (7).

In conjunction with these studies, crystallization of the waxes have been explored. Morphology of the waxes has been related to pour point depression by several researchers (9) and another has suggested that the response of the wax to a pour point depressant is a function of its crystal structure (10). A number of researchers have suggested it is the crystallization kinetics that may be the dominating factor. Holder (9) established that mixing different carbon number alkanes increased the extent of pour point depression and suggested that this occurred since the mixtures tended to crystallize more slowly. Reddy, (10) also suggested that additives have more time to interact with slow growing crystals and inhibit their growth.

Differential Scanning Calorimetry (DSC) is used in this study to determine if crystallization differences exist between high pour point oils that do and do not respond to rheology modifiers. Kawamura (11) and others have used the DSC to study waxes. Weslowski (12), using the DSC, noted that some lube oil waxes undergo phase transitions. After our work was completed (13), Redelius (14) used the DSC technique to characterize parafin wax in mineral oil products.

Experimental

Three waxy, high pour point crude oils were analyzed in this study. The three oils were selected since they respond quite differently to pour point depressants as seen in Table I. Crude A, a West African crude, responds extremely well to rheology modifiers, while crude C, a crude from the United States, shows no response to pour point depressants. These oils are characteristic of waxy oils that do and do not respond to rheology modifiers based on previous chemical analysis (6,7).

Sample Preparation. Each of the three crudes were separated into their saturate, aromatic, polar, and asphaltene fractions, using High Performance Liquid Chromatography (HPLC). Gel Permeation Chromatography and Gas Chromatography were used to verify the purity of the saturate fractions. Results of the GPC analyses for the crudes used in this study and several others were presented previously (6,7).

The HPLC separation was performed with a silica gel column equipped with a refractive index detector. The HPLC apparatus has been discussed elsewhere (15). The saturates acquired from the separation are all species which contain only carbon and hydrogen. The aromatic hydrocarbons contain at least one aromatic ring.

Nitrogen containing compounds, phenolic and carboxylic acids comprise the polar compounds. Asphaltenes, by definition, are a solubility classification, and for this analysis include all hexane insoluble material.

Samples treated with pour point depressants were heated along with the pour point depressant to 82°C in a pressurized cell to prevent vaporization of the light ends. Crudes A, B and C were treated with 2000 ppmw of Paradyne-85 (Exxon Chemicals) for subsequent DSC and pour point analysis.

A centrifuge was used to separate the portion of the crude that responds to pour point depression from the portion that does not respond to pour point depressants. Crude B was treated with the pour point depressant, and was transferred to a centrifuge tube at 10°C above the treated crude's pour point. The sample was centrifuged at this temperature, and the top liquid phase was decanted from the bottom wax phase. This liquid phase contained the waxes that were not treated by the rheology modifier, and was subjected to further analysis.

Pour Point Test. The pour point test is a standardized ASTM procedure (ASTM D97) for measuring the congealing temperature of a hydrocarbon fraction. This procedure recommends heating the sample to 46°C and then cooling it in 2.8°C (5°F) decrements to a temperature at which the sample shows no fluidity over a period of 5 seconds when subjected to a pressure head of 2.54 cm. The test is performed in standardized pour point tubes which automatically provide the required head. The pour point is reported as the next highest 2.8°C (5°F) increment above the recorded congealed value. A pour point is only accurate to +/- 2.8°C (5°F).

Since a temperature of 46°C was believed to be inadequate for the complete dissolution of all waxes, an alternate congealing temperature procedure was developed. In this modified procedure, the sample was heated to 82°C in a sealed rotating pressured cell to assure that all of the dissolved waxes were in solution. The container was then slowly cooled at a rate of 10°C/hr to 43°C and transferred to a standard pour point tube and steadily cooled at the same cooling rate until congealing occurred. A programmable, controlled temperature water bath was used for cooling.

Differential Scanning Calorimetry Analysis. Thermal measurements were made with a Perkin-Elmer DSC-2B calorimeter with a scanning auto-zero accessory. All studies were made on samples 2 to 4.5 mg. in weight. The samples were tested in hermetically sealed aluminum pans to prevent vaporization losses. Helium was used to purge the DSC head, and nitrogen was used in the dry box. All samples were loaded at ambient temperature. High instrument sensitivity was used for all scans.

Temperature calibration was performed using 99.9%(+) pure samples of benzene, indium, and mercury. Heats of crystallization were estimated with respect to an Indium standard.

Typical scans required that each sample be heated to 87°C to assure that all wax particles were in solution. Cooling curves were acquired at 40°C/min. using liquid nitrogen as the coolant. Similar results were also acquired with slower cooling rates (5°C/min), where the onset of nucleation/crystallization was only suppressed on the temperature scale (13).

Since the DSC-2B measures the rate of energy absorption or evolution by the sample, heats of transitions were determined. A Hewlett Packard 9816 computer equipped with a digitizer was used to integrate the areas under the cooling curves for each DSC scan.

Results and Discussion

The crudes are identified in Table I as being responsive or non-responsive to polymeric pour point depressant chemicals. Data is presented for Paradyne-85 as it is representative of several of the more effective pour point depressant chemicals. It should be noted that Crude C and the nonresponsive portion of crude B (denoted as b*) have never been affected by any of 49 pour point depressant chemicals that were available at the time of this study (5).

Differential Scanning Calorimetry curves are presented in Figure 1 for the crudes, in Figure 2 for the crudes treated with 2000 ppmw of Paradyne-85, and in Figure 3 for the saturates portion of the crude oils.

Nucleation temperatures and heats of crystallization are also noted on the figures. Non-isothermal cooling was studied because it is representative of the situation that a high pour point oil experiences during treatment with a pour point depressant. Besides, it has been documented that isothermal kinetics can differ from non-isothermal results (11). Additionally, initial microscopic studies indicated that isothermal crystallization proceeded at extremely slow rates, and the sensitivity of the DSC instrument could not measure quantitatively the energy transitions.

When analyzing the differences in responsive crudes (A and B) and non-responsive (C and b*) in Figure 1, the most striking difference is seen in the temperature span of crystallization. The responsive crudes crystallize over a much broader range of temperatures. Differences in nucleation temperature and heats of crystallization are apparent for each of the oils, but the data does not support a systematic trend between the treatable and non-treatable oils.

When treated with a pour point depressant chemical, the scans in Figure 2 show an immediate difference between the responsive and non-responsive oils. The responsive oils (A and B) show a phase transition at approximately -90°C. Again, nucleation temperatures and heats of crystallization of the treated crudes do not show a systematic trend between the successfully treated A and B, and the non-responsive b* and C. However, when compared with the untreated case in Figure 1, Crudes A and B show a decrease in the heats of crystallization when treated, whereas the non-responsive crudes do not.

In Figure 3, the saturates isolated from the responsive crudes A and B show a phase transition similar to that exhibited by the crudes after treatment. It appears that the additive unmasks the phase transition undergone by the saturates and that the other oil components suppress it. Comparison of the crudes in Figure 1 to the saturates in Figure 3 show that the saturates in the non-responsive crudes b* and C have higher heats of crystallization than the entire crude oil exhibits. This indicates that not all of the saturates

Table I
Response of Crude Oils to Pour Point Depressant Chemicals

Crude	Pour Point °C	(°F)	Treatment	Treated Pour Point °C	(°F)
A	21.1	(70)	2000 ppmw Paradyne-85	-17.8	(0)
B	35.0	(95)	"	18.3	(65)
b*	35.0	(95)	"	35.0	(95)
C	43.3	(110)	"	43.3	(110)

b* is the non-treatable portion of crude B isolated by the technique presented in reference 7 and described below.

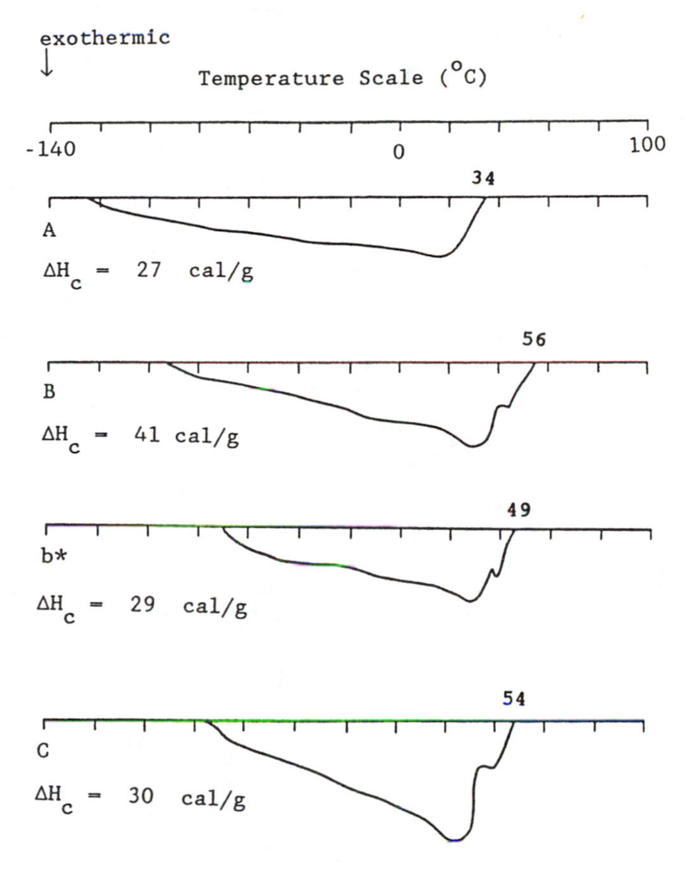

Figure 1. DSC scans of waxy, high pour point crude oils at 40°C/min. cooling rate.

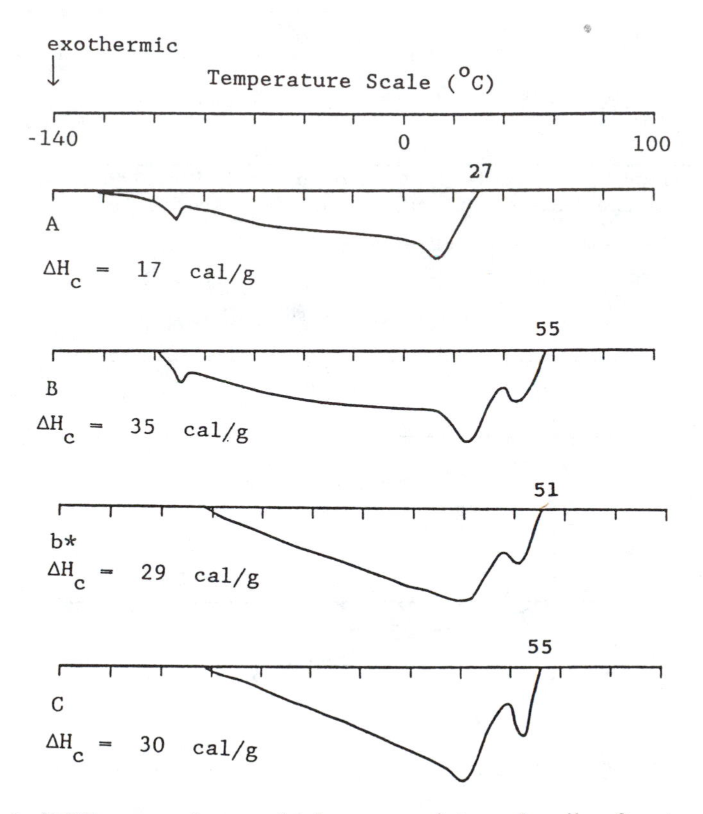

Figure 2. DSC scans of waxy, high pour point crude oils after treatment with 2000 ppmw of Paradyne-85 at 40 °C/min cooling rate.

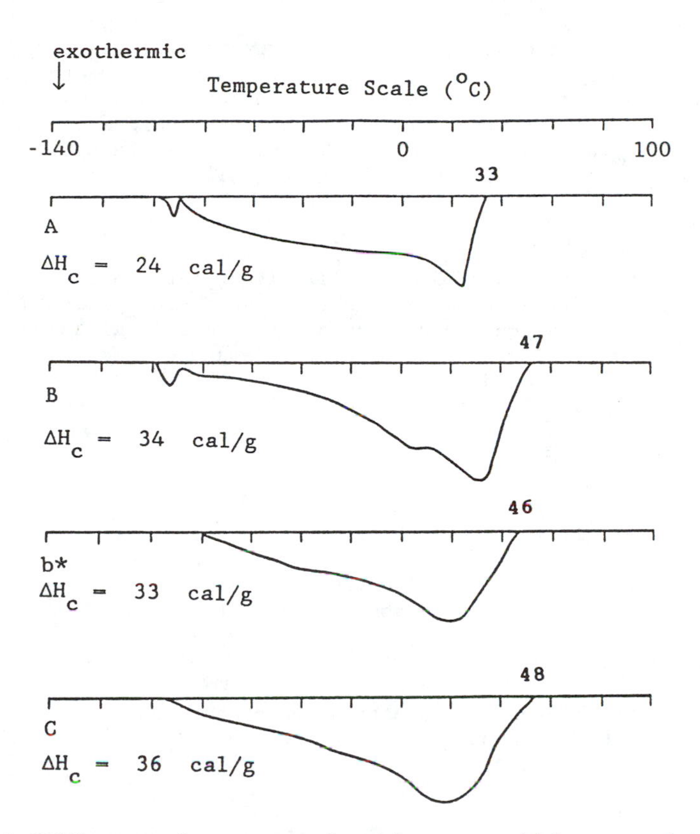

Figure 3. DSC scans of saturates isolated from waxy, high pour point crude oils at 40 °C/min cooling rate.

crystallize in the non-responsive crudes. In contrast, the responsive crudes A and B have higher heats of crystallization than do the saturates from these crudes, indicating that components other than the saturates also crystallize in these oils.

The results depict the complexity that exists in these systems and no simple picture describes it. Various cooling rates are needed where isothermal data may pin down one extreme end of the scale of interactions between the additive and the systems. Of course, one can invent an "exothermicity value" defined simply in cal./g./degree of crystallization undercooling to delineate responsive and non-responsive systems. Alternatively the value of pour point of the crudes (^{o}K) divided by their respective crystallization range may provide another practical way of doing it. Higher numbers would function as a guideline indicating non-responsive systems in which non-interacting additives fail to interact or suppress the onset of nucleation/crystallization and do not prolong the fluidity span of the system responsible for sustaining pourability.

Conclusions

Based upon the analysis of 3 crude oils it appears that crystallization differences exist between waxy crude oils that do and do not respond to a pour point depressant chemical. Although inference to all oils and pour point depressant chemicals can not be made based on only the 3 oils studied, the following characteristics are offered for consideration of responsive and non-responsive crude oils.

Responsive:

- Crude: Crystallizes over a broad temperature range
- Crude ΔH_C > Saturate ΔH_C
- Crude ΔH_C > Saturate ΔH_C > Treated Crude ΔH_C
- Saturates and treated crude show a phase transition

Non-responsive:

- Crude: Crystallizes over a narrow temperature range
- Crude ΔH_C = Crude + Additive ΔH_C < Saturate ΔH_C
- Show no phase transitions

Acknowledgments

Samples for this study were supplied by Gulf Oil Research and Development Company and the preparative HPLC was conducted by R. Ruberto. Technical discussions with C. Irani were greatly appreciated.

Literature Cited

1. Perkins, T.K.; Turner, J.B. Journal of Petroleum Technology, 1971, 23, 301.
2. Barry, E.G. Journal of the Institute of Petroleum, 1971, 57, 74.

3. Matlack, W.J., et. al. Proceedings SPE Rocky Mountain Regional Meeting, May 23-25, 1983, "Paraffin Deposition and Rheological Evaluation of High Wax Content Altamont Crude Oils".
4. Irani, C.A.; Zajac, J. SPE 55th Annual Fall Technical Conference and Exhibition, October 5-7, 1981, "Pipeline Transportation of High Pour Handil Crude".
5. Schuster, D.S.; Irani, C.A. SPE Rocky Mountain Regional Meeting, May 21-23, 1984, "Reducing Energy Costs for Heated Uintah Basin Crude Oil Transport".
6. Irani, C.A., et.al, "Understanding the Pour Point Depression Mechanism--I. HPLC and GPC Analysis of Crude Oils," ACS Division of Fuel Chemistry Preprints, 1985, 30, 158.
7. Schuster, D.S.; Irani, C.A. "Understanding the Pour Point Depression Mechanism--II. Microfiltration Analysis of Crude Oils, ACS Division of Fuel Chemistry Preprints, 1985, 30, 169.
8. Dooley, M.F.; Feldman, N.; Ryer, J. U.S. Patent 4210424, 1980.
9. Holder, G.A.; Winkler, J. Nature, 1965, 207, 719.
10. Reddy, S.R.; McMillan, M.L. SAE Fuels and Lubricants Meeting, October 19-22, 1981, "Understanding the Effectiveness of Diesel Fuel Flow Improvers".
11. Kawamura, K. JAOCS, 1980, 57, 48.
12. Wesolowski, Thermochimica Acta, 1981, 46, 21.
13. Schuster, D.S. M.S. Thesis, University of Pittsburgh, Pittsburgh, PA, 1984.
14. Redelius, P. Thermochimica Acta, 1985, 85, 331.
15. Suatoni, J.; Swab, R.E. Journal of Chromatography Science, 1976, 14, 535.

RECEIVED September 19, 1990

DEFORMATION-RELATED ORIENTATIONS

Chapter 19

Melt Rheology and Strain-Induced Crystallization of Polypropylene and Its High-Density Polyethylene Blends

Michael S. Chuu and Boh C. Tsai[1]

American National Can Company, Barrington Technical Center, 433 North Northwest Highway, Barrington, IL 60010

Melt elasticity and strain-induced crystallization (SIC) conducted under oscillatory shear for two different polypropylene resins: i.e. control-rheology PP and its virgin PP, having similar melt flow rates and polypropylene-high density polyethylene blends were investigated. It was found that the melt elasticity of the virgin PP was only slightly greater than that of the control-rheology PP at the low frequency side, but the induced crystallization rate of the virgin PP was much faster than that of the control-rheology PP. The high molecular weight tails of PP were concluded to play a critical role in determining the SIC rates. The rate was also dependent on the percent of strain imposed within the range of 1 and 10% studied. It increased with increasing strain. The addition of high-density polyethylene increased the melt elasticity of PP, and also slightly increased the quiescent crystallization temperature of PP. However, the SIC rate decreased with increasing HDPE content in the PP/HDPE blends.

The effect of molecular weight and molecular weight distribution on processing and the consequent

[1]Current address: Amoco Chemical Company, Naperville, IL 60566

0097–6156/91/0462–0312$06.00/0

properties of finished products has been reported and discussed quite extensively. The focal point has traditionally been on the effect of high molecular weight components on viscoelastic behavior. However, it is felt that the importance of the high molecular weight component is not only in its resultant viscoelastic behavior but also in its effect on strain-induced crystallization. This paper emphasizes that the contribution of the high molecular weight component should be, in many applications, assigned to its strain-induced cyrstallization behavior instead of its viscoelastic behavior, especially for semi-crystalline polymers.

Many papers have been published in the field of strain- or stress-induced crystallization. Polyethylene with high molecular weight was reported[1] to crystallize faster at a given rotational shear and temperature. Also, it was found[1] that the addition of a nucleating agent did not affect the rate of strain-induced crystallization of polypropylene random copolymer at shear rates above 5 sec^{-1}. Unique mechanical properties of polypropylene under flow-induced crystallization were also investigated in film[2] and stretched ribbon[3]. Of the papers published, the experimental techniques used to induce crystallization were mostly by rotational shear[1], drawing from melt[4] or solid state[5] and extrusion[6]. A patent[7] has also been issued for using SIC to produce enhanced mechanical properties of polybutene-1.

It is recognized that induced crystallization of polymers under uniaxial shear is expected to take place at a higher temperature than that under dynamic shear. However, both rotational and oscillatory techniques can be used to induce crystallization of semi-crystalline or crystalline polymers. In this paper, the oscillatory shear was applied to study strain-induced crystallization (SIC) of polypropylene and the effect of blending high elasticity HDPE on the SIC of PP.

Experimental

Sample Preparations. Two isotactic polypropylene resins used were the virgin PP, PP-I, and the control-rheology of the virgin PP, PP-II. Their characteristics are listed in Table I.

The isotactic PP used in the studies of PP/HDPE blends is a commercial grade, PP III, which has a melt flow rate of 1.60 g/10 mins. HDPE used is also a commercial grade and has a melt index (measured at 190°C and 2.16 kg) of 0.45 g/10 mins. PP-III/HDPE blends were mixed in a Brabender Mixer at 60 rpm and 220°C for 3 minutes under continuous nitrogen purge.

Table I. Characteristics of Polypropylene Resins

	PP-I	PP-II
Mn	94,500	92,500
Mw	460,000	427,000
MWD (Mw/Mn)	4.87	4.62
MFR *(g/10 min)	1.47	1.46

* ASTM D-1238; Measured at 230°C and 2.16kg

Rheological & Strain-Induced Crystallization Measurements. The dynamic mechanical spectrometer used was Rheometrics Mechanical Spectrometer RMS-7200 with a 10,000 gram-cm torque transducer and two 25mm parallel plates. Measurements of the storage moduli, G', and the loss moduli, G", for PP and PP/HDPE blends were conducted under nitrogen purge.

The SIC experiment was conducted by heating the PP to 200°C under nitrogen purge in the parallel plates for 5 minutes, then it was cooled down freely to the desired temperature at an estimated rate of 15°C/min. The oscillatory shear at a frequency of 1 rad/sec and 1mm gap was applied while the temperature was lowered and the storage modulus, G', loss modulus, G", and complex viscosity, η^*, were recorded. Measurements were stopped when G' reached about $1x10^7$ dynes/cm^2 due to slippage. The induction time was taken from the inflection point of G' vs. time plot when the temperature was at isotherm (as shown in Figure 1).

Quiescent Crystallization. Perkin-Elmer's Differential Scanning Calorimetry DSC-7 was used to measure the static crystallization of PP and PP/HDPE blends.

Results and Discussion

Polypropylene. Quiescent crystallizations of PP-I and PP-II were measured at 20°C/min heating and followed by immediate cooling and reheating at the temperature range between 40° and 210°C. Table II lists the transition temperatures and heat of transitions for PP-I and PP-II.

The on-set crystallization and remelt temperatures were only about 1-2°C greater for the virgin PP, PP-I, than for the control-rheology PP, PP-II. It is also noted that PP has a very wide range of supercooling, temperature difference between melting and crystallization. A small strain introduced could easily induce crystallization in this

Table II. Transition Temperatures & Heat of Transitions for PP-I and PP-II

PP-I	Peak (°C)	On Set (°C)	ΔH(j/g)
Heating	165.1	153.8	81.2
Cooling	109.4	114.4	-92.8
Reheating	162.0	154.2	90.9
PP-II			
Heating	166.2	151.7	80.1
Cooling	107.0	112.6	-89.1
Reheating	163.1	153.2	85.0

The melt elasticities vs. frequencies of the PP-I and PP-II were measured at 160°, 200°, and 240°C. They were then converted to form the master curves as shown in Figure 2.

A difference in elastic moduli of PP-I vs. PP-II was found in the low frequency side. The G' values of PP-I were the same as that of PP-II at high frequencies but were greater than that of PP-II at low frequencies. The greater G' at low frequencies was due to the contribution of melt elasticity from the high MW tails of PP-I.

Figure 3 shows the plots of G' vs. time at 130°C and 10% strain for PP-I and PP-II. It was found that the G' of the control-rheology PP-II increased under shear at a much later time than its virgin resin, PP-I. No quiescent crystallization peak was found when PP-I was subjected to the same thermal history with DSC and then held at 130°C or 134°C for 20 minutes. Therefore, the increase of G' isothermally at 130°C or higher was attributed to the strain-induced crystallization (SIC). The high molecular weight tails retained in the virgin PP, PP-I, played a critical role in determining the SIC rate. The high MW tails crystallized under shear first and could act as a nucleating site for the subsequent crystallization.

The induced crystallization of PP-I and PP-II under shear was also studied at 132° and 134°C at 10% strain. Table III summarizes the induction time, t_i, from the isothermal plots of G' vs. time, for PP-I and PP-II at these temperatures studied.

It was found the induction time increased with increasing temperature as expected for both PP-I and PP-II. In all cases, PP-I has a much shorter induction time than PP-II. The elimination of high MW tails prolonged the induced crystallization process of meta-stable state.

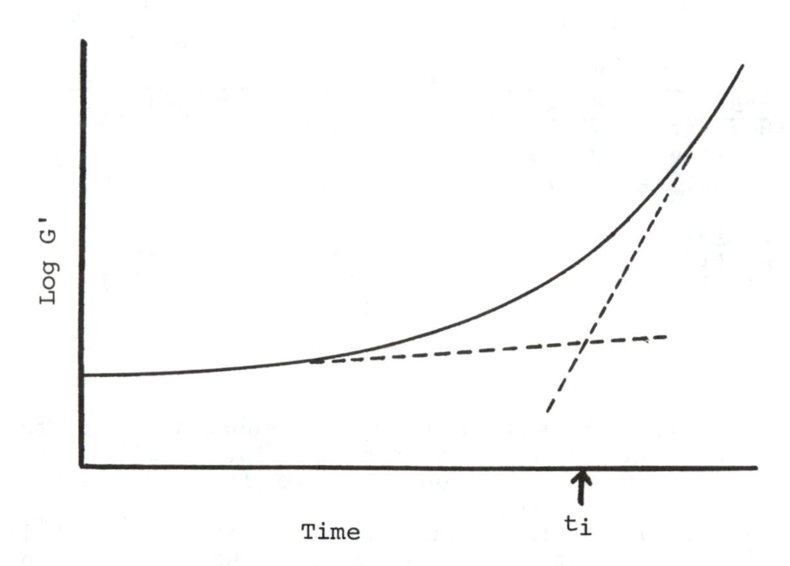

Figure 1. Schematic Plot of Log G' vs. Time For Induction Time, t_i, Measurement

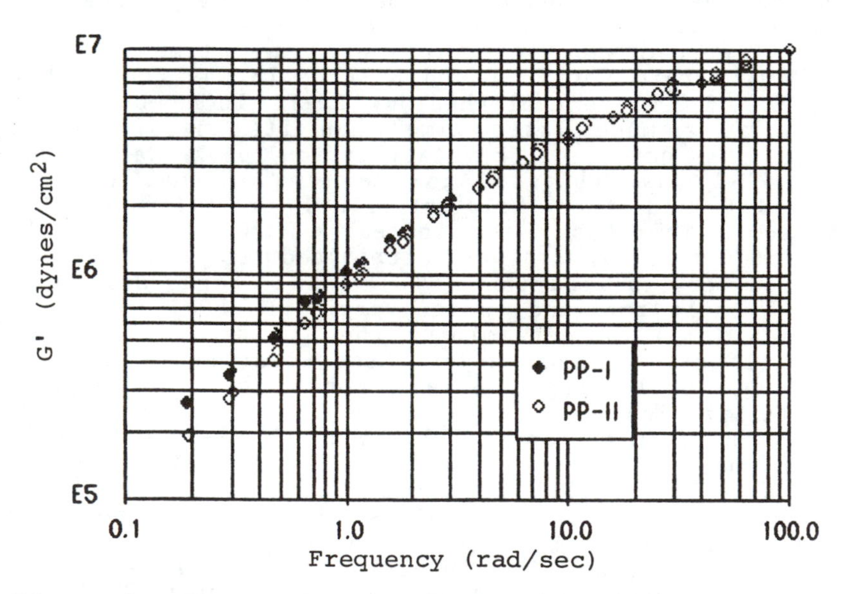

Figure 2. Master Curves of G' vs. Frequency at 160°C for PP-I and PP-II.

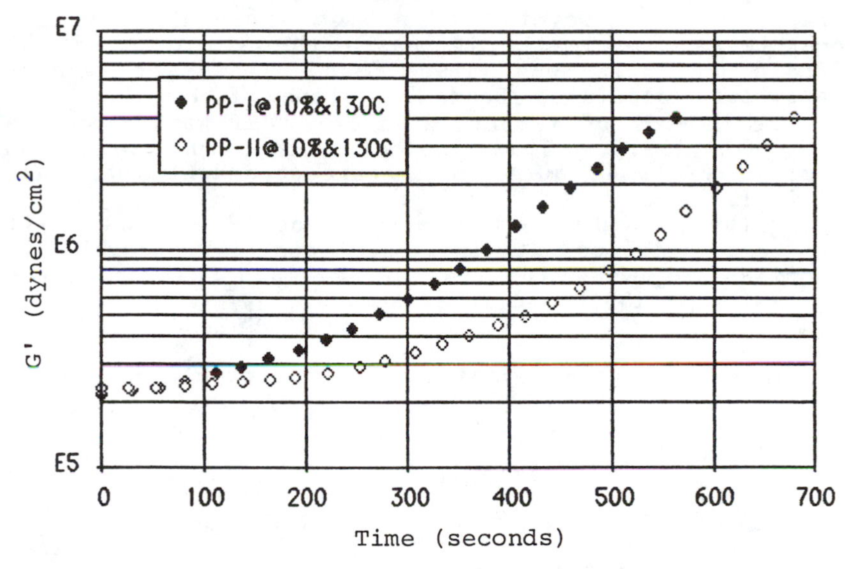

Figure 3. G' vs. Time at 130°C and 10% Strain for PP-I and PP-II.

Table III. Induction Time, t_i, (in seconds) of Strain-Induced Crystallization of PP-I and PP-II at 1 Rad/Sec and 10% Strain

Temperature (°C)	PP-I	PP-II
134	651	1291
132	346	726
130	218	302

PP-II. The high molecular weight components retained in the virgin PP only slightly increased the melt elasticity. However, the high molecular weight components drastically increased the SIC rate of the virgin PP. It was also found that the energy parameter calculated from the slope of log (t_i) vs. the reciprocal of temperatures for PP-I was less than that for PP-II. This indicates that the induction time of PP-II is more temperature-dependent than that of PP-I.

The strain dependence for both PP-I and PP-II was also studied at 1.6, 5.0 and 10%. Table IV compiles the induction time of PP-I and PP-II at these % strains at 130°C and 1 rad/sec.

Table IV. Strain Dependence of Induction Time, t_i, for PP-I and PP-II at 130°C

	Induction Time, t_i (seconds)	
% Strain	PP-I	PP-II
1.6	346	344
5.0	309	323
10	218	302

It was found that induction time decreased with increasing % strain imposed for both PP-I and PP-II. However, a more profound decrease was observed for the virgin PP, PP-I, as compared to PP-II. The strong dependence of induction time on strain of PP-I is a ramification of the long relaxation time of the high molecular weight tails. The deformed high molecular weight component again serves as a nucleation site to accelerate the crystallization under strain. The fact that the induction time of PP-II (control rheology PP) is much less dependent on strain probably is the main reason why it is widely used in fiber spinning for uniform cross-section area. The induction time of PP-I at low % strain, e.g. 1.6%, was approaching that of PP-II. This indicated that the difference of induced crystallization rates of PP-I and PP-II diminished when % strain was lowered and a static-state crystallization process was approached.

PP/HDPE Blends. The high molecular weight PP tails resulted in faster induced crystallization rate as discussed above besides greater elasticity as known in the literature. In order to separate the contribution of its induced crystallization rate from elasticity, the higher elasticity HDPE was blended to PP. Figure 4 shows the plots of G' vs. frequency at 240°C for PP-III/HDPE blends in 100/0, 80/20, 30/70 and 0/100 ratios.

It was found that the melt elasticities increased significantly with increasing HDPE content, especially at the low frequency side. In DSC quiescent crystallization, the crystallization peak temperature of HDPE is slightly greater than that of PP, i.e., 115.9 °C and 111.1°C for HDPE and PP, respectively. In the case of 80/20 and 30/70 blends of PP-III/HDPE, a sharp single crystallization peak was observed at about the crystallization peak temperature of HDPE for these blends in the DSC thermograms, i.e., 115.1° and 115.9°C, respectively. It was thought that the single DSC crystallization peak with the temperature close to that of HDPE component was due to the nucleation effect of HDPE in PP. However, peak crystallization temperature of the PP was only slightly increased by blending with HDPE.

In strain-induced crystallization, the crystallization kinetics are quite different for PP, HDPE, as well as the blends. The more supercooled PP began to show a more drastic change in crystallization rate as compared to HDPE. Figure 5 shows the SIC of η^* vs. time for 100/0, 80/20, 30/70 and 0/100 PP-III/HDPE blends at 136°C and 10% strain.

It was found that the induced crystallization of PP was at a much faster rate than that of HDPE. The SIC induction time of the PP/HDPE blends increased with increasing HDPE content. No induced crystallization was observed when the major phase of the blend changed from PP to HDPE. The effect of induced crystallization of PP was diluted with the addition of HDPE. To the contrary of quiescent crystallization, PP, which crystallizes first under shear, serves as a nucleating site for the subsequent crystallization of HDPE in the PP/HDPE blends. The contribution of higher elasticity by blending with HDPE does not facilitate the strain-induced crystallization process of PP.

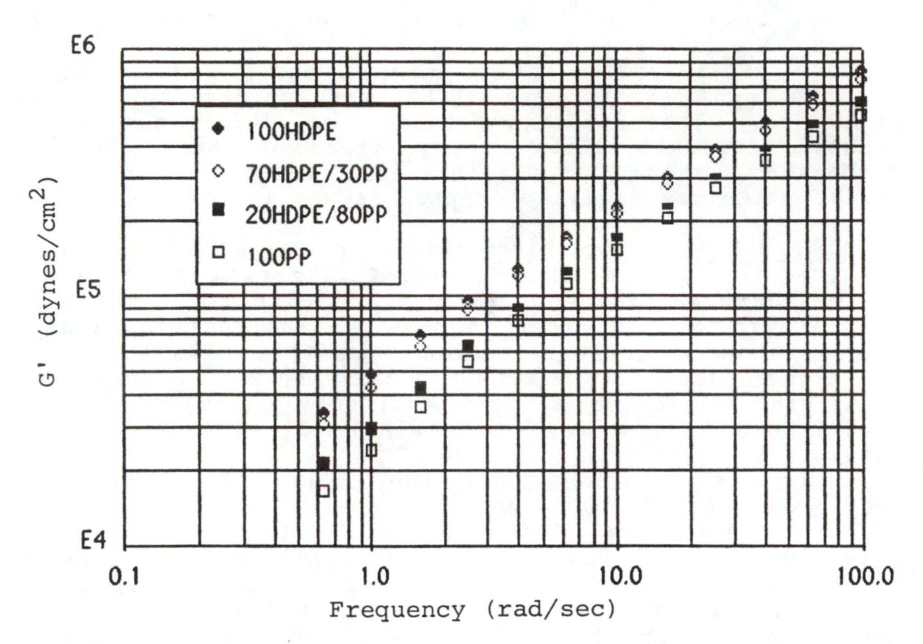

Figure 4. G' vs. Frequency for 100/0, 80/20, 30/70 and 0/100 PP-III/HDPE Blends at 240°C

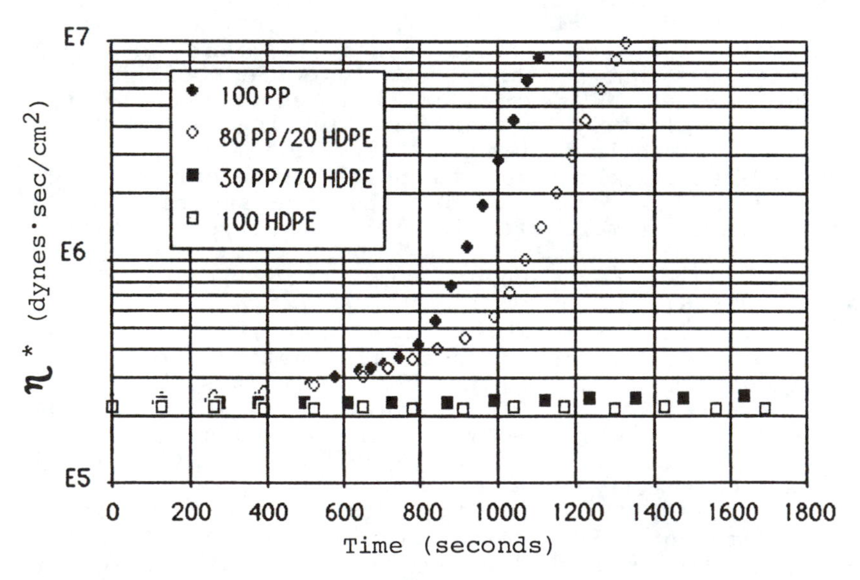

Figure 5. η^* vs. Time for 100/0, 80/20, 30/70 and 0/100 PP-III/HDPE Blends at 136°C and 10% Strain

Conclusion

The addition of high molecular weight (HMW) PP or HDPE slightly increases the melt elasticities of the PP or PP/HDPE blends. HDPE also acts as a nucleating site for the subsequent crystallization in the PP/HDPE blends. It results in a slight increase of the static crystallization temperature.

The situation is different in dynamic crystallization. The addition of HMW PP greatly increases the induced crystallization rate and increases the SIC temperature. However, the addition of HDPE dramatically decreases the induced crystallization rate and reduces the SIC temperature of the PP/HDPE blends.

In polymer converting process of semi-crystalline polymers, melt elasticity is important but should not be over-emphasized. It is believed that the strain- induced crystallization is playing a more critical role than melt elasticity in many applications.

Literature Cited

1. R. R. Lagasse and B. Maxwell, Polym. Eng. and Sci. 16 189 (1976).
2. P. G. Andersen and S. H. Carr, Polym. Eng. and Sci. 16 217 (1976).
3. D. T. F. Pals, P. Van Der Zea, and J. H. M. Albers, J. Macromol. Sci.-Phys. B6(4) 739 (1972).
4. T. Niikuni and R. S. Porter, J. Mater. Sci. 9 389 (1974).
5. S. M. Aharoni and J. P. Sibilia, Polym. Eng. and Sci. 19 450 (1979).
6. A. G. Kolbeck and D. R. Uhlmann, J. of Polym. Sci. 1 27 (1977).
7. K. Z. Hong, U.S. Patent 4,783,301 (1988).

RECEIVED November 6, 1990

Chapter 20

Dynamic Uniaxial Extensional Viscosity

Response in Spray Applications

David A. Soules[1], Gustav P. Dinga[2], and J. Edward Glass

Polymers and Coatings Department, North Dakota State University, Fargo, ND 58105

The importance of dynamic uniaxial extensional viscosities in the spray application of interior can coatings and aerosol adhesive formulations is addressed. Model systems are designed to approximate actual industrial formulations that cannot be examined with current dynamic uniaxial extensional viscometers. Increasing the cross-link density (and therefore the rigidity) of gel dispersions is observed to decrease the dynamic uniaxial extensional viscosities of thickened systems.

The addition of fillers to polymer melts and polymer solutions is a common practice for a variety of reasons: hiding and coloration, strength enhancement, rheology modification, and cost reduction. Polymer solutions in the absence of fillers exhibit increasing non-Newtonian rheology with both increasing molecular weight and concentration. This is related to overlapping and entanglement of macromolecular chains (Chapter 1, this volume). The entanglements and the aggregates in filled systems are disrupted with increasing shear rate, $\dot{\gamma}$, and the viscosity of the solutions, melts, or dispersions decreases. In melts or concentrated polymer solutions (*1*), particulate interactions are overshadowed by the medium viscosity and are generally neglected. Particle–particle interactions are important to the rheological properties of particle-loaded solutions at low polymer concentration. Three factors are primary in the shear deformation response of filled polymer solutions: the molecular weight of the polymer, the volume fraction of the filler, and the aspect ratio of the filler.

The addition of fillers has a more dramatic effect on extensional

---

[1]Current address: Phillips Petroleum Research Center, Bartlesville, OK 74006
[2]Current address: Department of Chemistry, Concordia College, Moorhead, MN 56560

0097–6156/91/0462–0322$06.00/0

viscosities of suspensions than on shear rate viscosities. Weinberger and Goddard (*2*) were among the first to observe the effect of rod-shaped glass particles on the extensional viscosity of Newtonian fluids. The addition of high-aspect-ratio fibers greatly enhanced the axial stress (~10-fold increase) relative to the unfilled solution. Alignment of the fibers appears to occur at extension rates less than 1 s^{-1}. The extensional viscosity of filled solutions that contain spherical beads is significantly different from that of fluids with needle-shaped fillers. Nicodemo, De Cindio, and Nicolais (*3*), using a tubeless syphon extensional viscometer, observed that the dynamic uniaxial extensional viscosity of the fluids decreases with increasing concentration of spherical particles (diameter between 40–50 microns). This behavior is in contrast to the shear viscosity of spherical dispersions and to the extensional viscosity of high-aspect-ratio particulates; the latter two viscosities increase with increasing particulate concentration.

This chapter will address a slightly different area: the importance of dynamic uniaxial extensional viscosities in spray application of fluids as a function of polymer structure and in filled systems with decreasing deformability due to increasing cross-linked densities.

Experimental Details

Acrylamide monomer was purchased from Aldrich and was recrystallized from ethanol. *N,N'*-Methylene bisacrylamide (BISAM) was electrophoresis-grade and was used as received. A hydrolyzed (23%) poly(acrylamide), an experimental polymer supplied by American Cyanamid, $M_v = 6 \times 10^6$, was used in preparing the semi-interpenetrating polymer networks (semi-IPNs). Potassium persulfate, purchased from Aldrich, was the free-radical initiator. The gels were prepared by dissolving the acrylamide, BISAM and potassium persulfate in water or an aqueous polymer solution (depending on whether a gel or an IPN was prepared) in a 500-mL, three-neck, round-bottom flask. The solutions were stirred for 30 min under an argon purge. The final solids at 12 h was 4 wt %. Four different BISAM concentrations were used: 1%, 2.5%, 5%, and 7.5% based on the total monomer weight. Two different linear polymers were used in preparing IPNs, the 23% hydrolyzed poly(acrylamide) and hydroxyethyl cellulose with two molecular weights, 9.5×10^5 and 3.0×10^5. The gels were diluted to 2%, dispersed with a high-speed mixer to promote a more-uniform gel particle size, and then blended with a linear-polymer solution to obtain a final gel concentration of 1%. The extensional viscosity data were obtained with a suction fiber viscometer (*4*).

Results and Discussion

Sprayability. Atomization of the fluid is the generally desired goal in spray applications. The critical breakup length of a filament from a spray nozzle has been described, over a decade ago, in terms of a dimensionless breakup length (*5*), L/2a, for HPAM-thickened fluids. Analysis of a filament's shape has been a standard technique for quantifying the dynamic surface tension (*6*) of surfactant solutions; recently this technique has been applied to estimate polymer dynamic uniaxial extensional viscosities (*7*).

In our initial study of low-viscosity industrial coatings applied by spray, dynamic uniaxial extensional viscosities of formulations with different spray characteristics could not be measured because of sensitivity limitations of the viscometer. In convergent-flow behavior in roll applications (*8*), high-molecular-weight (10^6) poly(oxyethylene) (POE) was added to roll coating formulations to overcome the sensitivity limitation. The formulation exhibiting the greater rib development exhibited the greater dynamic uniaxial extensional viscosity. With additional improvement in the instrument's sensitivity, this correlation also was observed in formulations to which POE was not added. The technique of adding POE and improved instrument sensitivity, however, did not facilitate the measurement of the dynamic uniaxial extensional viscosities of the interior can coatings applied by spray.

In a previous study of coal–water slurries (*9*), extension of such slurries over very short fiber distances provided a response under extensional deformation. One would expect that a greater dynamic uniaxial extensional viscosity would result in a more stable filament that would not mist. This was observed. The relative values of the two slurries were inversely related to differences in misting and combustion efficiencies (*10*). The dynamic uniaxial extensional viscosities, however, decreased with deformation rate, indicating a significant shear viscosity contribution in the response. To ascertain the dominance of extensional vs. shear viscosities in stabilizing the fiber and thus inhibiting breakup of the spray filament, thickened aqueous solutions were studied. The thickeners, described below, are comparable to those used to stabilize coal–water slurries.

The inverse relationship of a fluid's sprayability and dynamic uniaxial extensional viscosity is evident in thickened fluids containing water-soluble polymers of variable segmental and conformational flexibilities. Poly(oxyethylene), a segmentally and conformationally flexible polymer in aqueous solutions, contributes relatively little to the shear viscosity of aqueous solutions relative to fermentation carbohydrate polymers of comparable molecular weight, such as xanthan gum (XCPS) or scleroglucan (SGPS) (Figure 1). The reverse is true with respect to their contribution to the extensional viscosity of aqueous solutions (Figure 2). SGPS or XCPS solutions exhibit high shear viscosities but low dynamic uniaxial extensional viscosity and readily mist on exit from a spray gun (Figure 3). POE solutions, with notable dynamic uniaxial extensional viscosities even at low concentrations (1000 ppm), string rather than mist on exit from the spray nozzle (Figure 4). The antimisting phenomenon would appear to be related to the dynamic uniaxial extensional viscosities, as was the tendency to rib in roll applications.

Sprayability of Gel-Containing Formulations. As noted earlier, particulate structures can exert different influences in the rheological response of

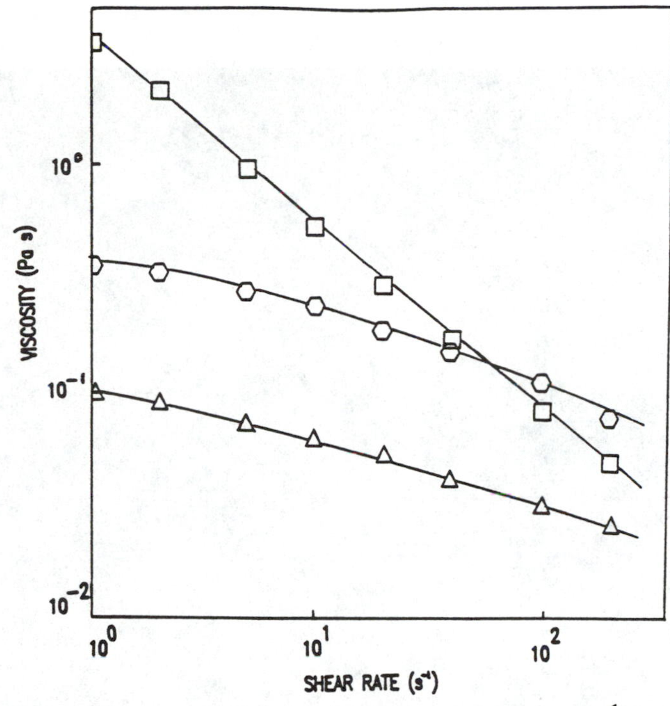

Figure 1. Dependence of viscosity (Pa s) on shear rate (s^{-1}) of 0.5 wt% aqueous solutions of poly(oxyethylene), $M_v = 6 \times 10^6$ (Δ); hydroxyethyl cellulose (HEC), $M_v = 9.5 \times 10^5$ (⬡); and *Xanthomonas campestris* polysaccharide (XCPS), $M_v = 2 \times 10^6$ (□). (Reproduced with permission from reference 1. Copyright 1985 Wiley.)

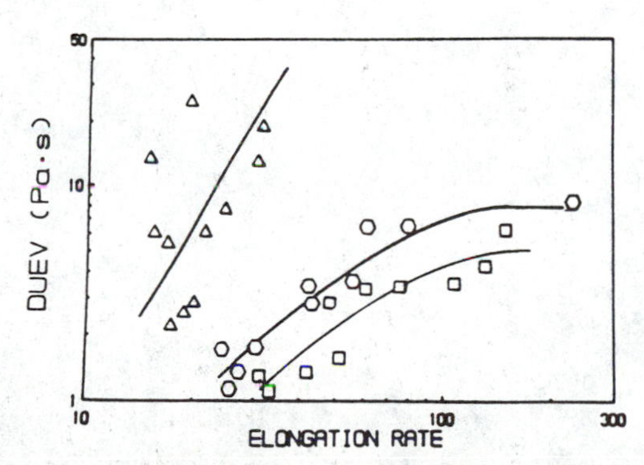

Figure 2. Dependence of viscosity (Pa s) on extension rate (s^{-1}) of 0.5 wt% aqueous solutions of poly(oxyethylene) [POE], $M_v = 6 \times 10^6$ (Δ); hydroxyethyl cellulose (HEC), $M_v = 9.5 \times 10^5$ (⬡); and *Xanthomonas campestris* polysaccharide (XCPS), $M_v = 2 \times 10^6$ (□). (Reproduced with permission from reference 1. Copyright 1985 Wiley.)

Figure 3. Spray misting behavior of XCPS, 0.5 wt % aqueous solution.

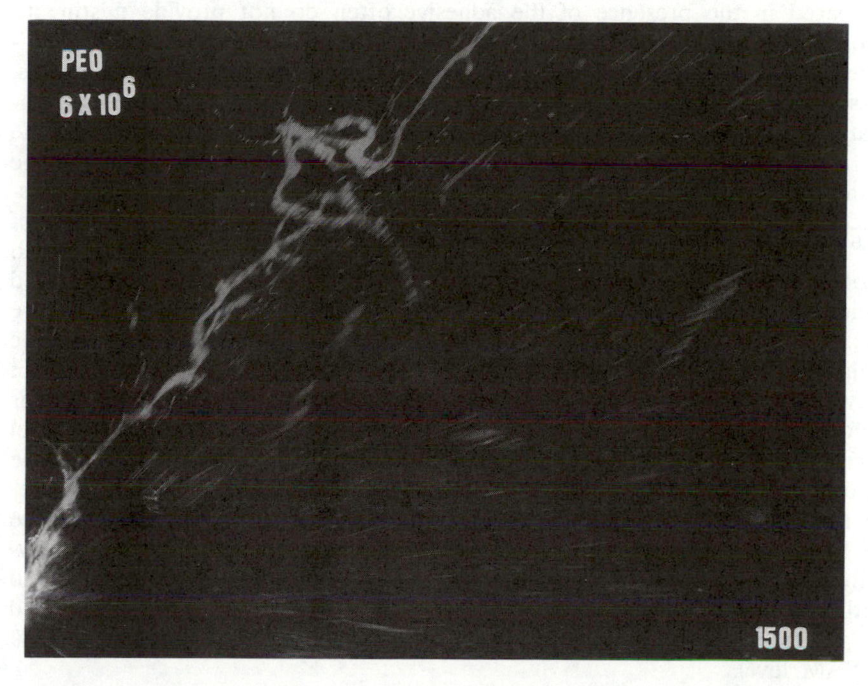

Figure 4. Spray cobwebbing behavior of POE, 0.15 wt % aqueous solution.

fluids and this may be related to various formulations used in different applications. For example, polymeric gels are added to high-solids coatings (*11*). The initial components of the coating are very low in molecular weight, and the formulations are very low in viscosity. The polymeric gels increase the low-shear-rate viscosity and thereby prevent sagging during oven curing (*12*). Antimisting of the formulation during spray application is not observed, presumably because of the spherical shape and rigidity of the gel.

Gels are used also for strength enhancement in adhesives. The gels tend to be lightly cross-linked to provide tactility to the adhesive. Many of the adhesives are applied by aerosol spray and the addition of the gels, prepared in the presence of the adhesive often do not provide misting to the aerosol adhesive. To model this system, a series of semi-interpenetrating network gels, prepared in the presence of linear polymers, was synthesized. At low cross-link (1–5%) levels, increasing the concentration of the difunctional monomer effects a smaller pore size (*13*). The semi-IPNs were prepared with acrylamide cross-linked with *N,N'*-methylene bisacrylamide. The free-radical polymerization was conducted in the presence of 0.3 wt % of the hydroxyethyl cellulose with molecular weight of 10^6. The gel was blended into 0.75 wt % of hydroxyethyl cellulose with molecular weight of 10^6. The extensional viscosity of any of the filled fluids is lower than that of the unfilled hydroxyethyl cellulose fluid (Figure 5). As the cross-link density of the semi-IPN is increased, the dynamic uniaxial extensional viscosities of the resulting filled fluids decrease. In view of the prior art, it would appear that the weakly cross-linked IPN is deformable. With increasing cross-link density, the rigidity of the gel increases and lowers the dynamic uniaxial extensional viscosity of the thickened dispersion, similar to the influence on the extensional viscosity of rigid glass beads (*3*). No evidence of polymer fragments on the surface of the IPN is evident in systems polymerized in the presence of hydroxyethyl cellulose of variable molecular weight. The semi-IPNs prepared without the carbohydrate polymer gave responses similar to those prepared in the presence of hydroxyethyl cellulose in systems using a 2.5 wt % BISAM level.

Hydrolyzed poly(acrylamide) (HPAM), unlike hydroxyethyl cellulose, is not prone to degradation in the presence of free radicals. Semi-IPNs were prepared in the presence of 0.02 wt % HPAM and then added to 0.01 HPAM solutions. HPAM polymers of $M_v = 1.6 \times 10^7$ have a much greater extensional viscosity at lower concentrations compared with carbohydrate solutions (*4*). The dynamic uniaxial extensional viscosities of 1 wt % semi-IPN slurries prepared without 0.02 wt % HPAM solutions are presented also in Figure 6. There appears to be a lower extension rate required to reach the plateau with decreasing cross-linker concentration.

Commercial Adhesive Hydrocarbon Formulations

The dynamic uniaxial extensional viscosities of commercial adhesive

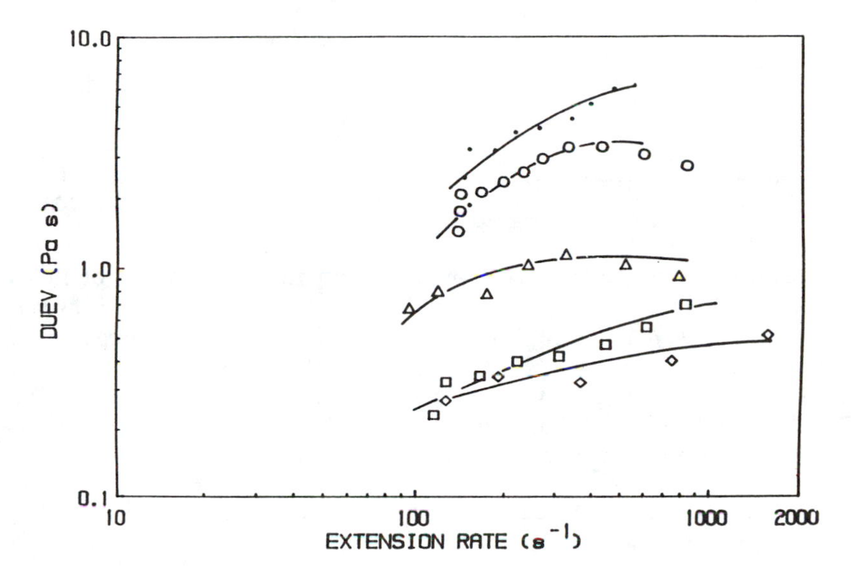

Figure 5. Dependence of dynamic uniaxial extensional viscosity (DUEV, Pa s) on extension rate (s^{-1}) for a series of fluids containing 1% acrylamide (AM) gels suspended in 0.75 wt % hydroxyethyl cellulose (HEC), $M_v = 9.5 \times 10^5$. The gels were prepared in the presence of 0.75 wt % HEC and with the following levels of *N,N'*-methylenebisacrylamide (BISAM): 1 wt %, ○; 2.5 wt %, Δ; 5 wt %, □; and 7.5 wt %, ◇. The DUEV of HEC in the absence of the AM/BISAM gel is noted by •.

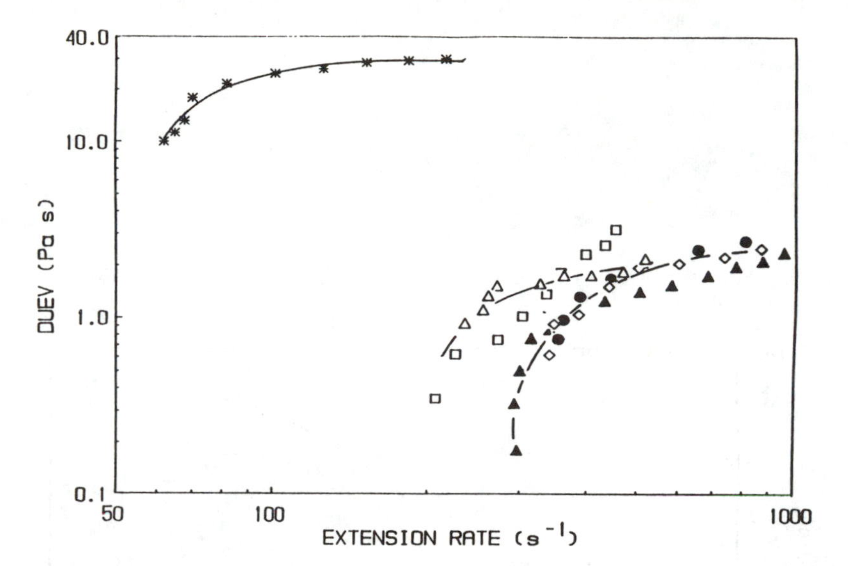

Figure 6. Dependence of DUEV on extension rate for a series of fluids containing 1% acrylamide (AM) gels suspended in 0.0001 wt % hydrolyzed (23%) acrylamide, $M_v = 1.6 \times 10^7$. The gels were prepared in the presence of 0.02 wt % HPAM and with the following levels of BISAM: 1 wt %, ●; 2.5 wt %, Δ; 5 wt %, □; and 7.5 wt %, ◇. The DUEV of HPAM in the absence of the AM/BISAM gel is noted by *. The closed symbols represent gel responses in the absence of HPAM.

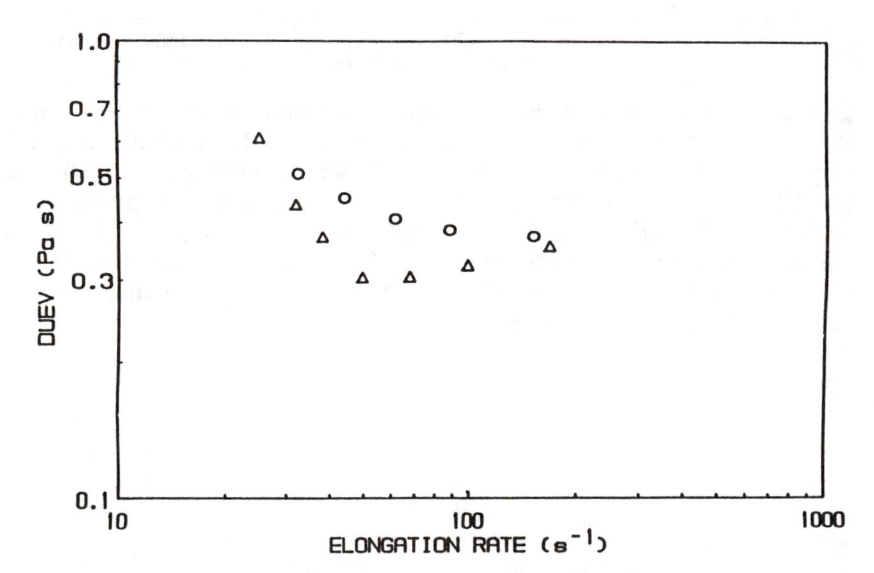

Figure 7. Dependence of DUEV on extension rate for a series of commercial gel (20 wt %) adhesive formulations. Key: ○, formulation that misted on spray; Δ, formulation that cobwebbed on spray.

formulations, supplied by the 3M Corporation, were examined. The "good" adhesive provided a mist during spray application from a pressurized container; the "poor" formulation did not mist. Like the coal–water slurries, the viscosity response of commercial formulations reflected a significant shear deformation response (Figure 7), and the responses were within experimental error. The hydrocarbon adhesives were in a poor solvent for the gel, possibly decreasing the difference between the gels under the measurement conditions. The model acrylamide gels and semi-IPNs were all in a good solvent. The commercial adhesive had a gel content much higher than the gel concentration studied with the aqueous model systems; the total solids content of the adhesive was ~20%. Our measurements of these hydrocarbon gels were made at 8 wt %. The extensional viscometer used in this study, as well as other techniques used in the past or in this period, is not capable of properly addressing the problem of complex formulations of this type. This is true of many commercial formulations utilizing polymers as rheological modifiers.

Acknowledgments

The financial support of this study by the 3M Corporation is gratefully acknowledged.

Literature Cited

1. Metzner, A. B. *J. Rheol.* **1985,** *29*(6), 739.
2. Weinberger, C. B.; Goddard, J. D. *Int. J. Multiphase Flow* **1974,** *1,* 465.
3. Nicodemo, L.; De Cindio, B.; Nicolais, L. *Polym. Eng. Sci.* **1975,** *15*(9), 679.
4. Soules, D. A.; Fernando, R. H.; Glass, J. E. *J. Rheol.* **1988,** *32*(2), 181–198.
5. Gordon, M.; Yerushalmi, J.; Shinnar, R. *Trans. Soc. Rheol.* **1973,** *17,* 303.
6. Adamson, A. W. *Physical Chemistry Surfaces,* 4th Edition; Interscience Publishers, Inc., 1982.
7. Schummer, P.; Tebel, R. H. *J. Non-Newtonian Fluid Mech.* **1983,** *12,* 331.
8. Fernando, R. H.; Glass, J. E. *J. Rheol.* **1988,** *32*(2), 199–213.
9. Fernando, Raymond H.; Lundberg, David J.; Glass, J. E. In *Polymers in Aqueous Media: Performance through Association;* Glass, J. E., Ed.; Advances in Chemistry 223; American Chemical Society: Washington, DC, 1989; Chap. 12.

10. Rakitsky, W. G.; Knell, E. W.; Murphy, T. J. *Proc. 11th Int. Conf. Slurry Technol.* **1986,** 137.

11. Lambourne, R. *Paint and Surface Coatings: Theory and Practice;* Halsted Press (Wiley): New York, 1988.

12. Bauer, D.; Briggs, L. M.; Dickie, R. A. *Ind. Engin. Chem., Prod. Res. Dev.* **1982,** *21*(4), 686.

13. Allen, R. C.; Saravis, C. A.; Maurer, H. R. *Gel Electrophoresis and Isoelectric Focusing of Proteins;* deGruyter: Berlin, 1984; Chap. 1.

RECEIVED March 11, 1990

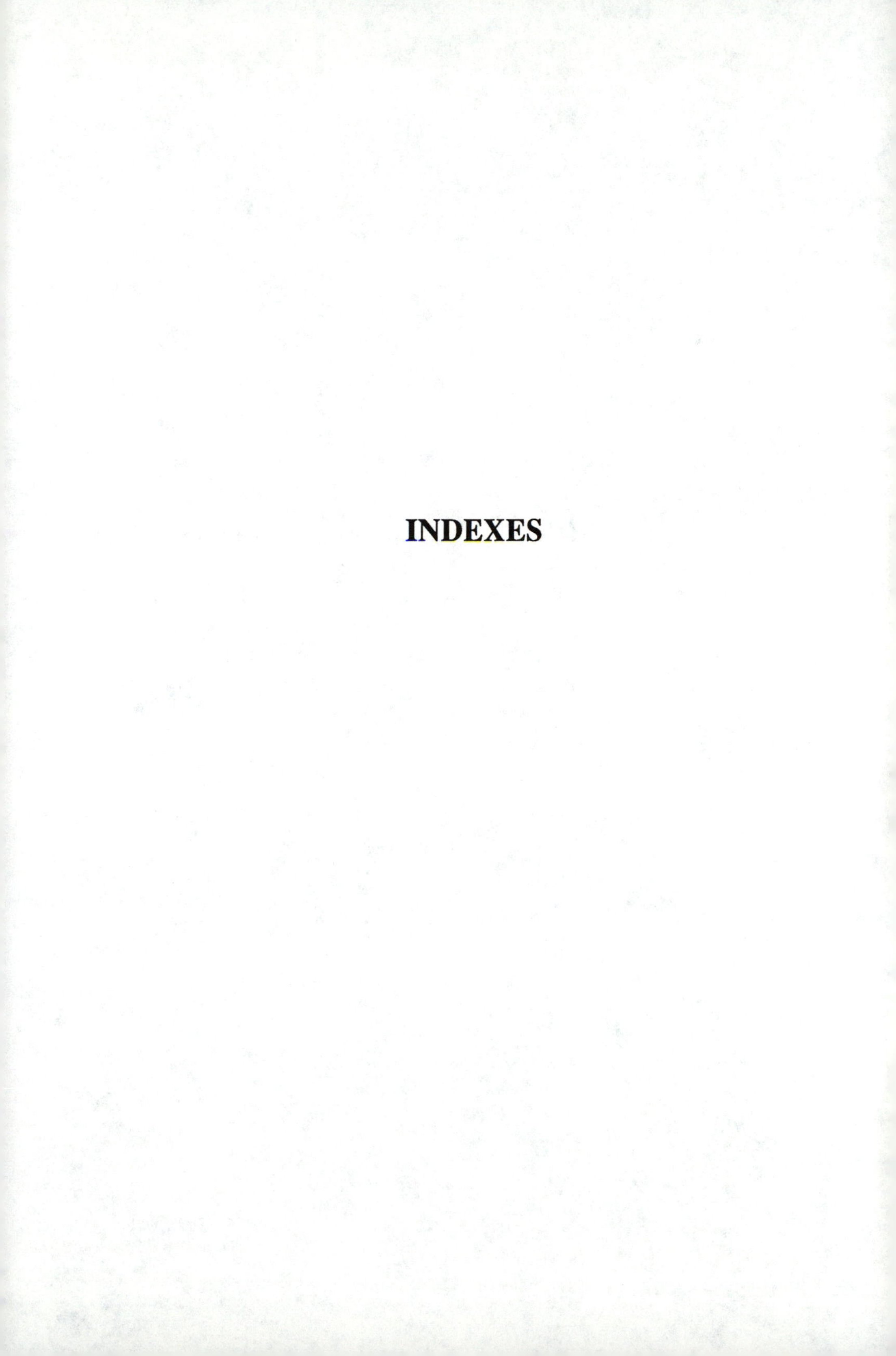

INDEXES

Author Index

Affiliation Index

Subject Index

R

S

Production: Paula M. Befard
Indexing: Deborah H. Steiner
Acquisition: A. Maureen Rouhi

Printed and bound by Maple Press, York, PA

Other ACS Books

Chemical Structure Software for Personal Computers
Edited by Daniel E. Meyer, Wendy A. Warr, and Richard A. Love
ACS Professional Reference Book; 107 pp;
clothbound, ISBN 0–8412–1538–3; paperback, ISBN 0–8412–1539–1

Personal Computers for Scientists: A Byte at a Time
By Glenn I. Ouchi
276 pp; clothbound, ISBN 0–8412–1000–4; paperback, ISBN 0–8412–1001–2

Biotechnology and Materials Science: Chemistry for the Future
Edited by Mary L. Good
160 pp; clothbound, ISBN 0–8412–1472–7; paperback, ISBN 0–8412–1473–5

Polymeric Materials: Chemistry for the Future
By Joseph Alper and Gordon L. Nelson
110 pp; clothbound, ISBN 0–8412–1622–3; paperback, ISBN 0–8412–1613–4

The Language of Biotechnology: A Dictionary of Terms
By John M. Walker and Michael Cox
ACS Professional Reference Book; 256 pp;
clothbound, ISBN 0–8412–1489–1; paperback, ISBN 0–8412–1490–5

Cancer: The Outlaw Cell, Second Edition
Edited by Richard E. LaFond
274 pp; clothbound, ISBN 0–8412–1419–0; paperback, ISBN 0–8412–1420–4

Practical Statistics for the Physical Sciences
By Larry L. Havlicek
ACS Professional Reference Book; 198 pp; clothbound; ISBN 0–8412–1453–0

The Basics of Technical Communicating
By B. Edward Cain
ACS Professional Reference Book; 198 pp;
clothbound, ISBN 0–8412–1451–4; paperback, ISBN 0–8412–1452–2

The ACS Style Guide: A Manual for Authors and Editors
Edited by Janet S. Dodd
264 pp; clothbound, ISBN 0–8412–0917–0; paperback, ISBN 0–8412–0943–X

Chemistry and Crime: From Sherlock Holmes to Today's Courtroom
Edited by Samuel M. Gerber
135 pp; clothbound, ISBN 0–8412–0784–4; paperback, ISBN 0–8412–0785–2